Desert Landforms

An Introduction to Systematic Geomorphology

This series consists of seven volumes published between 1968 and 1977.

1. *Humid Landforms*
   I. Douglas (1977)
2. *Desert Landforms*
   J. A. Mabbutt (1977)
3. *Landforms of Cold Climates*
   J. L. Davies (1969)
4. *Coasts*
   E.C.F. Bird (1968)
5. *Structural Landforms*
   C. R. Twidale (1970)
6. *Volcanoes*
   Cliff Ollier (1970)
7. *Karst*
   J. N. Jennings (1971)

Further details are available from the A.N.U. Press, P. O. Box 4, Canberra, A.C.T. 2600, Australia.

An Introduction to Systematic Geomorphology
Volume Two

Desert Landforms

J. A. Mabbutt

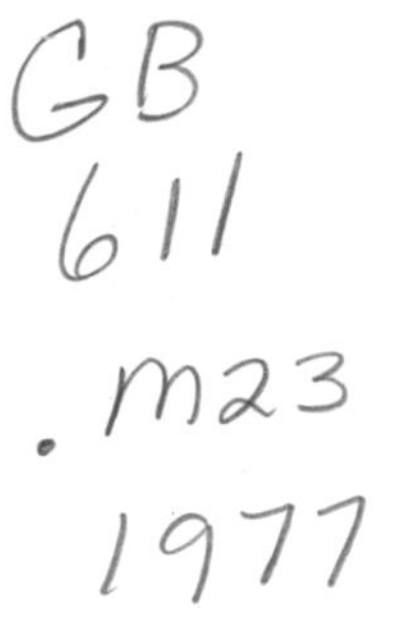

The MIT Press
Cambridge, Massachusetts

First MIT Press edition, 1977
First published in Australia 1977
Printed in Singapore for the Australian National University Press, Canberra

Library of Congress catalog card number: 77–76688
ISBN 0–262–13131–5

INTRODUCTION TO THE SERIES

This series is conceived as a systematic geomorphology at university level. It will have a role also in high school education and it is hoped the books will appeal as well to many in the community at large who find an interest in the why and wherefore of the natural scenery around them.

The point of view adopted by the authors is that the central themes of geomorphology are the characterisation, origin, and evolution of landforms. The study of processes that make landscapes is properly a part of geomorphology, but within the present framework process will be dealt with only in so far as it elucidates the nature and history of the landforms under discussion. Certain other fields such as submarine geomorphology and a survey of general principles and methods are also not covered in the volumes as yet planned. Some knowledge of the elements of geology is presumed.

Four volumes will approach landforms as parts of systems in which the interacting processes are almost completely motored by solar energy. In humid climates (Volume One) rivers dominate the systems. Fluvial action, operating differently in some ways, is largely responsible for the landscapes of deserts and savanas also (Volume Two), though winds can become preponderant in some deserts. In cold climates, snow, glacier ice, and ground ice come to the fore in morphogenesis (Volume Three). On coasts (Volume Four), waves, currents, and wind are the prime agents in the complex of processes fashioning the edge of the land.

Three further volumes will consider the parts played passively by the attributes of the earth's crust and actively by process deriving energy from its interior. Under structural landforms (Volume Five), features immediately consequent on earth movements and those resulting from tectonic and lithologic guidance of denudation are considered. Landforms directly the product of volcanic activity and those created by erosion working on volcanic materials are sufficiently distinctive to warrant separate treatment (Volume Six). Though karst is undoubtedly delimited lithologically, it is fashioned by a special combination of processes centred on solution so that the seventh volume partakes also of the character of the first group of volumes.

J. N. Jennings
General Editor

PREFACE

This book discusses desert landforms and landforming processes in a range of physiographic settings, from upland desert to sand desert. The approach was chosen partly because the aim is to emphasise landforms first, and processes in relation to them. Moreover it serves a theme that recurs often in the text, namely that the geomorphic response to desert climatic controls is profoundly influenced and compartmented by structure and major relief. In retrospect the choice was fortunate, since by this treatment of its subject this book complements the larger *Geomorphology in Deserts* (1973) by Cooke and Warren which appeared in the interim and which is structured around systems of geomorphic processes.

The Australian deserts receive fuller treatment in this book than their relative extent would warrant, but that is where I have worked during the past twenty years, and this treatment may help to remedy a neglect abroad. The examples have been chosen and discussions are, one hopes, wide-ranging enough to establish general principles concerning desert landforms, to account for the development and interplay of ideas about their origins, to direct and encourage readers towards further sources, and above all to entice them into the desert to see and enjoy for themselves.

I am very grateful to a group of friends who helpfully criticised early drafts of chapters: Fred Bell, Jim Bowler, Bill Bull, Karl Butzer, Ron Cooke, Cliff Ollier, Asher Schick, Marjorie Sullivan, Dan Yaalon, Aaron Yair and Tony Young. I also thank those who provided photographs and allowed the use of published illustrations as a basis for the line diagrams, not only those acknowledged in the text but others who offered materials which could not be used, or who helped procure materials. The Editor of the CSIRO Division of Land Use Research, Margaret Mills and the staff photographers Len Leslie and Jack Cavanagh assisted generously with photographs. The book could not have been completed without the dedicated help of Joy Gallagher, David Johnson, Kevin Maynard and Vi Threader of my own staff, and that of Keith Mitchell of The Australian National University, who drew the final line diagrams. Most of all, I am deeply indebted to Joe Jennings for his advice and help at all stages, for his continued encouragement, and for his personal kindness.

I thank the authors, photographers and other sources, as acknowledged in the captions to the figures and plates, for permission to use their materials.

In addition, permission from publishers to reproduce material is acknowledged as follows: Academic Press for Fig. 53A; Advertiser Newspapers, Adelaide for Plate 37; American Geographical Society for Figs. 74 and 91; American Journal of Science for Figs. 27A, 29, 30 and Plate 16; Association of American Geographers for Figs. 93, 95 and Plates 13 and 26; Association de Géographes Français for *part* of Fig. 61; Australian Department of National Resources for Fig. 52 and Plate 47; B. T. Batsford Ltd for Fig. 71a; Cambridge University Press for Figs. 7, 8 and 13B; Centre National de la Recherche Scientifique for Fig. 88b; Chapman and Hall for Figs. 65, 68a, 70, 71b and 73; Columbia University Press for Fig. 37; CSIRO Australia Editorial and Publications Service for Figs. 31 and 79; Elsevier for Fig. 69; Ferd. Dümmlers Verlag for Figs. 40a, 40c and 96; Gauthier-Villars for Fig. 85; Gebrüder Borntraeger for Figs. 5, 19, 22, 33, 68b, 76, 77, 84, 88a and Plates 6, 31, 34a, 35 and 36; Geographical Society of New South Wales for Fig. 59; Geological Society of America for Fig. 39; Geological Society of Australia for Fig. 20Bc; Geological Society of South Africa for Fig. 10; Gustav Fischer Verlag for Fig. 3; Institute of British Geographers for Plate 43; Institut Fundamentale d'Afrique Noir for Fig. 72; Institut Geographique National for Figs. 68c, 68d, 68e and Plates 44 and 49; Lenz-Verlag Giessen for Fig. 12; Librairie Armand Colin for Figs. 40b and 87B; Methuen & Co. for Fig. 13A; Office de la Recherche Scientifique et Technique Outre-Mer (ORSTOM) for Figs. 35 and 87A; Princeton University Press for Figs. 80 and 94; Royal Geographical Society for Figs. 9B, 54, 63 and 82; Royal Geographical Society of Australasia (South Australian Branch) for Fig. 50; Royal Society of South Australia for Fig. 55c; Scottish Academic Press for Fig. 9A; Society of Economic Palaeontologists and Mineralogists for Fig. 69c; Société d'Édition d'Enseignement Supérieur (SEDES) for Fig. 92; South African Geographical Society for Fig. 18; South African Government Printer for Plate 23; Swets and Zeitlinger for Figs. 43 and 44c; UNESCO for Fig. 1; United States Department of the Navy for Fig. 15; United States Geological Survey for Figs. 6, 23, 25, 26,

28, 47, 53B, 56, 57, 58, 61 (part), 86 and Plates 3, 18, 25 and 41; University of California Press for Fig. 17; University of Chicago Press for Figs. 24, 27B, 67, 78, and Plates 34b and 50; University of Dakar, Department of Geography for Fig. 75; University of Massachusetts for Fig. 55; Georg Westermann for Plate 28; Wissenschaftliche Verlagsgesellschaft for Fig. 90; John Wiley & Sons for Fig. 81.

Sydney J. M.
July 1976

CONTENTS

FIGURES

PLATES

TABLES

I

DESERT ENVIRONMENTS

Deserts are areas in which vegetation is sparse or absent, where the land surface is left exposed to the atmosphere and the associated physical forces (Pl. 1). These circumstances lend a distinctive character to the effectiveness and operation of desert land-forming processes.

Many factors can limit plant growth; for example a bare rock outcrop constitutes a lithological desert, but the extensive deserts of the world are climatically determined, through lack of moisture or low temperatures. This book deals with deserts caused primarily by aridity, in lower and middle latitudes. For reasons of space the intensely cold polar deserts have been omitted, but many features of cold dry regions, which commonly exhibit a remarkable simi-

1. Desert hillslope illustrating the inadequacy of the sparse vegetation as protection to the land surface. Photo N. H. Speck, CSIRO, Division of Land Use Research.

larity to those of warmer deserts, are treated elsewhere in this series as *Landforms of Cold Climates* (Davies, 1969).

EXTENT AND TYPES OF ARID DESERTS

Generally, these are areas where rainfall is small in amount and infrequent and irregular in occurrence. More specifically, arid climates are defined as those under which potential moisture losses from evaporation and transpiration exceed incoming precipitation. Early classifications of climate measured the degree of this deficiency by indices of aridity based on empirical relationships between precipitation and temperature as a major determinant of evaporation. A typical index, chosen because it has been used in considering relations between climate and landforms, is that of Martonne (1942),

$$I = \frac{P}{T + 10}$$

where P is the mean annual precipitation in mm, and
T the mean annual temperature in ° C.

The limit of dry climates was set at the index value 10 and the division between semiarid and arid climates at 5.

Later classifications sought improved estimates of evaporation and transpiration. That of Thornthwaite (1948) used a moisture index

$$\frac{s - 0.6d}{e} \times 100$$

where s is the sum of monthly surpluses of precipitation above estimated potential evaporation
d is the sum of monthly deficits
e is the estimated annual evaporation based on mean monthly values of temperature with adjustment for season of rainfall and including a factor for soil-moisture storage.

Areas with a moisture index of between −20 and −40 were defined as semiarid, and those with below −40 as arid.

The differences between such indices are probably not important geomorphologically, mainly because our understanding of the link between aridity and landforms remains at best general. The forms of the indices do, however, demonstrate that no single rainfall value can delimit the desert lands; in Australia, for example, the accepted boundary of the arid zone accords closely with the 250 mm isohyet along its cooler southern margin, whereas localities in the tropical northwest receiving above 500 mm rainfall must still be regarded as arid.

Distribution and controls of arid deserts

The map of arid deserts in Fig. 1 is based on Meigs (1953), who supplemented the semiarid and arid types of Thornthwaite by an extreme-arid climate with a moisture index below −57, generally with less than 25 mm annual rainfall, with no seasonal rhythm of precipitation and in which at least twelve consecutive rainless months have been recorded.

Five desert provinces are shown, in each of which an arid core reaches to the west coast between latitudes 10° and 30°, and from which a dry belt extends inlands and polewards, locally to more than 55°. This pattern reflects two major controls of aridity and two associated minor factors. The first, or *zonal* control is the large-scale subsidence and divergence of stable air in the anticyclonic belts of low latitudes. On the west coasts of the continents this control is commonly reinforced by an *oceanic* effect due to cool oceanic currents and the upwelling of cold water inshore. The second major control of aridity is that of remoteness from maritime sources of precipitation, or *continentality*. This reinforces zonality within the broad continents of low latitudes, such as northern Africa and Australia, but becomes the major determinant of deserts in middle latitudes. Both major controls of aridity, but particularly continentality, may be reinforced regionally by an *orographic* effect, where air subsides in dry rainshadows in the lee of mountain barriers.

Zonal or low-latitude deserts are most extensive in the Sahara-Arabia-India belt and in Australia. In North and South America and to a lesser degree in southern Africa they are confined by mountains to coastal deserts in which oceanic effects are prominent (Meigs, 1966). Continental or mid-latitude deserts have their greatest extent within the broad mass of Asia and a lesser development in North America, and in both areas aridity is heightened by orographic effects in intermont basins. The only mid-latitude desert in the southern hemisphere is the Patagonian Desert, where

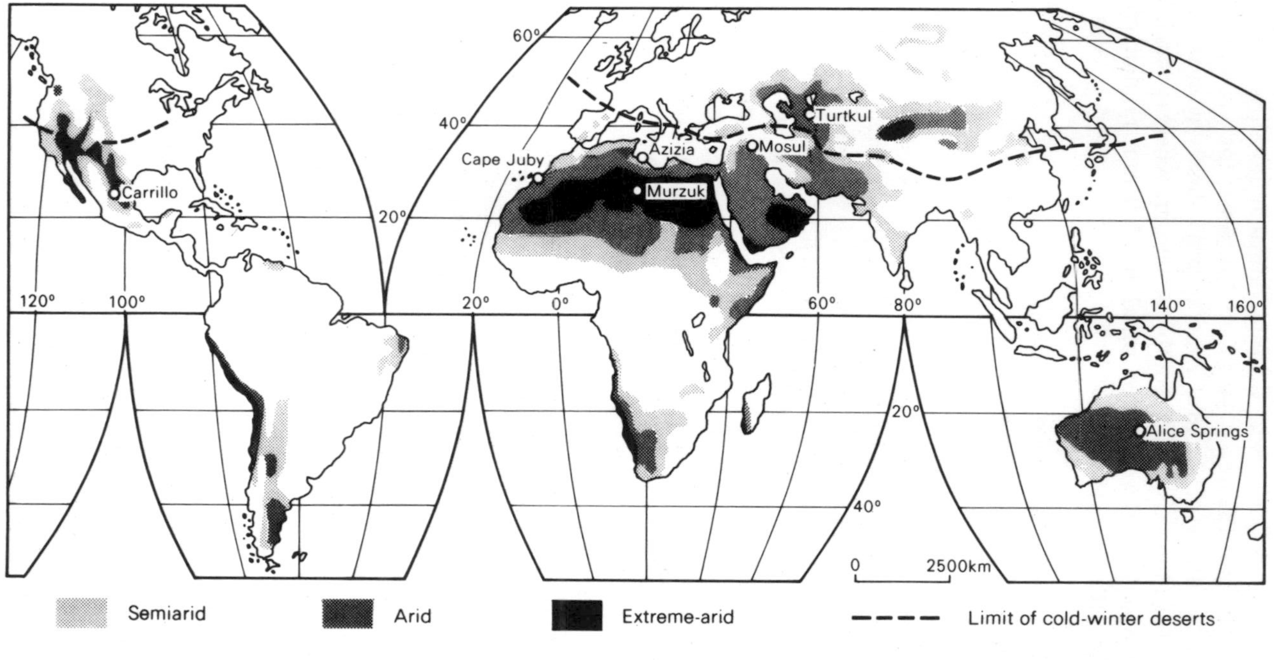

1. *World distribution of dry climates from Meigs, 1953. The southern limit of the cold-winter deserts is drawn where the mean temperature of the warmest month falls below 0 °C.*

dry climates extend to the east coast between latitudes 40° and 52°S in the lee of the southern Andes.

With the exception of the Takla Makan Desert in central Asia, extreme aridity is confined to the low-latitude deserts — to the inner Sahara and southern Arabia and to the west-coastal deserts of South America and southern Africa. No part of Australia attains extreme aridity.

Warm and cold-winter deserts

Apart from differences in degree of aridity, there is a major distinction between the warm deserts and those with such cold winters that frost becomes geomorphologically significant. The limit of the cold-winter deserts in Fig. 1 is drawn where the mean temperature of the coldest month falls below 0°C, as indicative of a severe winter regime with frozen ground (Meigs, 1953). The division corresponds in large measure with that between low-latitude and mid-latitude deserts, but some desert uplands in low latitudes are also subject to strong frost action.

Climatic characteristics associated with aridity

The stable descending airmasses which are responsible for low precipitation in deserts introduce secondary climatic characteristics which may also be geomorphologically significant, and these are briefly illustrated by the example of Alice Springs, Australia (Fig. 2), in a low-latitude continental desert (Slatyer, 1962).

With a mean annual rainfall of 275 mm, Alice Springs is only moderately arid, but its summer rainfall maximum is characteristic of all but the poleward margins of these deserts. An average rainfall of 34 mm per rain day reflects the occasional incidence of heavy rainfalls associated with the penetration of moist equatorial air with the passage of tropical depressions. Variability of rainfall is high, at between 35 and 40 per cent, as is general in deserts.

With little cloud, Alice Springs receives 3350 out of a possible 4100 hours of sunshine, and high incoming and outgoing radiation are expressed in diurnal temperature ranges generally in excess of 25°C, greater than the annual range. Day temperatures are high, typically between 35° and 40°C in the warmer months. The geomorphologically significant ground temperatures fall outside these limits and are characteristically between 20° and 25°C above screen temperatures at midday and between 5° and 10°C lower at

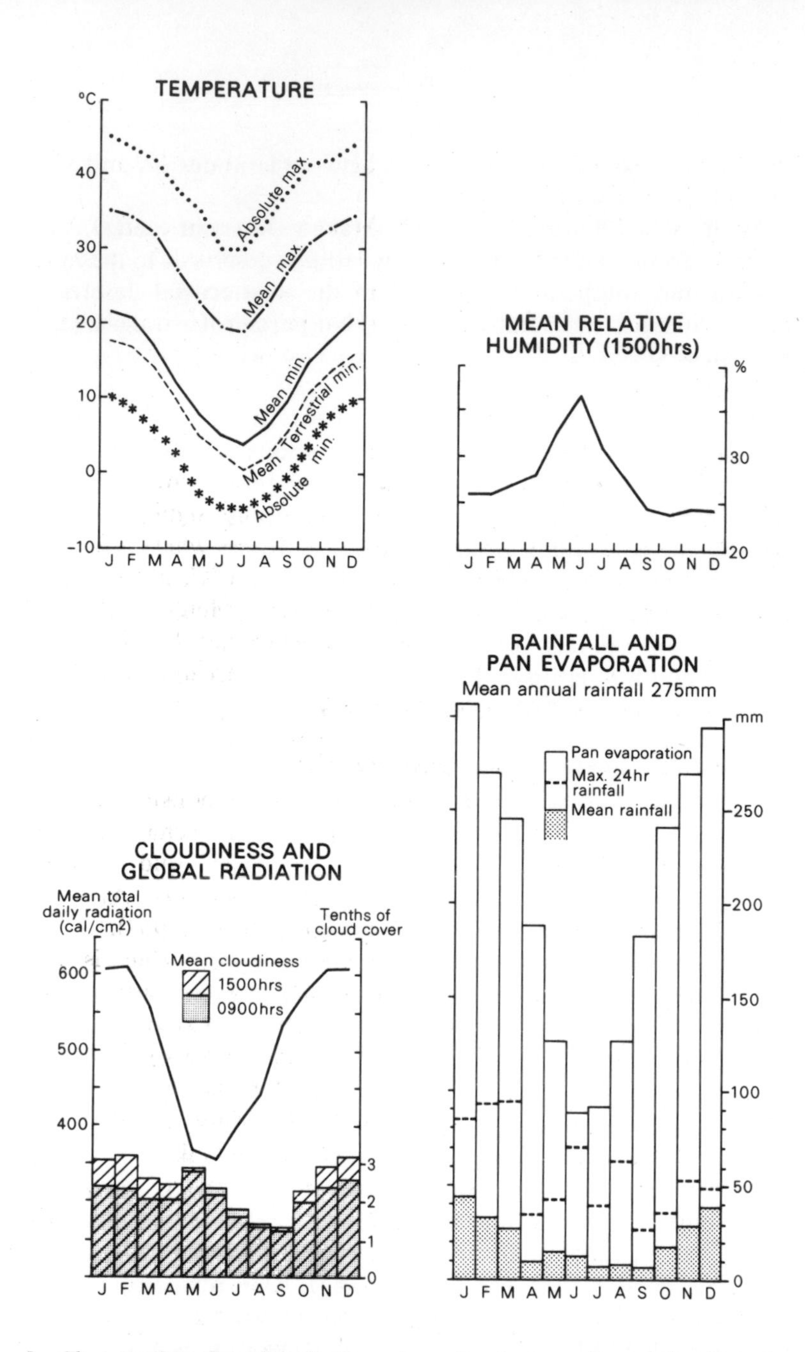

2. *Climatic data for Alice Springs, Australia, in a continental low-latitude desert of moderate aridity. From Slatyer, 1962 and data supplied by Bureau of Meteorology, Australia.*

night. With its continental setting reinforced by an appreciable altitude (570 m), Alice Springs records between 20 and 30 frosts annually in a frost season of 100 days, but not severe enough to be geomorphologically important.

Dryness of the atmosphere is shown by afternoon values of relative humidity of between 25 and 40 per cent at Alice Springs. The large moisture deficit, also associated with marked drying of the ground, is reflected in a pan evaporation of 2800 mm and a biologically more relevant potential evapotranspiration of between 1250 and 1500 mm. Dew falls are relatively unimportant so far inland and occur mainly during winter rainy periods.

As in most central sectors of the low-latitude deserts, winds tend to be lighter and more variable at Alice Springs than in the regions to north and south (Brookfield, 1970a). The strongest winds are associated with steep pressure gradients on the margins of thermal lows of the inter-tropical convergence zone, particularly in summer. However, even average gradients can produce geomorphologically effective winds over sparsely vegetated desert plains, especially where turbulence is generated over heated surfaces.

Type of arid desert climate

Degree of aridity, cold-season temperatures, and extent of maritime influences distinguish the other main desert climatic types in Table I, in which are listed those climatic characteristics likely to have geomorphological significance. Climate diagrams showing temperature and moisture regimes of typical stations are illustrated in Fig. 3.

TABLE I **Types of arid desert climate**

Warm Deserts	Insignificant Frost Action
extreme-arid continental	*example* Murzuk (Fig. 3) extremely low, variable rainfall without a regular season; extreme moisture deficit, with pan evaporation to 4000 mm and evapotranspiration of 2000 mm; high day temperatures and large diurnal ranges throughout year; variable winds

extreme-arid
coastal

example Cape Juby
extremely low, highly variable rainfall; relative humidity generally above 50 %, significant dewfall to 50 mm, and frequent fog in littoral tract; moderate cloud cover and moderated temperatures with very small diurnal and annual ranges; evaporation as low as 850 mm; strong unidirectional winds throughout year; abundant atmospheric salts

arid continental
central & equatorward
margin

example Alice Springs (Fig. 2)

[semiarid transition to savana with increasingly dependable wet season]

arid continental
mid-latitude margin

example Azizia
very low, moderately variable rainfall with a more dependable and effective winter frontal element and a less dependable summer thunderstorm component; relative humidity 30–40 % and moderately large moisture deficit with pan evaporation 1500–2000 mm; diurnal and annual temperature ranges comparable, and frosts common in winter; moderately to strongly windy, especially in summer
[semiarid transition to Mediterranean climate with increasing winter rainfall and some upland storage of precipitation as snow]

arid-upland

example Carrillo (1102 m)
higher, more dependable rainfall than in adjacent desert lowlands, with moisture deficit reduced by lower temperatures; increased diurnal temperature

	ranges, with night frosts throughout the year; rather windy
Transitional	
low to mid-latitude moderately continental	*example* Mosul very low, highly variable rainfall, and very large moisture deficit; spring or summer rainfall, often of high intensity from local convective storms; high day temperatures in summer, but annual temperature range exceeds diurnal range; long frost season, but significant frost action restricted to higher ground
Cold-Winter Deserts	Significant Frost Action
mid-latitude continental	*example* Turtkul low to very low, highly variable rainfall influenced by topography, mainly in late spring and early summer and locally of high intensity in convective storms; snowfall may be important at higher elevations; cloudier than low-latitude deserts but receiving 60–70% possible insolation; annual temperature ranges to 45°C, far in excess of diurnal ranges, with high day temperatures in summer and one or two winter months with mean temperature below 0°C; night frosts throughout year on higher ground; moderate moisture deficit with evapotranspiration reduced to 500 mm by lower temperatures; strong winds in winter and spring [semiarid transition to steppe]

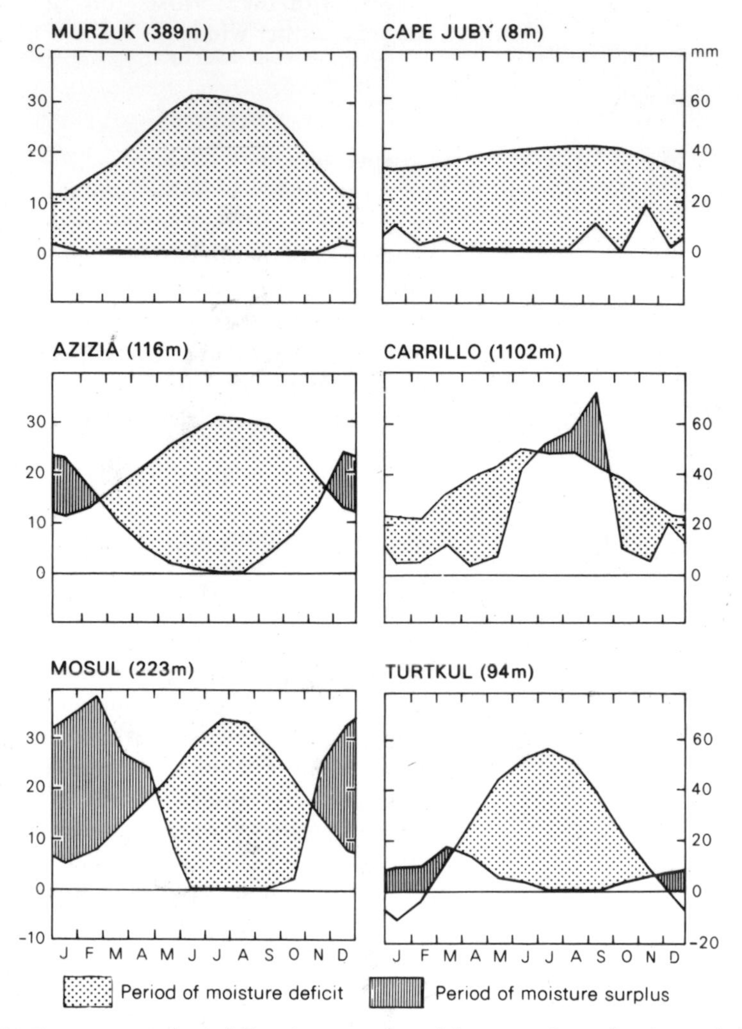

3. Temperate and rainfall regimes at selected desert stations, showing periods of moisture deficit and moisture surplus, from Walter and Lieth, 1960. Location of stations is shown in Fig. 1.

CLIMATIC BASES FOR A GEOMORPHIC SUBDIVISION OF DESERTS

It can be anticipated that aridity and the associated climatic characteristics will profoundly and *directly* affect the operation and balance of geomorphic processes: for instance that the depth and degree of rock weathering will be limited where surfaces are strongly heated under a dry atmosphere, with mechanical breakdown significant; that infrequent but locally intense rains on waste-strewn surfaces will lead to flash runoff heavily charged with sediment; that stream flooding may be torrential in immediate response to rainfall, but short-lived because of high evaporation losses; and that hot dry turbulent air passing over dry waste-strewn surfaces will be particularly effective in deflation or sand drift.

In addition, the desert climate also influences landforming processes *indirectly* through its influence on the vulnerability of desert land surfaces exercised through vegetation and soil.

The geomorphologically significant components of the desert vegetation are the perennial xerophytes. These comprise woody shrubs and small trees, which survive by means of adaptations reducing transpiration, increasing osmosis, or enabling water to be stored, and perennial grasses or bulbous or rhyzomaceous plants that die back during dry periods and shoot again from the base during the next rains.

These adaptations generally reduce the effectiveness of the plants as ground protection. Stunted forms and narrow leaves intercept little of the rainfall; the extensive root systems that are necessary leave the canopy sparse, with ground cover commonly as low as 10 per cent, and the open spacing of the plants offers little hindrance to runoff and sheet erosion; the lack of shallow roots leaves the near-surface soil unbound and readily washed away.

A cover of desert shrubs of the order of only 30 per cent can protect the surface against wind erosion, and an even sparser shrub cover may trap sediment, particularly drifting sand which mounds up about the plant roots. However, it is this component of the desert vegetation that has suffered most from man and his grazing animals, particularly in the Old World.

The vulnerability of scantily-vegetated surfaces is increased by the structural weakness of many desert soils. A small contribution of plant matter from a low biomass is rapidly mineralised under high temperatures and dry oxidising conditions; hence there is little litter or organic horizon to protect a surface which tends to crust under raindrop impact and to respond rapidly in runoff, with heightened erosional effects. A soil structure weakened by lack of humus may further deteriorate with the concentration of salts where leaching is restricted by aridity. Alkaline soils with thin brittle topsoils above contrasting clay subsoils are particularly liable to sheet erosion; the puffy surfaces of salt-impregnated soils are subject to deflation; and the flocculation of fine particles in saline environments robs clay soils of cohesion and increases their erodibility. Marked expansion and contraction of saline fine-textured soils, results in cracking and surface weakening, and similar effects are produced in soils with expansive clays such as montmorillonite, which form under restricted desert weathering. The widespread calcareous soils also tend to be powdery and readily deflated.

These general relationships suggest that a climatic division of the arid deserts could also serve as a basis for systematic geomorphic treatment. The cold-winter deserts should be differentiated by higher yields of rock waste from uplands subject to frost action, by more prominent inheritance of glacial or periglacial forms in the piedmonts, and by continuing deflation of comminuted sediments under the prevailing cold dry conditions. Among the warm deserts, the coastal deserts stand apart by virtue of their high atmospheric humidity, abundant oceanic salts, and moderated temperature ranges, giving distinctive weathering regimes, and also through their strong unidirectional winds. Arid and extreme-arid types of the continental warm deserts should be contrasted by a greater frequency and range of water action in the former and by more mobile sand surfaces due to lack of vegetation in the latter. Finally, the warm upland deserts should also form a distinctive geomorphic type in consequence of the vertical zonation of climate, with moderated aridity and significant frost action at higher levels, and more abundant erosional and depositional evidence of Quaternary cold periods.

Despite their geomorphological significance, however, these differences alone do not constitute an adequate framework for the description of desert landforms. In the main they are differences of degree only and are gradational rather than sharply limited. More important, however, within broad limits set by climate, the regional compartmentation of landforms and the spheres of effective action of the formative processes are determined by non-climatic factors, particularly by geologic structure and relief. For example the fate of runoff locally and hence its geomorphological role is determined by slope and by surface permeability, whilst the location of a dunefield may be determined by lithological controls of sand source rather than by degree of aridity or wind regime.

The structure and patterns of desert vegetation and local contrasts in soil stability are also strongly influenced by non-climatic controls; particularly by topography and the texture of surface deposits as determinants of the movement and storage of surface water. In the desert of central Australia, for example, hillslopes are devoid of soil, but rocky surfaces allow deeper infiltration of water and support a sparse cover of mixed trees and shrubs; piedmont lowlands receiving run-on characteristically have medium-textured soils which here carry open tall shrubland; flats subject to occasional flooding have clay soils supporting perennial tussock grasses; salt-affected areas near drainage terminals bear halophytic shrub-steppe; whilst wind-sorted sands in areas remote from stream flooding support xerophytic hummock grasses which effectively stabilise these surfaces.

For these reasons a geomorphologically significant classification of desert environments must include consideration of geologic structure and relief as well as climatic divisions.

MORPHO-STRUCTURAL TYPES OF DESERT

The world deserts can be grouped into two broad types on the basis of major geologic structure and resulting gross relief, namely the deserts of the shields and platforms and the mountain-and-basin deserts of tectonically more active zones, particularly the young folded belts (Dresch, 1964).

Shield and platform deserts

These mainly comprise the Gondwana fragments of Africa, Arabia, Australia and India. Typical landforms include broad plains on granitic rocks of the exposed shields, as in Western Australia (Pl. 2), and tablelands and basin lowlands on subhorizontal platform strata, as over much of the Sahara. Bolder structural relief is restricted to ancient orogenic belts and areas of younger igneous activity, for instance the volcanic massifs of the central Sahara, and to rift zones as in Sinai and east Africa. These are of limited extent, however, and rarely is relief sufficient

2. Shield desert in Western Australia, with a relict lateritic surface and younger granitic plains and tors below the breakaway. Photo N. H. Speck, CSIRO, Division of Land Use Research.

to moderate aridity or to introduce forms due to current frost action.

Similarity of relief in these deserts stems from a common long and stable tectonic and geomorphic history, under which ancient landforms have survived extensively. Extensive planation of the shields under savana regimes in the earlier Cainozoic, as indicated by remnants of laterite, was followed by stripping of weathered layers from higher parts and the extension of depositional surfaces on lower ground, associated with a disorganisation of drainage with progressive increase in aridity. On this trend were superimposed the effects of Quaternary wetter and drier phases, the former associated with increased fluvial activity in the piedmonts and the latter with the extension of sand dunes in the remote lowlands.

Gently sloping erosional plains are well developed on the shields, for the island hills yield only small amounts of sandy detritus which are swept beyond the piedmont plains into the lowlands. The structural tablelands of the platforms are commonly reinforced by resistant weathering crusts. The lower parts, beyond the limits of active drainage, are occupied by aeolian sands which cover between a quarter and a half of the shield and platform deserts, as sandplain or dunefields.

Mountain-and-basin deserts

These deserts are dominated by the juxtaposition of mountain and plain as a result of earth movements, commonly with sufficient relief to generate climatic contrasts (Pl. 3). Seasonal floods from the uplands, locally fed by snowfields or glaciers, carry abundant debris from steep slopes in fractured rock, its breakdown commonly accelerated by frost. At the sharp junction of mountain and plain, often tectonically determined, coarse alluvium is deposited as gravel fans at valley outlets. Further downslope are plains of finer alluvium, partly saline, and playa lakes. The basin sediments may attain great thicknesses where subsidence of the floor has continued.

These tectonic and relief components also tend to be surface

3. Mountain-and-basin desert, Death Valley, California, showing fans and playa within the mountain rim. Photo W. B. Hamilton.

and groundwater hydrologic units; the playas form evaporative terminals and salts concentrate in the lowest parts of the systems. Wind-blown sands are extensive only where relief opens out and the plains extend beyond the limits of streamfloods, as in the Takla Makan basin of central Asia.

The shield and platform deserts largely correspond with the climatic types of extreme-arid and arid continental warm deserts, and the homogeneity of their relief is emphasised by uniformity of climate. The mountain-and-basin deserts include the cold-winter mid-latitude deserts, but also extend into low latitudes along the Pacific mountain belt of the Americas. Differences of structure and relief and the associated climatic contrasts give greater regional variety in this desert type. For instance the Andean deserts include the extreme-arid coastal deserts of Peru and northern Chile, the montane deserts of the Pacific slope leading to the *paramos* of the Bolivian *altiplano,* the tropical semiarid Argentinian flank, and the cold Patagonian Desert in the rainshadow of the southern Andes.

The North American deserts have a more broken and compartmented relief, including the fault-block topography of the Great Basin, the intermont Colorado plateaux, and the high plains east of the Cordillera. Relief is less than in the Andes, but is nevertheless generally adequate to introduce orographic contrasts in rainfall, and there are snowfields on the main watersheds. Over much of this region aridity is moderate, being tempered by altitude. The range of arid climates is widened by the longitudinal extent of the deserts, supplemented by the increasingly severe winters towards the interior of the continent.

The Asian deserts include high mountain ranges comparable to the Andes and complexes of lower mountain and basins similar to the North American deserts, although on a broader scale. In the west the climates are transitional to the low-latitude deserts (e.g., Mosul in Fig. 3), but winters become increasingly severe eastwards as latitude, continentality, and altitude increase. The severity of frost shattering is seen in the general mantles of waste on the hillslopes. Their landforms include inheritances from drier past periglacial regimes, such as the loess mantles on the southern foothills of many of the basins. These mountain-girt continental deserts are naturally areas of interior drainage and their lowlands are entirely depositional.

DESERT PHYSIOGRAPHIC SETTINGS

Within the gross morpho-structural classes, each desert region consists of a sequence of physiographic settings from upland to plain. These are made up of characteristic landform groupings as shown in Pl. 4, and constitute spheres of action of related complexes of geomorphological processes. There are erosional and depositional surfaces, wind-dominated and stream-dominated settings, and areas of removal and of surface concentration of salts. The dynamics of the processes within a given setting may be strongly influenced by climate, but the processes themselves and the physical framework within which they operate are mainly determined by relief and fundamentally by geologic structure. Each physiographic setting is functionally related to those upslope and downslope in an overlapping sequence. In mountain-and-basin deserts the sequence of physiographic settings tends to be geographically compressed, whereas in the shield and platform deserts it opens out, and in Australia it has been possible to map desert physiographic settings at a broad scale, as shown in Fig. 4. The major desert physiographic settings provide the bases for the succeeding chapters of this book.

Desert uplands are the sectors of predominant erosion in which bedrock is extensively exposed and where geologic controls of relief are best observed; accordingly, *rock weathering* is conveniently treated in this setting, as are *hillslopes* and *drainage on desert watersheds. Desert piedmonts* tend to be sharply separated from the uplands by break of gradient, but the uplands nevertheless exercise a dominating influence through control of runoff and the supply of detritus: erosional or depositional forms may prevail here, depending on climate, lithology and the vigour of relief. The desert lowlands show a wider range of forms reflecting the influence of materials and climate on progressive changes in the balance of erosion, transport and deposition with distance from the piedmont. *Stony deserts* comprise structural plateaux and stony plains formed at lower levels. These surfaces exhibit many similar features and pose common problems, for example the nature of desert pavements. *Desert rivers and floodplains* constitute the main elements of riverine deserts; the regimes and forms of ephemeral channels and the status of desert drainage systems are appropriately reviewed in this context. At or beyond the drainage terminals are the *desert*

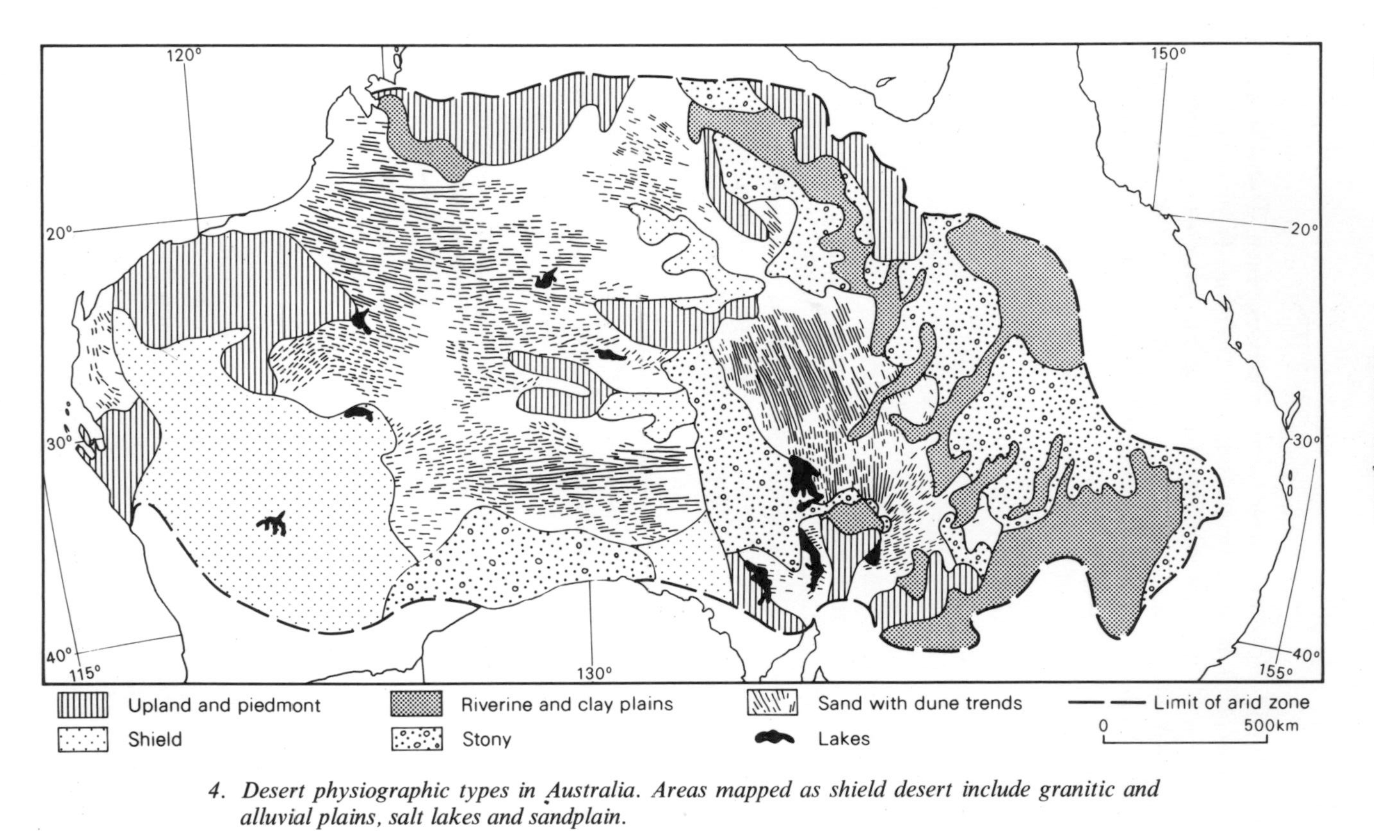

4. Desert physiographic types in Australia. Areas mapped as shield desert include granitic and alluvial plains, salt lakes and sandplain.

lake basins, tectonically sited or isolated by disorganisation of drainage. Their surface forms need to be interpreted in the light of the ephemeral hydrologic regimes and their related salinity and sediment budgets. Forms due to wind action on exposed soft lake sediments are also important here. Beyond the limits of active drainage are the *sand deserts*, which present problems concerning the movement of sand by the wind and the nature and controls of dune form.

4. Stony desert in central Australia, with stony tablelands (hamadas) and plains (regs) at lower levels. Photo by the author, CSIRO, Division of Land Use Research.

II

ROCK WEATHERING IN DESERTS

The sharp outlines and rocky slopes of desert uplands are shared to some degree by mountains fashioned under other climates, for such forms, with their clear expression of underlying geology, occur wherever the removal of waste keeps pace with preparatory rock weathering. What distinguishes desert uplands is the wider range of rock types over which these conditions are fulfilled and the extreme degree of exploitation of lithologic contrasts (Pl. 5). Rectilinear facets, angular breaks of profile and a relatively abrupt change of slope at the hill foot reflect this structural control, which may also be expressed in detail in bizarre weathering forms. Upland deserts with their extensive rock exposures are the appropriate settings for consideration of arid weathering of rock *in situ*. The weathering of bedrock and of rock waste in other desert environments, for example on stony desert plains or in depressions

5. *Exploitation of lithologic contrasts in sandstone and shale terrain, central Australia. Photo R. A. Perry, CSIRO, Division of Land Use Research.*

with saline groundwater, comprises similar processes but with differences of emphasis and effect which will be discussed in the appropriate settings.

MECHANICAL WEATHERING

Insolation weathering

The widespread and apparently little-altered debris on desert hillslopes fostered an impression that rock breakdown was rapid and encouraged an explanation in physical terms judged to be appropriate to a hot desert setting. A source of disruptive stresses which appeared obvious to those who had experienced the power of a desert sun was the expansion and contraction of bare rock on heating and night-time cooling. It was known that ground temperature in summer may exceed screen temperatures by as much as 20°C and that around sunset the ground cools more rapidly than does the air above. It was also recognised that these changes would be concentrated in a near-surface layer because of low thermal conductivity. This so-called *insolation weathering* was judged to be effective in various ways. Rock spalling was considered to be due to the shearing normal to temperature gradients of layers subjected to alternate tension and compression relative to the insulated rock beneath; the fissuring of rocks and boulders normal to the surface was seen as the expression of tensional stresses set up by unequal heating and expansion; granular disintegration was attributed to differential heating of mineral grains of various colours with different coefficients of expansion, as in heterogeneous crystalline rocks.

But some field evidence cast doubt on initial assumptions concerning the general efficacy of *thermoclastis* or rock fracture caused by solar heating. Barton (1916, 1938) noted that formerly polished surfaces on granite monuments were more roughened by weathering in the slightly more humid atmosphere of Lower Egypt than in hotter and drier Upper Egypt, that marked changes were restricted to levels below or close to former ground surfaces and within reach of soil moisture, and that weathering was greater on relatively shaded north sides than on faces directly exposed to the sun. The importance of selective rock weathering in shaded desert sites (*Schattenverwitterung*) had previously been remarked upon by Walther (1900). It is generally the case that rock outcrops, angular debris and the margins of the finest incipient clefts are stained by weathering or impregnated by salts, and that the faces of mineral grains released in rock disintegration are dulled by chemical change. If thermal fracturing does occur it must therefore be infrequent relative to superficial chemical weathering.

Calculation of the stresses in rocks heated by intense solar radiation also suggests that thermoclastis may be uncommon in fresh rock. Maximum surface temperatures on dark rocks are known to attain 78–79°C (Peel, 1974; Fig. 5), and the penetration of the diurnal heat wave under such strong heating may be between 50 and 100 cm depending on thermal conductivity, or about twice that in a dune sand. Temperature gradients of between 0.5 and 1°C per cm have been measured in near-surface rock layers (Roth, 1965; Peel, 1974). Linear dilatation of a granitic rock under these conditions may be as much as 0.025 per cent in a diurnal cycle, although differential expansion within or between near-surface layers at any time will be less than this. Since the compressive and shear strengths of rocks are greater than their tensional strengths in the approximate ratios of 8:3:1, failure is most likely during cooling, and Birot (1968) has considered the extreme case of hot rocks chilled by 30°C in a desert thunder shower. On a granite outcrop the tensional forces at the surface could amount to 10 MPa,* which might conceivably cause fissuring normal to the surface, but the stresses parallel to the surface at the base of the chilled layer would fall well below the shear strength and are most unlikely to cause flaking of fresh rock. However, they could come close to the limit of rupture in fine-grained materials such as flint or chalcedony in which the coefficient of expansion is 50 per cent greater than in granite. The forces involved in granular disintegration of heterogeneous rocks are so complex that it is possible only to generalise on theoretical grounds that the most favourable circumstances are those in which a rock with grains of contrasted size, composition and colour such as a porphyritic biotite-granite is subject to strong and repeated heating to some depth. Birot concluded that the envisaged mechanisms of thermoclastis appear to be marginally capable of fracturing fresh rock only under the most favourable conditions.

These indications have been reinforced by laboratory tests.

*1 Megapascal (M Pa) = 10 kg per sq cm.

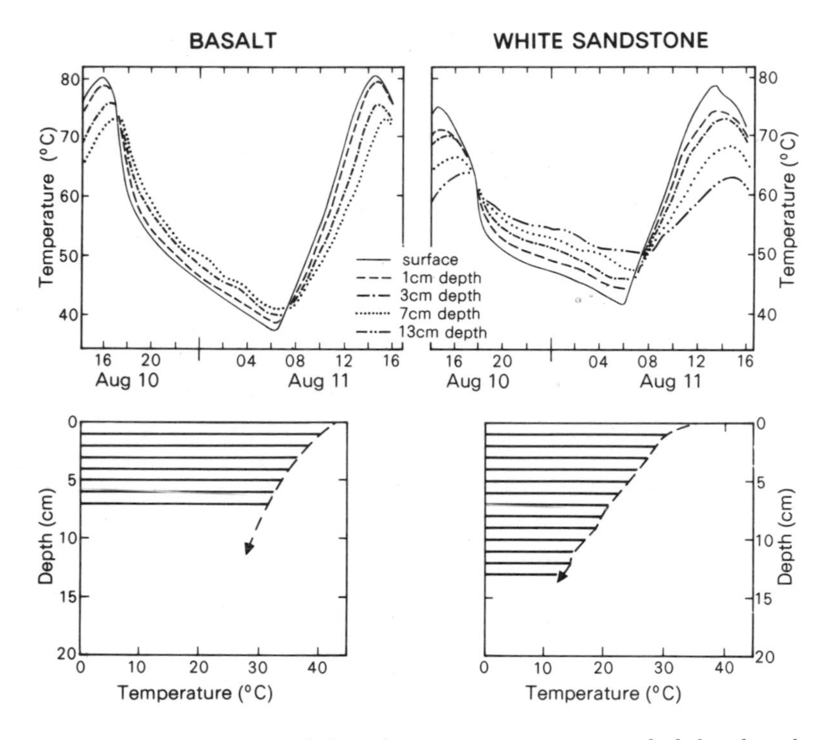

5. Temperature curves and diurnal temperature ranges in a dark basalt and a white sandstone in the Tibesti, August 1961. From Peel, 1974.

Tarr (1915) heated granite samples to 250°C with no sign of physical damage nor loss of compressive strength, and Blackwelder (1933) chilled heated samples of basalt and obsidian through at least 200°C, also without apparent effect. One problem facing such experiments is that of incorporating the time factor and the effect of accumulated stress after thousands of cycles of heating and cooling. Griggs (1936) went some way in this in simulating 250 years of diurnal cycles with a temperature range of 110°C. Under dry conditions he could discern no microscopic effects in a polished sample of fresh granite, and his findings were widely interpreted as precluding natural thermoclastis in fresh rock.

Nevertheless, in apparently unweathered and chemically inert rocks, fractures are found which invite explanation by thermoclastis, notably boulders and cobbles split by radial or multiple transverse fractures described as *Kernsprung* (Walther, 1900) or *diffission* (Hume, 1925). Reiche (1950) has noted that these are commonly dark fine-grained lavas and has suggested that such disruption may have followed the release of remanent cooling stresses, but flint (Walther, 1900) and limestone (Williams, 1970; Pl. 6) are also represented and the phenomenon is typical of many fine-grained materials that naturally resist granular disintegration. Ollier (1963) has attributed the crazing or blocky fracturing of quartzite in central Australia to thermoclastis. Alternative explanations such as lightning strikes and grass fires have been offered, but these forms have continued to keep alive the hypothesis of insolation weathering and have drawn attention to possible shortcomings of the laboratory experiments. For example, heating tests were necessarily on unconfined rock samples free to expand and contract as a whole. More important, the phenomenon of rock fatigue may have been overlooked. Repeated stressing of a rock at a frequency below that of its relaxation period leads to accumulated forces which are relieved by progressive dislocation along microscopic undulating slip planes. Thermoclastis could occur

6. Cleaved limestone boulder in reg pavement, Biskra, Algeria. Photo G. E. Williams.

in a boulder so weakened without the tensional strength of the fresh rock being exceeded (Bertouille, 1972).

Ollier (1965) has described as *dirt cracking* the wedging open of cracks in heat-expanded rocks by fillings of fallen soil or dust, which could lead to an accumulation of stress.

Fracture due to unloading

Rock fracture through the release of confining pressures with the removal of superincumbent rock (Twidale, 1971) occurs independently of climate, but its effects are marked in desert uplands because of the extensive outcrop of rocks which tend to remain physically cohesive to the surface. On a large scale, unloading has contributed to the domal outlines of hills in granitic and similar massive rocks, as in Ayers Rock and Mt Olga in central Australia (Ollier and Tuddenham, 1962), through the sheeting of unweathered slabs several metres thick, beyond the range of significant diurnal heating. In sedimentary strata unloading may result in topographic jointing parallel with the surface, which may in turn induce exfoliation and so influence upland forms (Bradley, 1963). The slow rate of denudation in deserts may allow stresses to persist and find release on a relatively small scale, and some minor fractures as in *Kernsprung* may result in this way. For example Ollier and Tuddenham (1962) describe the subsequent shearing and slicing off, flush with the surface, of conglomeratic boulders left projecting above sheeting planes on Mt Olga in central Australia.

PHYSICO-CHEMICAL WEATHERING

Distinction between physical and chemical weathering is nowhere more artificial than in the desert, where strong surface heating and a dry atmosphere limit the penetration of moisture into the the rock and foster its upward return through capillarity. Under these conditions hydration or more complex chemical alteration is restricted to a relatively thin layer in which superficial enrichment may occur through deposition from ascending weathering solution. The resultant changes commonly involve an increase in volume in a rock layer that remains sufficiently coherent to transmit the attendant physical stresses. Where rock breakdown results, the term *physico-chemical weathering* best describes the complex of contributory processes.

Hydration weathering

When Griggs (1936) switched from dry air cooling to a fine water spray, granite samples that had resisted the equivalent of 250 years of extreme thermal strain crumbled in the equivalent of a further 2.5 years. A cause of this rapid granular disintegration was probably hydration, or increase in moisture content through molecular adsorption: this involves expansion and stresses comparable with those caused by surface heating, which it may effectively reinforce. Hydration is particularly effective along the boundaries of mineral grains, but also penetrates in silicates such as the biotites and feldspars in which the structure and pronounced cleavage favour surface adsorption, and these factors explain the vulnerability of granite to hydration weathering. It may also be facilitated by the strong heating of rock surfaces in deserts, for Birot (1968) found that the porosity of crystalline rock was increased threefold after prolonged heating and cooling, presumably due to microfissures formed along grain boundaries. He demonstrated with the use of dyes that this mainly affected a thin surface layer, favouring the shedding of expanded flakes of hydrated rock. The process has not been reproduced experimentally, but flakes of otherwise unaltered granite shed from Saharan inselbergs have been found to contain significantly more moisture than the rock in place.

The extent to which colloidal silica is vulnerable to hydration weathering is unknown, but it remains a possible explanation of fractured boulders of flint, chalcedony and silcrete on desert pavements.

The expansive effects of hydration become increasingly important where rock material is reduced to colloid dimensions, as in products of incipient weathering in minute fissures in otherwise sound parent rock. One important hygroscopic colloid is ferric hydroxide from the weathering of ferromagnesian minerals such as biotite. Birot (1968) found that a sample of moderately porous sandstone impregnated with ferric hydroxide crumbled after between three and four months of repeated hydration from exposure to alternate humid and dry atmospheres at room temperature. Arid weathering, with restricted moisture and a non-acid environment, favours the development of 2:1 lattice clays such as montmorillonite which undergo marked expansion on hydration as a result of inter-layer adsorption (Reiche, 1950), and where soil or dust fillings in cracks contain such expansive clays they may contribute to further breakdown on wetting and drying.

Salt weathering

Rock surfaces in deserts are commonly impregnated with soluble salts as void fillings or as efflorescences. These may be oceanic (cyclic) or from groundwater in nearby playas, products of rock-weathering, or connate salt in sedimentary rocks of marine origin, and they include the mobile chlorides, nitrates, sulphates and carbonates of weatherable sodium, calcium, magnesium and potassium. The disintegration of surface rock by salt action in deserts (*Salzsprengung*) was first given publicity by Walther (1891). Two main processes are involved, namely crystal growth and the expansion of hygroscopic salts on hydration.

The effects of crystal growth had come to be discounted since they involve crystallisation from supersaturated solutions (cf. Mortensen, 1933; Wilhelmy, 1958). However, Evans (1969/70) has recently cited experimental and theoretical evidence that rocks may be shattered by the initial process of crystallisation of interstitial salts such as sodium chloride and gypsum. Crystal growth depends on the maintenance of a film of solution at the salt-rock interface by surface tension.

> Continued crystallisation from this film is possible only if the solute diffuses inward, which requires that the effective normal stress decreases inward (toward the centre of the area of 'contact'). Since the solubility of calcite, quartz and probably other common minerals increases with pressure, crystallisation can occur against pressure only in a supersaturated solution. The force against which crystallisation can occur increases with the degree of supersaturation. With 1% supersaturation, calcite might crystallise against 10 bars pressure (Evans, 1969/70: 172).

Evans noted that this is of the same order as the tensile strength of rocks, which is commonly between 2 and 20 MPa. Crystals have in fact been known to grow against pressures of up to 5 MPa, and higher pressures are theoretically possible (Goudie, Cooke and Evans, 1970). Supersaturation is favoured under desert conditions where surface heating and drying winds cause excessive evaporation, and crystallisation forces are therefore likely to reinforce thermal stresses. The forces may be cumulative with repeated solution and recrystallisation under high supersaturation, and since salts crystallise preferentially around nuclei, an initial crystal once formed tends to enlarge, so localising and intensifying the disruptive effect.

Disintegration of rocks through the hydration of impregnating salts was particularly stressed by Mortensen (1933). Crystallisation generally occurs in the anhydrous form under high temperatures and low humidity, in cracks and voids at the rock surface, and moisture is later adsorbed, particularly after rain or dewfall. The process is repetitive; as humidity drops the salts again become partially dehydrated, leaving additional voids which are in turn filled by anhydrous salts from solution, and the whole is again hydrated. The disruptive forces depend in the first instance on the hydration pressure of the salts, which can exceed 100 MPa; as the rock yields, the important factor is expansion on hydration, which ranges between 25 and 300 per cent. Crystals have been observed to hydrate against 6.3 MPa (Goudie, Cooke and Evans, 1970).

For salt-weathering in general, Goudie, Cooke and Evans (1970) and Goudie (1974) have demonstrated experimentally the great efficacy of Na_2SO_4 and $MgSO_4$ and of $CaCl_2$ in the weathering of sandstone, and the increased importance of Na_2CO_3 and $NaNO_3$ and the reduced significance of $MgSO_4$ in the weathering of chalk. $CaSO_4$ (gypsum) is unimportant owing to its low solubility. Evans (1969/70) has concluded that hydration is probably most important in the case of gypsum, which has a coefficient of expansion on hydration five times that of calcite. With chlorides, crystal growth is probably more important than hydration and this form of weathering is a feature of coastal deserts (as of non-arid coastal environments) and apparently also of the Egyptian deserts, where chlorides predominate. Salt-weathering is accentuated in combination with freezing, which stimulates crystallisation, and is therefore marked in polar deserts such as Antarctica (Wellman and Wilson, 1965).

Salt-weathering is most effective in porous rocks with high water absorption capacity and with low cohesion, as in weakly-cemented sandstones. Crystalline rocks are subject to granular parting along grain boundaries. Surface flaking or *exsudation* (Jutson, 1918) occurs where crystallisation takes place preferentially at a shallow depth. Jutson (1934) attributed great importance to both forms of breakdown in the hollowing and retreat of breakaways in disaggregated pallid granite in Western Australia. The processes are most effective where repeated crystallisation or hydration are favoured, as in seepage zones and other sites with lingering humidity, and may result in basal sapping and in various forms of cavernous weathering because of the susceptibility of shaded surfaces to attack. Rocks embedded in horizons of salt-enrichment in desert soils are naturally subject to intense salt-weathering. Sections in alluvial regs in the Negev, Israel, contain innumerable fractured flint and limestone boulders, identical with *Kernsprung*, at levels of gypsum and lime accumulation at 20–50 cm depth (Yaalon, 1970). The gravels below are virtually unweathered, and the fractured boulders are less common above the gypseous horizon. Seepage zones of saline groundwater are another environment of marked salt-weathering, and Beaumont (1968) has described the disintegration of gravels on lower parts of alluvial fans in the Great Kavir through this mechanism, whilst those upslope remain unaffected.

Cooke and Smalley (1968) have proposed an additional effect in salt-weathering, namely the strong heating of dried-out rock surfaces impregnated by salt crystals. Since the salts have higher coefficients of thermal expansion than most rocks, in some cases by a factor of three or more, differential expansion may extend cracks initiated by the process of crystallisation. Subsequent experiments, however, suggest that this process may be relatively ineffective (Goudie, 1974).

Since periods of alternate wetting and thorough drying are necessary for crystallisation, salt-weathering is relatively important in hot moderate deserts where moisture is regularly forthcoming; it is apparently less effective in the most arid inland deserts such as northern Chile (Mortensen, 1933). It reaches its maximum efficacy in the coastal deserts, where high salinity and atmospheric humidity are favourably combined.

CHEMICAL WEATHERING

Although the reactions comprised in the chemical weathering of rocks in deserts may not differ fundamentally from those under moister climates, the arid environment imposes constraints which radically alter the relative effectiveness and range of the chemical changes, with important consequences for rates and patterns of weathering. The chief such environmental limitation is a lack of water in the weathering zone, its shallow penetration and capillary return through excessive evaporation, and hence its short-lived contact with the weathering front. Another is the reduced aggressiveness of weathering solutions that tend to be neutral to alkaline in reaction, with little reinforcement by organic acids and, in warm deserts at least, a lack of CO_2. As a result, the chemical weathering of desert rock surfaces is slowed and tends to be shallow and partial only.

These limitations apply particularly to the more complex reactions such as hydrolysis involved in advanced clay-weathering of the common silicate minerals. The rocks tend to crumble at the surface before decomposing and the resulting waste is generally coarse-textured, with weatherable minerals such as the feldspars, and lacking in clay. Hydrolysis and hydration mainly affect iron and manganese which in a warm dry oxidising environment are left extensively as reddish or dark stains and varnishes on rock surfaces.

Solution of susceptible rocks tends to be localised, for example in the tiny vermicular groovings developed on limestone boulders in desert pavements. In the absence of leaching the solutes accumulate in place through evaporation. Where they impregnate the weathering rock they facilitate various forms of mechanical breakdown as described earlier, and in waste mantles they may form cements in crusts or subsurface pans.

The alkaline weathering environments allow the persistence and production of swelling clay minerals such as montmorillonite, with important consequences for soil heaving and associated microtopography and patterned ground to be discussed in Chapter VI. They may also favour the mobilisation of silica, as an acid in solution or in colloidal form, and its eventual concentration in siliceous crusts.

Weathering crusts

With upward capillary movement and evaporation of weathering solutions, salts are redeposited at desert rock surfaces. They include those soluble compounds which in a humid region would be leached from the system, and the role of such salts in physico-chemical weathering has been discussed above. Their presence can also foster further chemical weathering by retaining moisture at the atmosphere-rock contact through deliquescence and by facilitating exchange-reactions. However, they remain subject to repeated solution and mobilisation and to eventual removal with the weakened rock.

Other surface impregnations harden and protect the rock face, notably encrustation with oxides of manganese and iron (German: *Schutzrinde*). These salts move in colloidal suspension or by ionic diffusion through films of surface water and are concentrated near the surface as hydroxide gels which are relatively stable under strongly oxidising desert conditions and in the relative absence of organic acids. Birot (1968) has shown experimentally how in a dry atmosphere and at temperatures above 40°C a surface enrichment of iron hydroxide is transformed into a hard red crust which grows into the rock surface. On outcrops and large boulders, where the source is the rock itself, the amount of encrustation varies with rock type and is greatest on compact rocks rich in ferromagnesian minerals. It is commonly marked along joints and bedding planes. Tricart and Cailleux (1969) have claimed that the thickness of such crusts increases with rainfall, and state that crusts on sandstones on the savana boundary of the southern Sahara, with 500 mm annual rainfall, are between 5 and 10 mm thick, that they range between 0.2 and 0.4 mm in the southern desert margin with 200 mm rainfall, and that they are thinner still in the Sahara proper. The reddish coatings are normally associated with a deeper impregnation of the rock by oxides of iron and manganese and by silica, forming a hardened layer a centimetre or two thick. Beneath this the rock is commonly bleached and porous to comparable depth through loss of the minerals concentrated in the indurated outer horizon. This phenomenon of *case hardening* is most noticeable in sandstone and in coarse-grained igneous rocks liable to leaching and interstitial recementation.

Cobbles resting on the desert soil are subject to stronger varnishing where the oxidised coatings originate externally, from soil solutions. These desert varnishes are characteristic of stony desert pavements and are discussed in Chapter VI.

PATTERNS OF DESERT WEATHERING

Inasmuch as the effectiveness of the different weathering processes is modified under aridity, the relative resistance of certain rock types in deserts differs from that in humid regions. Physical strength rather than chemical stability determines resistance to denudation and in consequence massive crystalline rocks, particularly the basic varieties, tend to be more prominent in desert relief than in humid areas. Limestone and calcareous sandstone, normally subject to solution, are also more resistant in a dry environment.

Most hillslopes in the warm deserts are an alternation of bare rock and bedrock surfaces thinly covered with rock debris. In the absence of frost-shattering, constructional slopes of debris accumulation are poorly developed. On the salient outcrops, rocks that are physically resistant by virtue of hardness, compactness, massiveness, cementation or crystalline cohesion remain intact to the surface and preserve the details of bedding, jointing and lithology. The debris cover is in general angular and coarse and lacks the fine matrix of secondary alteration products, indicating that initial weathering is by *disintegration* rather than by advanced chemical decomposition of the rock in place. Breakdown takes various forms depending on the nature of the rock, including the detachment of joint blocks or bedding slabs, the shedding from rounded massive faces of shells ranging in dimension from thick plates to flakes, the fissuring of compact blocks or boulders, and the smoothing of surfaces by the release of mineral grains through disintegration.

The clarity of surface expression of small differences in bedrock owes much to the slow rate and to the selectiveness of weathering in deserts.

Rates of weathering in deserts

The predictable slowness of desert weathering because of deficient moisture was early demonstrated by the good state of preservation of historic monuments. Barton (1916) calculated that in Lower Egypt the average weathering of dressed granite faces between 4000 and 5000 years old had been between 0.5 and 1.0 mm per 1000 y on shaded sides and less on southern faces. It was negligible in the drier air of Upper Egypt. Weathering was five to ten times greater on surfaces subsequently covered by moist and presumably saline soil.

Polished crystalline rocks are unnaturally resistant to weathering, and higher rates are reported for fissile, weakly cemented or porous rocks. Emery (1960) estimated an annual surface loss of 0.2 mm over the last 1000 years on soft limestones on the Great Pyramid. The rates predictably increase with rainfall, and Mortensen (1956) and Schumm and Chorley (1966) report annual recessions of between 0.2

and 0.6 mm on hard sandstone cliffs of the Grand Canyon, Colorado under a semiarid regime (250 mm annual rainfall). Weathering is probably considerably more rapid in coastal deserts, and Yair and Gerson (1974) have recently reported recession rates of between 100 and 400 mm per 1000 y in sandstones on coastal terraces of the Sinai Peninsula, and between 100 and 200 mm on granite.

On available evidence the arid zones are revealed as areas of slow rock weathering, as is suggested by the following comparative figures for average recession rates of hard rock faces (Young, 1972):

	Years required for retreat of 1 m
Humid tropical	10^{2-3}
Humid temperate or subtropical	10^{3}
Semiarid	10^{3-4}
Arid	10^{7}

Selectivity of desert weathering

With slow rock breakdown, slope waste tends not to accumulate and much bedrock is left exposed to weathering which is markedly differential because of sensitivity to site conditions. Joints and other planes of weakness which allow the deeper penetration of moisture are exploited, as are recesses which collect water and waste and tend to stay moist, being shaded; salients, on the other hand, remain little weathered and tend to increase in relative relief because they are naturally water-shedding. Waste-mantled surfaces are more prone to weathering than bare rock, but the mantles are commonly loose and shallow and do not store moisture well, and on steep slopes they are likely to be removed by erosion before advanced weathering can occur within or beneath them.

Basal steepening. The desert hillfoot is often a zone of waste accumulation and influent seepage, and weathering of the underlying bedrock may be accentuated accordingly. Selective erosion at the piedmont junction has commonly been attributed to such localised weathering (cf. Mabbutt, 1965a). It is particularly important on crystalline rocks which are physically resistant in the dry atmosphere above ground but are subject to chemical weathering where mantled by moisture-storing waste, and the steepened or flared margins and piedmont angles of desert granitic domes may be explained by differential etching of rock weathered at and below ground level (Twidale, 1962; Mabbutt, 1966a). On a smaller scale it produces miniature pediments or ground-level platforms around joint blocks (Ollier, 1965).

Basal sapping. Concentrated weathering near the base of rock faces reflects the availability and capillary range of soil moisture near ground level and the location of seepage zones. Upper surfaces and higher faces, on the other hand, remain relatively dry and free from attack, often with well-developed hardened crusts. This situation can lead to basal sapping of rock walls, particularly where they are partially shaded, as has been invoked to explain widening of the now fossil wadi canyons of the Tibesti plateaux (Peel, 1941).

In a similar way, outliers may be shaped into *mushroom rocks,* often with a pronounced basal overhang as much as a metre or so above ground level. It was perhaps inevitable that these, like the steepened bases of desert cliffs, should initially have been attributed to undercutting by wind and sand blast, but Walther (1900) early recognised that many were the result of differential weathering, and illustrated mushroom rocks (*Pilzenfelsen*) with overhanging canopies formed by hard weathering crusts. They were cited as an example of shadow weathering, with wind action confined to the deflation of the fine weathering products. Bryan (1925a, 1927) demonstrated conclusively that although wind was locally operative, undermining could be severe on lee faces, and demonstrated the effects of differential weathering, rainwash and sapping. He noted that mushroom rocks formed on sandstone blocks in which the bedding had remained horizontal and in which marginal undermining could occur in softer layers beneath weathering crusts, whereas *pedestal rocks* occurred where bedding was vertical and where the face of the rock retreated evenly under weathering attack.

Cavernous weathering. The characteristic combination of hardened crust and weakened underlayer, and the critical importance of shade and shelter for weathering under aridity, account for the widespread occurrence of cavernous weathering in deserts (Blackwelder, 1929). It is particularly important on coarse-grained igneous and sedimentary rocks that are prone to surface encrustation and granular disintegration. The hollows, termed *niches* by Bryan (1922) but more generally known as *tafoni,* may be initiated at any point where a crust has been breached by the weathering out of a large grain; however, their further development is favoured on shaded faces or undersides of cliffs or joint blocks. Any such protected recess maintains a relatively humid and equable micro-climate protected from desiccation by wind and extreme heating; within it hydration and related weathering can proceed through granular disintegration, possibly with some small-scale desquamation. Growth of the hollow tends to be upwards and inwards beneath the hard crust, away from the entrance and the desert climate outside, and the hardened outer layer tends to form a canopy-like overhang and side walls, and sometimes a lower lip. The fine products of weathering are apparently swept clear by wind, rain or wash. Spherical or ellipsoidal hollows several metres in dimension may develop in this way, especially in massive rocks such as granite (Pl. 7), and large joint blocks may be completely hollowed out by such 'inside-out weathering' (cf. German *Hohlblockbildung*). The growth and coalescence of tafoni may eventually detach

7. Granite tafoni, central Australia. The joint blocks with hardened crusts may have been corestones in a weathered mantle. Photo by K. Fitchett, CSIRO, Division of Land Use Research.

parts of the carapace, leaving only isolated plates. Crusting may then resume on the newly exposed surfaces, or the block may collapse entirely.

Cavernous weathering tends to be influenced by rock structures at various scales, through initial control of encrustation and through variations in resistance to weathering. On Ayers Rock in central Australia for example, cavernous weathering tends to be shallower but more extensive on the almost vertical bedding planes as on the northeast flank; in contrast, a line of deep cylindrical caves with prominent overhangs has formed near the base of the southern face where a joint zone of preferred weathering runs at right angles to the bedding, into the mass (Ollier and Tuddenham, 1962). The siting of the innumerable weathering caves on nearby Mt Olga has been controlled by the gently-dipping bedding planes.

Patterns of smaller pitting known as *honeycomb weathering*, characteristic of sandstone faces, exemplify a finer scale of struc-

8. Honeycomb weathering of feldspathic sandstone with an indurated crust about 5 cm thick, central Australia.

tural control (Pl. 8). As the small tafoni develop they etch out a network of hardened walls with a spacing of a few centimetres, and these become increasingly indurated as they are narrowed in relief. The outer surfaces of the networks represent the initial rock surface with its incomplete or pitted encrustation. Commonly the honeycombing is regular and reflects control by and cementation of minor bedding and jointing in the parent rock. Honeycomb weathering is generally shallow; it exploits the softened subsurface layer and can eventually cause the detachment of the perforated crust.

Tafoni are not characteristic of extreme deserts because severe aridity does not favour hydration and the other modes of weathering involved, nor the formation of hardened crusts. They are pronounced in coastal deserts and in similar locations where they are reinforced by salt-weathering (Schattner, 1961).

9. 'Gnamma hole' in granite, central Australia. The dome shows thick detached plates and thin exfoliating layers of weathered granite. Rock waste and vegetation aid biochemical weathering in the joint cleft. Photo by the author, CSIRO, Division of Land Use Research.

Weathering pits. A form of selective weathering on flattish rock surfaces is the weathering pit, for any small depression that will hold rainwater tends to become enlarged by hydration and related weathering on periodic filling. *Gnammas* on granite hills in arid Australia have originated this way (Twidale and Corbin, 1963). The commonest type, where the granite is indurated or has a pronounced laminar structure, is a shallow pan up to a metre or so in diameter and 15–30 cm deep, with a flat bottom and an overhanging rim (Pl. 9). Others include deeper, circular pits in more homogeneous granite, and canoe-shaped forms elongated along joints.

III

DESERT HILLSLOPES

Certain general features of hillslopes in warm deserts indicate the operation of common controls under the influence of climate.

On a broad scale, there is a striking lack of slopes of intermediate angle (Pl. 10). The hills rise abruptly, island-like, from plains of gentle declivity, a characteristic which suggested to early investigators that the deserts were abandoned seabeds, or which led to erroneous views of the hills as projections through depositional surfaces in landscapes drowned in their own detritus. This relative absence of moderate hillslopes is confirmed by sampling of slope angles and is shown for example in Figs. 6 and 7 and in the bipolarity of slope values in Fig. 8.

Reasons for the gentleness of the footslopes and their abrupt piedmont junctions with the hills will be discussed in Chapter V. The relatively steep desert hillslopes comprise two main elements, namely the steepest, rugged cliffy segments and straighter debris-mantled segments. The prominence of the former reflects the wide extent of rock outcrop in upland deserts due to the slow ingress of

10. Western Macdonnell Ranges, central Australia, with a characteristic lack of slopes of intermediate angle. Photo by the author, CSIRO, Division of Land Use Research.

desert weathering and the relative ease of removal of its finer products across little-vegetated slopes; even relatively weak rocks which would be soil-covered under a moister climate here stand out in relief. The outcropping rocks stand at steep angles limited only by structural strength, which is greatest in deserts, since the dry conditions leave the rock relatively sound to the surface.

The debris-mantled segments are generally only thinly covered; constructional debris slopes or screes are mainly limited to the cold-winter deserts subject to frost shattering. This again indicates the ready removal from the hillslope of the products of slow weathering, particularly by wash. Relationships between debris size and limiting angles on the relatively steep mantled segments are discussed below; what is of concern here is the insignificant degree of slackening of such slopes towards the foot. This partly reflects the complete removal from the hill of those secondary weathering products which under conditions of more advanced rock breakdown and closer vegetation might be formed by sheetwash into concave colluvial aprons; partly it indicates the relative unimportance of slow mass movement such as creep, owing to the thinness of the waste covers, the dry environment, and the lack of clay from secondary weathering required to absorb and store moisture and lubricate flowage.

In closer detail the profiles of desert hillslopes exhibit considerable irregularity, in which ledges and faces alternate with

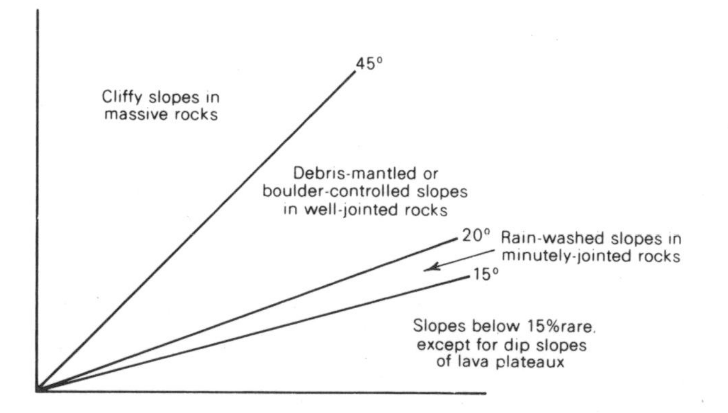

6. Slope elements on crystalline rocks in southwest Arizona. From Bryan, 1922.

short debris-covered sectors. These contrasts are exploited and enhanced by selective desert weathering as discussed in the previous chapter. They result from the paucity of the supply of debris in relation to potential rates of removal, and the related absence of those slow mass movements which under wetter regimes contribute to the overall smoothing and rounding of profiles.

The balance of desert hillslope processes, under which removal waits upon slow weathering, was perhaps first recognised by Gilbert (1877) and has been aptly termed *weathering-limited.* It favours the expression of structure in slope form at all scales, and this has been allowed for here in treating desert hillslopes and the related processes under three geologic settings: on crystalline rocks, on escarpments of layered rocks of differing hardness, and on weak rocks alone.

HILLSLOPES IN CRYSTALLINE ROCKS

Boulder-controlled slopes or slope-controlled boulders?

In an early study, Bryan (1922) distinguished three main hillslope elements on igneous rocks in southwestern Arizona (Fig. 6). Dominant was the *debris-mantled slope*, typically of granite (Pl. 11) and with straight profiles estimated by Bryan to be mainly between 20° and 35°, although ranging to 45°. Above that grade was the bare *cliffy slope*, more commonly on volcanic rock in the area studied, and below 20° was the relatively uncommon *rainwashed slope*. Hillslopes gentler than 15° were observed to be rare, consistent with the marked separation of upland and plain in such settings.

Bryan commented that the debris mantles were not significantly derived from cliffy slopes above, which were minor and often absent, but were mainly joint blocks that had weathered out *in situ*. These 'woolsack boulders' were mainly between 20 and 300 cm in diameter, as initially determined by joint spacing, and formed a layer only one or two boulders deep. They appeared not to be undergoing mass movement because of the absence of boulder accumulation at the foot of the slopes, and the main export of waste was by wash transport of intervening sand and grit supplied from granular disintegration of the boulders. Bryan noted that the boulders were weathered and suggested that any that might slide or roll to the foot of the slope after undermining would be likely to disintegrate in the process. It was held to be fundamental to this slope regime that granite tends to break down directly from the

11. Granite hillslope, central Australia, straight overall but in detail an alternation of bare rock and joint blocks, many in situ. *Abrupt piedmont angle. Photo by the author, CSIRO, Division of Land Use Research.*

stable slope boulders to transportable sand and grit, without intermediate ranges. In contrast, the few rainwashed slopes were on closely jointed rocks yielding a range of medium to fine slope debris, all of which could be directly removed by wash processes.

In his explanation of the straight debris-mantled slopes, Bryan adopted an earlier conclusion by Lawson (1915) that slope angle varies with the average size of superincumbent boulders since it is adjusted to their angle of repose, and described the relationship as *boulder-controlled*. He considered that control was exercised through the protective role of the debris mantle. Local oversteepening would result in removal of boulders and cause increased weathering and wash on the exposed bedrock, tending to diminish that part of the slope and renew the boulder cover; conversely, an extra accumulation of debris on a slackened gradient would protect the underlying rock until recession of the sector below had restored the angle of repose. Inasmuch as boulder size was controlled by joint spacing, the angle of a debris-mantled slope should remain unchanged during its retreat, independent of relief, and the straightness and narrow range of gradients of the hillslopes studied was a measure of the uniformity of joint spacing in the granite.

Later surveys in the same area (Melton, 1965a) have shown that the debris-mantled granite slopes are neither so straight nor so uniformly steep as Bryan had considered (Fig. 7a), and that the relationship between the size of the superincumbent blocks and the slope angle is less close than might be expected from a dependent relationship (cf. Fig. 8). It was also evident that the debris cover,

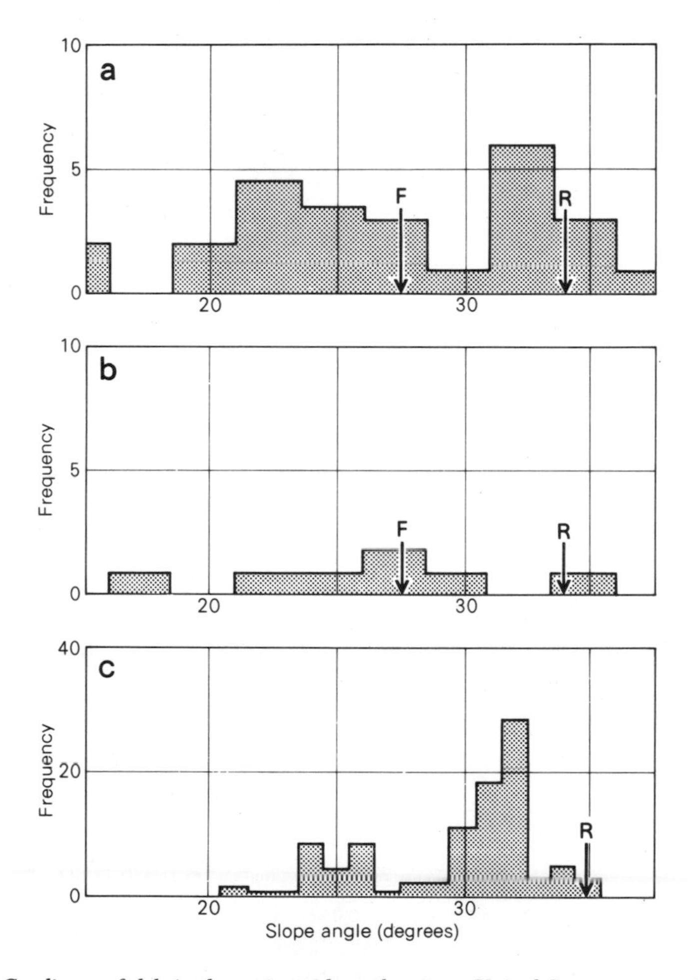

7. Gradients of debris slopes in arid southwestern United States. a. granites in south Arizona; b. volcanics in south Arizona; c. mainly volcanics in New Mexico. From Carson and Kirkby, 1972.

which ranged from stable to unstable, did not protect the bedrock completely. These observations indicate that the bedrock slope may in fact be set by the usual factors of rock resistance, including joint spacing, and the vigour of erosion at the foot. The approximate uniformity of the slopes probably reflects the homogeneity of the parent granite and the ineffectiveness under desert conditions of processes tending to modify the slope angle. Hence boulder size and slope angle may be associated through a common control. This could arise from removal by wash of smaller fragments to an approximate limit set by slope angle, leaving original, joint-determined blocks prominent.

It is possible that slope angle may control boulder size, rather than the reverse. This is suggested by the bimodality of slope values in Fig. 7a, with the possibility of two limiting angles. The upper limit of debris-mantled slopes at about 36°, below that claimed by Bryan, approximates to the angle of repose of granite blocks on

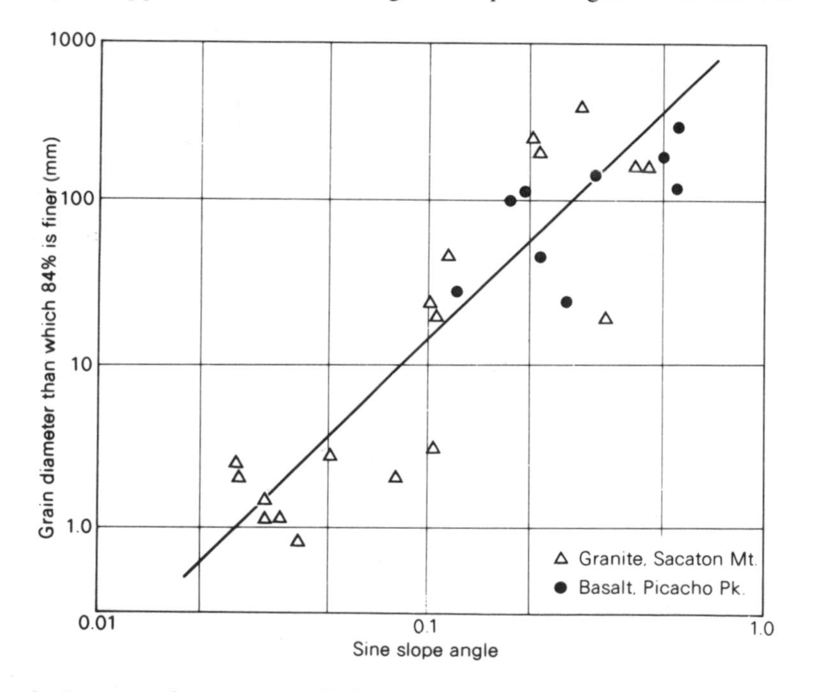

8. Increase of grain size with slope angle in southeast Arizona; points to upper R denote hillslopes, those to lower L are on pediments. From Carson and Kirkby, 1972.

granite (R in Fig. 7a), whereas the angle of sliding friction (28.5°) at which blocks come to rest (F in Fig. 7a) separates the steeper, surficially unstable debris slopes from the stable class.

The mechanism of boulder control proposed by Bryan is in any case suspect since, as discussed earlier, granite in an arid setting should weather at least as rapidly beneath a debris mantle as where it is exposed. It is possible in some sites that the slope blocks may have originated as corestones in weathered slope mantles inherited from more humid climates antecedent to the desert regime (Oberlander, 1971).

Debris mantles on granitic rocks in hot deserts are rarely continuous or cohesive enough for the internal shear strength of the mantle to be a limiting factor as suggested by Carson (1971) for the Laramie Mountains of Wyoming. In detail the slopes consist typically of bedrock steps, minor ledges with rock debris, and protuberant joint blocks in place; only from a distance do they appear rectilinear, at angles set by the frequency and size of these components. Thicker debris mantles occur only where relief is great enough to bring frost-weathering in summit zones, where fossil periglacial mantles survive (Weischet, 1969), or below exceptionally prominent cliffs of finely-jointed rock in mountain deserts, as in southern Sinai, where they attain the dimensions of true scree.

Debris trains. Granitic debris slopes on the east coast of the Sinai Peninsula bear numerous debris trains which head at re-entrants in the rock faces above. They consist of ill-defined bouldery channels flanked by prominent paired or multiple boulder levees up to 3 m high, on the crest of which are perched exceptionally large blocks as much as 1.5 m across. In the middle and lower slopes the outer levees link and outline tongue-shaped lobes 5-10 m across, with a pronounced inner depression marking the end of the channel. Boulders near the terminals show markedly imbricate packing directed upslope. Inner levees may link with remnants of intermediate lobes breached by the channel. Interstitial sand and silt have been removed from the surface and carried into a vesicular horizon at 5-20 cm depth, suggestive of sieve deposition in the end-stages of a debris flow. Several generations of debris trains can be recognised by differences in colour due to the degree of subsequent weathering. The process appears to be a major one in the transport of debris to and across the slopes, and it is possible that debris flows and stream flows alternate along the same courses.

In the Henkries Valley in Namaqualand, South Africa, comparable trains of diorite boulders traverse debris slopes in gneissic rocks and locally reach the slope foot, where they have built small boulder cones out into the valley floor. The travelled boulders have taken on a black varnish in contrast to the greyish gneissic debris, producing a vertical striping of the slopes (Pl. 12).

12. Debris trains of dark granodiorite boulders form stripes on a straight slope of granite-gneiss with scanty debris cover. Henkries Valley, Namaqualand, South Africa.

CLIFFED ESCARPMENTS IN LAYERED ROCKS

Wherever a resistant caprock protects a softer layer a cliffed or faceted hillslope tends to develop with great clarity in an arid setting. The hillslope components defined by Wood (1942) and propounded by King (1957) as an almost universal model assemblage (Fig. 9A) are in fact most characteristic of the desert regime, where they achieve expression over the widest range of structural settings under arid conditions of rock weathering and removal. For this reason they are appropriately considered in some detail here.

Schumm and Chorley (1966) have distinguished between simple scarps formed by a single hard layer, compound scarps with

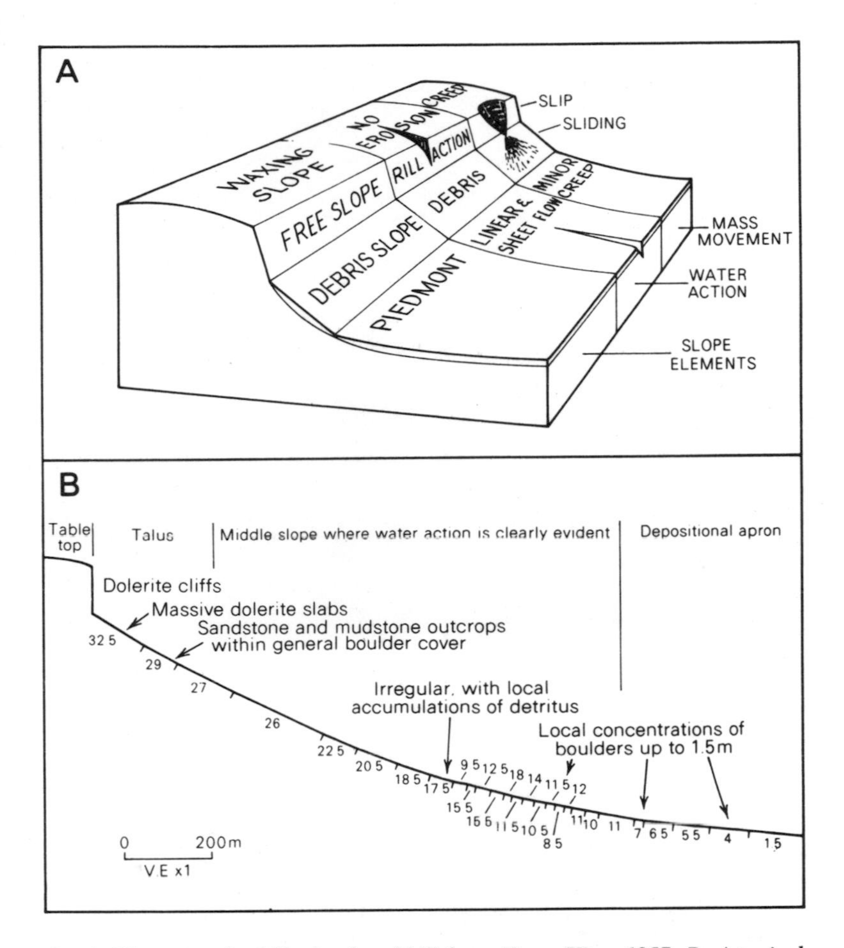

9. *A. Elements of a fully developed hillslope. From King, 1957. B. A typical hillslope profile on a dolerite-capped mesa in the South African Karroo. From Robinson, 1966.*

a resistant caprock above a soft bed, and complex forms where several harder and softer layers alternate. This account will deal mainly with the compound scarp (Pl. 13), and will treat the other forms as variants.

Summit forms

In the general absence of soil creep on caprocks, the summit behind the cliff top tends to be rounded by rainsplash and slopewash

13. Eagle Mesa, Monument Valley, Arizona-Utah, U.S.A., a compound scarp of massive sandstone above weak shale and sandstone. The straight debris slope has little cover. Photo F. Ahnert.

combined with weathering at the upper cliff angle, and since these processes are more effective on the steeper cliff than on a flattish structural crest, summit rounding is usually restricted to a small segment above the face. In extremely arid settings, such as the Gilf Kebir of the central Libyan Desert, modification of the cliff top may be negligible (Peel, 1941).

The tendency to rounding is increased where the capping dips towards the scarp (*Achterstufen*) and where the area draining to the face is thereby increased, where a caprock is porous and weakly-cemented and subject to granular disintegration rather than to fracture along joints, or where a slowly-receding massive caprock develops curvilinear pressure-release jointing (Bradley, 1963). Conversely, rounding is limited where the cliff recedes by failure along joint planes. The summit profile reflects the balance of these opposing tendencies; for example Ahnert (1960) and Schumm and Chorley (1966) have demonstrated that escarpment crests in massive sandstones of the Colorado plateaux are rounded on simple scarps where no soft basal layer is exposed to erosion and where the cliff is less subject to recession through undermining.

Cliff recession

With summit rounding relatively ineffective, and under the weathering-limited conditions prevailing on the hillslopes, a a resistant caprock tends to form a particularly prominent cliff face in an arid setting. The slowness and selectiveness of desert weathering combined with the general competence of transport removing debris from the slope ensure the dominance of the cliff during the evolution of the slope, and it is not surprising that

14. Sandstone tower, western Tibesti. A simple scarp with no basal weak horizon. Photo R. F. Peel.

desert landscapes with flat-lying resistant rocks have provided classic models for the evolution of landscapes by the parallel recession of escarpments, for instance in extreme-arid Libya (Peel, 1941) and in semiarid southern Africa (King, 1947).

Where the resistant bed extends the cliff to the slope foot, recession may be the result of basal weathering or sapping by seepage in the hard rock, as in the sandstone walls of wadis in the Gilf Kebir of the central Libyan Desert (Peel, 1941) and will be extremely slow. The sandstone towers of the Tibesti (Pl. 14) are a spectacular example of evolution of this type. In a compound escarpment, cliff recession will probably be dominated by mass movements resulting from undermining through attack on a softer underlayer, or from failure in the underlayer itself.

Major events in the recession of cliffs of this type include rock falls and slides of joint-bounded columns and projecting slabs of rock. These may result from tensional failure where rock strength has been reduced by weathering or where the face is rendered unstable through undermining by sapping or by erosion at the base. Falls of up to 15,000 tonnes of calcareous Navajo Sandstone in Zion Canyon, Utah have been ascribed to these causes (Robinson, 1970). Where recession of the caprock is slowed by virtue of its massiveness or cohesion the cliff may extend across a weak underlayer until collapse follows. In parts of the southwestern United States, Koons (1955) has noted that where there has been recent rock sliding caused in this way the cliff is fresh and extends only to the base of the caprock; conversely, faces extending into the underlayer are more weathered. Rotational slumping occurs where a thick soft underlayer is subject to failure through oversteepening; for example the Toreva blocks of the Black Mesa region of northeastern Arizona consist of masses of Mesaverde Sandstone up to 600 m long that have rotated backwards in moving above the expansive Mancos Shale (Reiche, 1937).

Events of such magnitude and infrequency raise the question whether they are characteristic of present conditions or whether they relate to a past wetter climate. Evidence that some forms due to major mass movements along the escarpments of the Colorado plateaux are fossil will be discussed in Chapter X. The general paucity of debris below cliffs of this region has been claimed as an indication of their current inactivity (Ahnert, 1960) save for localised recession on favoured *Achterstufen* (Mortensen, 1953), but

Schumm and Chorley (1966) have assembled numerous accounts which show that rockfalls and landslides are still wide-spread and that the lack of debris reflects competent weathering and removal. This is also attested by the general freshness of rock faces in the area. Active movements appear commonly to be associated with freeze-thaw in the winter and spring, as in the case of Threatening Rock (Schumm and Chorley, 1964), and findings in this cool-temperate montane desert may not apply to all arid regions; nevertheless they are a salutary reminder that climatic change need not necessarily be invoked to explain these catastrophic geomorphic events.

Debris slopes

Commonly the cliff above supplies only part of the mantle on a desert debris slope. True screes are restricted in warm deserts and may be limited to an uppermost straight sector. Fair (1948) and Robinson (1966) describe short screes of up to 34° below the dolerite *kranse* of kopjes or mesas in the semiarid Karroo of South Africa, and relate these slopes to the angle of repose of fallen blocks of 1–1.5 m diameter (Fig. 9B). This is slightly less steep than scree of comparable texture in wetter climates, where binding vegetation and soil confer an added stability.

It is a corollary of the above that much of the mantle lower on desert debris slopes is yielded from the underlying rock itself. The desert conditions of restricted debris supply and competent removal of fine secondary waste from the slope and at its foot satisfy conditions postulated by Bakker and Le Heux (1952) for the development of a Richter denudation slope, that is a thinly mantled rocky slope at the talus angle. Ledges and other structural breaks on debris slopes attest to the essentially erosional forms of such slopes in desert settings. They resemble the bouldery slopes on igneous rock described earlier, but the bedrock here is generally less uniformly massive, and a smooth concave profile replaces a straight slope to the extent that the finer components of the debris mantle become subject to transport by wash.

The equilibrium of a debris slope across a layer of relatively weak rock rests on the relationship between the yield of debris from the cliff and from the slope itself, and its subsequent weathering and removal, with implications for the erosion and recession of the lower slope component and hence for the undermining and further retreat of the cliff. This balance, quantified rather ponderously by Schumm and Chorley (1966) as the *talus weathering ratio*, is particularly significant in desert conditions because of the vital protective role of slope debris where vegetation is scanty. The debris cover controls the relative importance of areal wash processes and the close linear erosion to which soft rocks are subject under aridity. This balance is a delicate one under a dry climate because cliff recession tends to be slow and removal by wash to be relatively effective; hence debris slopes are commonly only thinly mantled.

The range of equilibrium conditions has been presented in idealised mathematical models by Carson and Kirkby (1972, 352-5). One extreme case is where the yield of primary debris to the debris slope far exceeds in bulk the products of secondary breakdown removed, when the cliff fails to survive. This does not apply here, but the joint-block slopes in massive igneous rocks discussed earlier form an example. The opposite extreme is that in which primary debris is insignificant, for example where a massive caprock breaks down entirely into granules or where rockfalls disintegrate on impact. The cliff then develops to dominate the slope in a simple escarpment, as do some massive but weakly-cemented calcareous sandstones in the Colorado plateaux (Schumm and Chorley, 1966). Compound escarpments constitute an intermediate range of forms in which the debris mantle may range from sparse to a complete cover.

Sparse debris mantles occur where the caprock is compact and cohesive and hence resistant to joint fracture, where it is inherently subject to granular disintegration, or where debris from the face weathers rapidly. Structural stability conferred by dip of the caprock into the scarp may reinforce these effects. Under these conditions the soft-rock slope is usually dissected by parallel gullies which head near the base of the cliff, and talus is confined to their heads. Where a caprock is thick relative to the height of the scarp the narrow debris slope is steepened to an alternation of short spurs and talus cones. The base of the caprock is exposed and the cliff may extend across the underlayer to a limit set by failure of the face through shearing and landsliding (Koons, 1955). Where the capping is relatively thin and recession more rapid the spurs on the debris slope lengthen and begin to take on the appearance of badlands. The longer spurs are encroached upon by wash slopes at the base and their profiles are markedly concave.

Circumstances which give a complete mantle of waste include a well-jointed caprock of reasonable thickness, and debris resistant to further weathering and comminution. These conditions are fulfilled on debris slopes below silcrete duricrusts in arid Australia (Pl. 15), where the hard patinated silcrete boulders spall only very slowly, and a similar role is played by flints from escarpment-capping limestones in the Negev of Israel. These debris mantles form a smooth armour which also checks gullying by dispersing runoff. Colluvial armours on the duricrust-capped mesas in central Australia range up to 1.5 m thick and consist of silcrete boulders haphazardly cemented in a sandy matrix derived from slopewash and commonly further indurated by impregnation with limonite and clay. They are much more resistant than the pallid sandstone and shale which they cover. Where they remain intact they extend to cover the base of the caprock, and on such hillslopes the cliff is typically weathered and softened in outline. Where a gully breaches

15. Silcrete-capped hamada tableland, central Australia. Debris slopes are protected by a mantle of cemented boulders which cap talus flatirons when detached by linear erosion. Photo by the author, CSIRO, Division of Land Use Research.

the mantle it tends to flare out headwards in the soft underlying material and to expose the base of the caprock; the cliff is thereby rejuvenated and debris begins to collect afresh on the rock slopes of the gully head. Where the former mantle is detached it commonly preserves an upstanding triangular facet or *talus flatiron* (Koons, 1955) with a steep infacing scarp and an outer slope which represents the former debris-slope profile. The detachment of talus flatirons indicates a replacement of dominant slopewash by linear erosion, and evidence that this may result from a change of climate will be discussed in Chapter X.

The prevalence of unconfined wash on debris-mantled slopes is reflected in the common occurrence of *wash terracettes* fronted by convex lines of boulders one, or at most two deep, behind which is a flattish sandy tread up to 1 m across. The terracettes follow the contour, although they rarely extend for more then a few metres, and the whole debris slope may be crudely stepped in this way. They form by the accumulation of sand carried by slopewash behind a few conveniently adjacent boulders, and grow out and extend and link laterally as suitable protective obstructions occur. Minor, shorter-lived steps also form upslope from shrubs.

Where the debris is underlain by argillaceous rock or by a clay subsoil, talus creep may become significant where freeze-thaw action occurs. Schumm (1967) reported boulder creep of between 2 and 7 cm per annum over a period of seven years at high elevations in western Colorado, over shallow lithosols on little-vegetated shale slopes between 20 and 40°, the amount varying with the slope angle. These movements are less than those reported for periglacial talus creep in high latitudes, but considerably greater than those on vegetated slopes under a humid climate.

Slow mass movement of hillslope debris appears to be relatively unimportant in warm deserts. After the boulders fall or roll to an initial resting place, or weather out on the slope itself, their further travel results mainly from sporadic undermining as adjacent fines are removed by slopewash. The relationship between slope angle and boulder size weakens in the lower part of the debris slope as wash processes become dominant, and such correspondence as does remain reflects the increasing comminution of debris with age and distance travelled rather than selective transport, for the sorting of slope materials is always very poor (cf. Dury, 1970).

Debris slopes with expansive clays such as montmorillonite are subject to strong swelling and heaving on wetting, followed by shrinkage and cracking on drying. These movements involve the stone mantle, and a downslope component can result in the formation of steps with lobate stony risers and relatively stone-free treads with cracking soils at the centre. They may occur singly or in festoons across the slope. In Australia they are fairly widespread on debris slopes of moderate steepness, and on sandstone slopes in the Barrier Range of western New South Wales they may exceed 10 m wide, with risers more than 50 cm high. They are equivalent to the stony gilgai to be described in Chapter VI, but are distorted so that only the downslope arc of the puff is developed, with a steep stony front. The affinity with gilgai is shown by well-structured, relatively stone-free subsoils that have been thrust to the surface along the riser arc. They have been named *desert gilgai* (Ollier, 1966), but *sorted stone step* seems to be more appropriate as it expresses the analogy with similar periglacial forms. The steps form part of a wide range of patterned ground associated with swelling clays in the Australian arid zone.

HILLSLOPES IN WEAK ROCKS

Loss of a caprock

Removal of a caprock brings with it changes in the form of the remaining slope (Fig. 10). With the cessation of supply of cliff

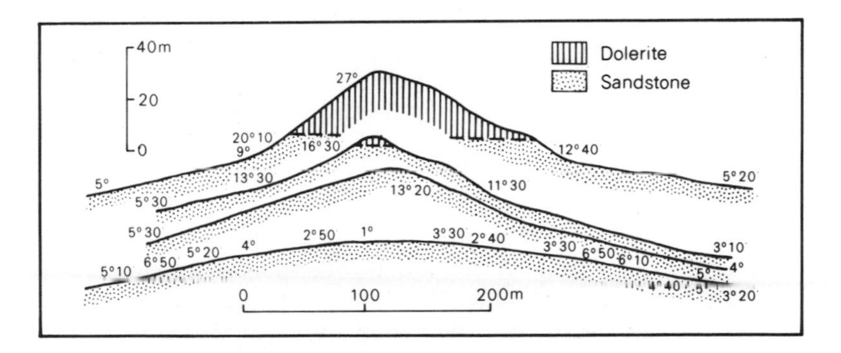

10. Comparative profiles indicative of successive stages in the lowering of a dolerite-capped sandstone divide in the semiarid interior of Natal, South Africa, showing summit rounding and slope reduction following removal of the caprock. From Fair, 1947.

debris and the progressive comminution of any surviving mantle the declivity of the upper sector of the former debris slope becomes adjusted to a lessened equilibrium angle. Regrading of the unprotected summit through rainsplash and creep introduces a convexity of profile which may be more or less sharply rounded depending on the permeability and physical resistance of the bedrock and the relative importance of slopewash. At the same time, the concave basal slope encroaches upwards as runoff becomes increasingly competent to transport all weathered material on the slope. The upward extension of concavity may be localised in the headward recession of gullies on midslope, as the reduction of the mantle brings an increasing liability to linear erosion, with a transition to badland forms.

Badlands

These extremely closely dissected landscapes on weak impervious rocks devoid of vegetation are a wilderness of steep smooth slopes, knife-edged or sharply rounded crests, and steep narrow valleys. Badlands evolve so rapidly that they can persist only in areas of regenerating relief. For this reason they are not widespread, but are probably more important in the evolution of desert landscapes than their restricted extent might indicate.

Among the best-known badlands are those in the semiarid Great Plains of North America, developed in flat-lying shales at receding escarpments which form steps between major erosional levels. For instance Badlands National Monument in South Dakota occupies a 'break of the plains' 5 km wide known as 'The Wall' between a Lower Prairie drained to the White River in the south and an Upper Prairie graded northwards to the South Fork of the Cheyenne River. Another well-known tract of badlands is the *ghor* of the Jordan rift valley, the Biblical wilderness, eroded in lacustrine clays.

In the absence of vegetation, little-consolidated impervious rock is very vulnerable to erosion by rainsplash and slopewash in occasional intense rainfalls. The efficacy of these processes was confirmed in measurements by Schumm (1956a, 1956b) at Badlands National Monument and elsewhere in the United States which revealed annual rates of lowering of between 1 and 2 cm on steep hillsides in shale. The preponderance of slopewash and related processes in the initial erosion of weak rock has been confirmed by measurements in the small upland catchment of Coyote Arroyo in New Mexico, shown in Table II. Nevertheless the intense runoff generates ultrafine channel networks in the soft materials, and drainage densities in badlands typically range between 125 and 350 km per km^2, or between 10 and 20 times the average under humid conditions.

Schumm found that erosion on badland slopes varied with the shear component as expressed in the sine of the slope angle, but that it did not increase with distance below the crest, indicating the attainment of a limit set by the capillary resistance and roughness of the shale surface. Strahler (1950) has termed this a *high-cohesion slope*. It is expressed as a very straight and steep midslope profile which commonly attains between 40 and 50°.

Uniform erosion over a slope implies parallel retreat, and Schumm confirmed this by measurements on badlands over a few years. In turn, crests should become increasingly narrowed. Most badlands are narrow-crested, and knife-edge forms do occur, but a degree of summit rounding is common and there are some markedly convex divides. As Gilbert (1877) noted long ago this poses a problem where parallel recession through slopewash is the dominant slope evolution. Recent measurements in the Canadian badlands have indicated remarkably uniform rates of erosion, independent of slope, as though the forms, having attained an equilibrium relationship, maintain themselves despite the rapid rates of erosion (Campbell, 1974).

Schumm (1956b) has invoked surficial rock creep as a main cause of rounding, as suggested by contrasted forms on different lithologies at Badlands National Monument (Pl. 16). The Brule Formation is a compact shale subject mainly to slopewash. It is extremely finely dissected and forms straight very steep slopes (>45° maximum) with close parallel rills and very sharp crests. In contrast, the less-cemented Chadron Formation, with an element of expansive clay, has loose crumbly surfaces which allow greater infiltration and surface saturation and which are consequently more vulnerable to creep than to slopewash or rilling. It forms rounded 'haystack' hills with convex slopes that average only about 27° where steepest. The density of drainage is half that on the Brule Formation, and even where rills do form in summer thunderstorms they tend later to be smoothed out by subsequent mass movement. The Chadron shale is more resistant to erosion than the Brule

16. Escarpment of Badlands National Monument, South Dakota, with steep cliffs in Brule Formation and convex forms on the underlying Chadron Formation. Photo S. A. Schumm.

Formation, despite its loose surface, for Schumm determined that the average rate of surface lowering through creep was only half that by slopewash.

Seasonal changes where shale slopes in montane desert are subject to winter freezing also express the differing roles of creep and slopewash. On footslopes in Mancos Shale in western Colorado, runoff from light spring rainfalls may be less than 1 per cent because the surface has been loosened and cracked by frost during the winter. However, the slopes become increasingly compacted under the heavier summer rainfalls and runoff may then attain between 20 and 40 per cent (Schumm and Lusby, 1963). Parallel rills form on the slopes, only to be obliterated by frost heaving in the following winter and by creep during the thaw. Schumm (1964) recorded rock creep of between 5 and 10 cm on the crests and upper slopes during winter and early spring, a considerably greater rate than creep on similar rocks in humid temperate regions and approaching that of talus creep under cold climates.

However, processes other than creep can produce summit convexities on weak rocks, notably rainsplash as observed on badlands at Lusk, Wyoming (Mosley, 1973). Deflation of shale splinters from dried crests may also contribute to rounding.

Young (1972) has stressed similarities between badlands and areas of accelerated erosion where the removal of vegetation has left the soil loose and unprotected. In the North American badlands the intact prairie sod plays the role of caprock above the retreating scarp, but once this turf cover is undermined grasses cannot recolonise the rapidly eroding slopes and can be seen to advance only on the more stable flats below. Other sloping turf remnants preserve spur facets, like debris mantles on talus flatirons, as evidence that the escarpment itself formerly had a smooth concave profile. Their isolation suggests the replacement of wash slopes by forms due to aggressive linear erosion, and again raises the question whether such changes may not have resulted from fluctuations between wetter and drier climates (Smith, 1958).

Dust-mantled landscapes

Forms comparable in some respects with badlands arise where bedrock has been thoroughly and finely weathered or buried beneath transported mantles of fine texture.

The coast ranges of the desert of northern Chile exemplify the former case. They constitute a special environment by virtue of an association of high atmospheric humidity and abundant salt under virtually rainless conditions. Most rock types have been severely affected by salt-weathering, outcrops have been reduced, and the rounded crests and hillslopes have been smoothed in a mantle of weathered detritus with an admixture of salt crystals and aeolian dust. The extremely rare falls of rain produce little runoff on these surfaces — rather a form of transport akin to mudflows in short-lived channels. There are few slope gullies and the local valleys appear fossil under the prevailing extreme aridity. Wind is the dominant transportive agent and slopes exposed to the coastal breezes are extensively covered with blown sand. Below the rare cliffs are bedded screes consisting of alternate layers of sand and rock waste.

Similar rounded landscapes have also been described from the extreme desert about 100 km further inland (Mortensen, 1927). Here the slopes are covered by up to 50 cm of salt-impregnated loam with bedrock splinters, regarded as a product of rock weathering in place, and the uppermost 10 cm has been reduced to a drab brown dust which bears a thin silt crust (*Staubhaut*). In the absence of sand the wind is generally powerless to corrade the crusted surfaces and there is little evidence of mass movement; despite the aridity the dominant imprint is that of running water, the convex slopes are closely rilled and local valleys have been kept clean of detritus.

These smooth mantled landscapes strongly resemble loess-covered hills, as in the northern Negev of Israel.

Badlands as relief models

As rapidly evolving fine-textured landscapes, badlands constitute useful models which exemplify in miniature and at an accelerated rate the interactions between process and form on unvegetated slopes. Schumm (1956b) has used them to contrast the creep-dominated profiles of the Chadron Formation, as a simulation of humid-temperate landscapes, with the wash-dominated forms of the Brule Formation, as characteristic of desert hillslopes. Where creep predominates, summits are broadly rounded, slopes are convexo-concave and diminish with reducing relief as the area of dominant creep extends downslope, and the texture of drainage is moderate; in contrast, wash-dominated uplands show narrow crests, steep straight slopes which maintain their angle as relief is destroyed, and very high drainage density, all features which are held typical of arid regions. This exemplification of the effects of differing relative importance of creep and wash may point to an explanation of the differences between the forms of arid and humid hillslopes (Holmes, 1955).

IV

DRAINAGE ON DESERT WATERSHEDS

The characteristic topographic break between desert upland and plain is paralleled by contrasts in their drainage systems. In the hills are connected, close-branching drainage systems, with clean-scoured rocky channels in head tracts, passing down-valley into gravelly floors; on the nearby plains are partly disintegrated sparser networks of sand-filled channels which break down progressively with distance from the upland front. Slope and surface characteristics are so fundamental to the differences in runoff which underlie these drainage contrasts that it is appropriate to treat the watersheds and their lower-order channels separately, as part of the complex of upland desert forms.

RUNOFF FROM DESERT WATERSHEDS

Initial runoff from a desert hillside may be expected to resemble that described as Horton overland flow (Horton, 1933), the result of rainfall intensity in excess of the infiltration capacity of rock or thin debris mantles, or more locally to take the form of saturation overland flow (Kirkby and Chorley, 1967) where thin soils have become saturated. Detrital covers on desert hillslopes are generally too thin and patchy for interflow through them to be important. In general, runoff in deserts is more strongly dependent on rainfall intensity than on duration, suggesting that infiltration rate is a more important determinant than degree of saturation. Desert rainstorms commonly include bursts of intense rainfall beyond the rate of watershed infiltration, and with a sparse and stunted vegetation and lack of litter interception and detention storage are diminished.

Upland streamfloods

These effects are expressed in the rapid rise and cessation of streamfloods in desert uplands and foothills. Their impressiveness is indicated by the many local names to describe them, for example *avenida* in Chile, *foum* in the Algerian Sahara, and *seil* in the Egyptian deserts. Hume (1925) gives this early account of floods in the mountain wadis of the Sinai Peninsula:

> Storm-clouds first gather heavily on the mountain summits; then when the storm breaks, all the fury of the elements seems to be let loose. It displays itself in the continual flash of the lightning, the roll of the thunder, the roar of torrents, and the splash of the rain on the bare rock surfaces. In a few moments a whole mountain-side may be foaming with innumerable cascades, while the rocks rolling down the boulder-strewn valleys descend with the roar of an avalanche. Yet in a very short space of time the scene may change: an isolated waterfall forming a small stream quickly swallowed up by the gravel of the valley, and patches of sand bound together where the waters have passed, being the sole evidences remaining of the sudden and destructive water-floods (pp.86–7).

The relative infrequency and short duration of the flows, their remoteness and the hazard and technical difficulty of recording them account for the few measurements of streamfloods in desert uplands. One upland drainage system that has been monitored is that of Nahal Yael near the head of the Gulf of Aqaba, not far from the region described by Hume (Schick, 1970; Schick and Sharon, 1974). It is a 4th-order catchment with about one hundred head channels, with an area of 0.6 km^2 and a drainage density of about 25 km per km^2 (Fig. 11a). It is in rugged country of schist, granite and amphibolite, with about 170 m relief, and the unvegetated boulder-strewn slopes typically attain around 30°. The average annual rainfall is 30 mm, and during the period 1966–73 ten hydrologic events were recorded on a small network of gauges, with an extreme maximum discharge of 14–17 m^3/s. The hydrograph in Fig. 11b is schematic and composite, embodying features typical of several of these events. It shows the characteristic abrupt flood rise, often as a wall of water, averaging 5 minutes but exceeding 20 minutes in the largest floods, the short duration of peak flow (defined as above half the maximum discharge) which had a median value of 10 minutes or about 5 per cent of the flood duration, and the slower falling stage until the last trickle has disappeared into the channel gravels. Peak velocities have been estimated as 1–2 m/s for the majority of the floods. Most flows lasted between 1 and 4 hours, but the largest events lasted almost 24 hours.

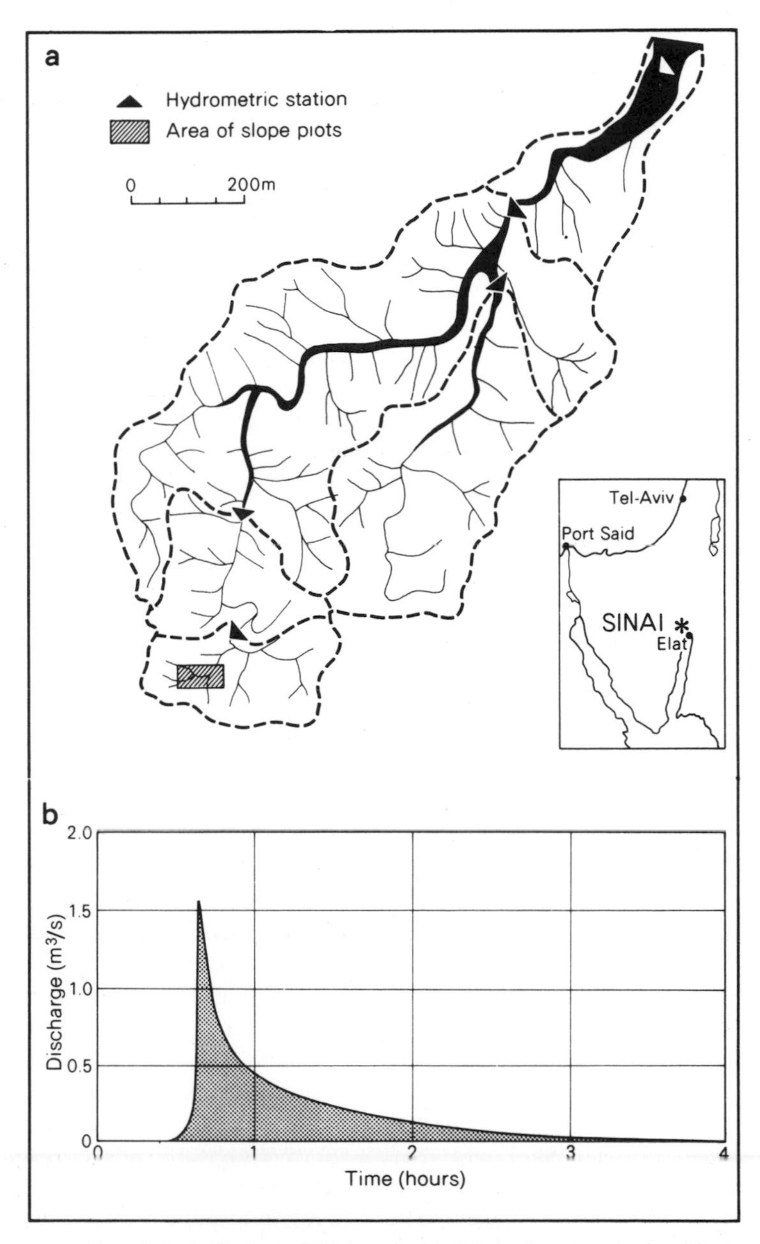

11. a. The Nahal Yael catchment, Sinai Peninsula, Israel; b. schematic hydrograph typical of larger flows in the Nahal Yael. Based on data in Schick, 1971.

Rainfall required to initiate flow in the main channel of Nahal Yael was found to be 7.5 mm at a mean intensity of 0.5 mm/min, a value close to that of 5 mm with an intensity of 0.5 mm/min cited as the threshold for streamflow in the Hoggar, central Sahara (Dubief, 1953a, b). At the extreme, catastrophic floods in the canyon outlets of the White Mountains of California and Nevada were associated with falls of more than 100 mm with intensities greater than 1 mm/min (Kesseli and Beaty, 1959). These threshold values are commonly attained in conventional 'cloudbursts' and in high-intensity cells in frontal rains in deserts. In mountain-and-basin deserts such rainfalls may be supplemented by snowmelt floods in spring or by winter floods where rain falls on snow or frozen ground (Kesseli and Beaty, 1959), generally with a slower but more sustained flood response. In the upland valleys of moderate deserts more than one such flow may occur each year, whereas in extreme deserts several years may intervene between floods which last for less than an hour. Nevertheless records for the western Sahara suggest that few wadis are likely to escape flooding for periods greater than 5–10 years, even in severe desert settings (Capot-Rey, 1953).

Controls of runoff

A study of runoff on hillslopes of the Nahal Yael watershed (Yair and Klein, 1973) has confirmed that for each discharge event in the head channels several minor episodes of runoff may occur on the hillslopes without resulting in streamflow. This is also reflected in differences between runoff coefficients of 40–50 per cent claimed for head catchments of Nahal Yael (Schick, 1970) and those of 1.5–4.5 per cent measured on the slopes and incorporating sub-flow events (Yair and Klein, 1973). Only in the larger rainfalls does slope runoff approach the higher values.

The minor runoff episodes which fail to reach the slope foot express a discontinuity of operation inherent in all desert drainage systems. Failure of runoff to persist across the hillslope may be in part due to evaporation losses, but the major initial losses result from absorption and depression storage in dry rock debris. The former can be large on porous rocks; for example it was the main component in an initial infiltration of up to 10 mm per minute on debris slopes of coarse sandstone in eastern Sinai, where the corresponding rate was only 2 mm per minute on bare crystalline rock and between 3 and 5 mm on a range of finer and less porous rock debris (Yair and Lavee, 1974). The slopes gave correspondingly longer and shorter delay periods in runoff response. Such measurements underline the important control over runoff exercised by rock debris in deserts in the general absence of soil and close vegetation.

This control also affects the runoff coefficient through its determination of surface roughness and consequent impedance of overland flow, as shown by differences between three hillslopes studied in the Nahal Yael catchment (Fig. 12), summarised in Table III (Yair and Klein, 1973). Runoff was greatest from the slope with the finest debris and hence with the smoothest surface, despite its gentler gradient. To the extent that steeper slopes tend to have coarser debris therefore, runoff may be inversely related to hillslope angle in deserts, in contrast to conditions on vegetated hillslopes in humid regions. This is further illustrated by runoff yields on an experimental flood farm at Avdat in the northern Negev in Israel (Evenari *et al.*, 1971), where the hillsides bear mantles of limestone rubble and loess. Removal of stone, as practised by Nabatean irrigators fifteen centuries earlier, was found to increase runoff from moderate storms by smoothing the surfaces and permitting crusting by rain impact on the exposed silty loess. Of all hillslope surfaces, smooth rock outcrops have been found to yield the highest runoff (Yair, 1974).

Even in the larger rainfalls, hydrologic discontinuity on the hillslopes may result in only part of the watershed contributing to streamflow. The area contributing will vary with slope properties and importantly with the magnitude and intensity of the storm. Using simulated rainfalls, Yair and Lavee (1974) determined it to be between 15 and 35 per cent on steep debris-mantled slopes in Sinai, and found the main contributing areas to be the rocky channels and adjacent outcrops, with low roughness and absorption. Runoff was strongly dependent on rainfall intensity, and long-continued light rain gave low runoff yields, confirming infiltration rates to be more important than saturation. The contributing areas are comparable in relative extent with the partial-area contributions to storm runoff under humid conditions, where the stream channels and margins have also been shown to be the main yielding surfaces, although here the degree of soil saturation becomes a dominant control (Dunne and Black, 1970).

In the larger hydrologic events of desert uplands, with storm

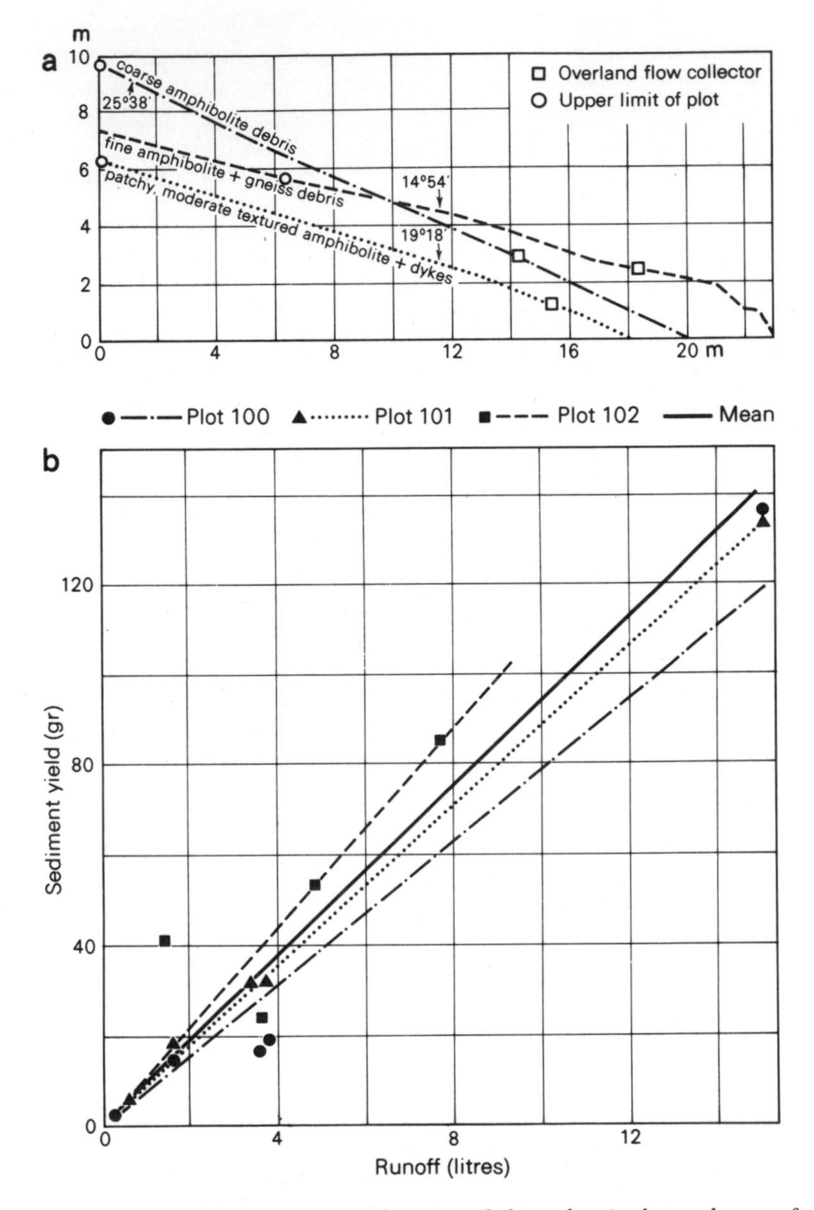

12. a. Profile and debris mantles of monitored slope plots in the catchment of Nahal Yael; b. relationship between volume of runoff and sediment yield on plots. From Yair and Klein, 1973.

rainfalls attaining 50 mm and above, the percentage of the watershed contributing to streamflow increases and with it the runoff coefficient. The larger flash floods in the small tributary catchments of the Nahal Yael (0.05 km^2) have represented runoffs greater than 50 per cent, with yields as high as 10 $m^3/s/km^2$. For the devastating floods in the Arava watershed near the south end of the Dead Sea rift valley in 1966, peak yields of between 20 and 30 $m^3/s/km^2$ have been estimated, with peak velocities as high as 7 m/s and surges attaining 12 m/s (Schick, 1971). An upper limit to the magnitude of such floods is set by the dimensions of intense desert rainstorms and the maximum effects are predictable at the outlets of upland catchments of around 25 km^2.

In an 8-year period only the three largest flows in Nahal Yael reached the gravel fan at the piedmont exit. At this point two of the flows had diminished to about 10 per cent of their maximum discharges, and only in a single extreme event did flow in the master channel persist to link with an adjoining upland drainage system (Schick, 1971; Schick and Sharon, 1974). In the majority of the smaller flows, runoff values of between 35 and 60 per cent in the small head catchments were reduced to between 5 and 10 per cent for the watershed as a whole if estimated for the diminished discharges down-valley. This illustrates the lack of meaning of average runoff values for desert regions given without reference to size of watershed. Downstream diminution in discharge, well-known in desert lowlands, is also a feature of the moderate and smaller-sized upland floods and is largely due to abstraction through infiltration into the gravelly channel alluvium, which increases in amount with stream order. Yair and Klein (1973) measured infiltration rates as high as 30 mm/min into channel sediments, or between 10 and 50 times more rapid than that on adjoining rocky hillsides. Rapid infiltration into the stream bed in the van of the flow may contribute to the 'wall-of-water' onset of the larger upland floods.

SEDIMENT TRANSPORT

On desert hillslopes

High sediment concentrations in runoff from desert uplands reflect the vulnerability of unvegetated surfaces to water erosion, particularly by rainsplash and overland flow, and additionally by gullying. Sparsity of vegetation is the essential control, leaving the surface exposed to raindrop impact, which erodes directly by dislodgement and indirectly by breaking down the surface layer. The bare surfaces with loose dry waste are equally vulnerable to erosion by wash during flash runoff. Both sets of processes are most effective under arid conditions, although the greatest erosion is produced with more frequent rainfall in semiarid climates as shown in Fig. 13A and 13B. The effects of such erosion are seen in earth pillars and perched stones and in 'pedestalling' around perennial shrubs. The preponderance of slopewash in initial erosion is demonstrated in Table II by measurements from Coyote Arroyo, New Mexico, where it accounted for more than 97 per cent of the sediment yield.

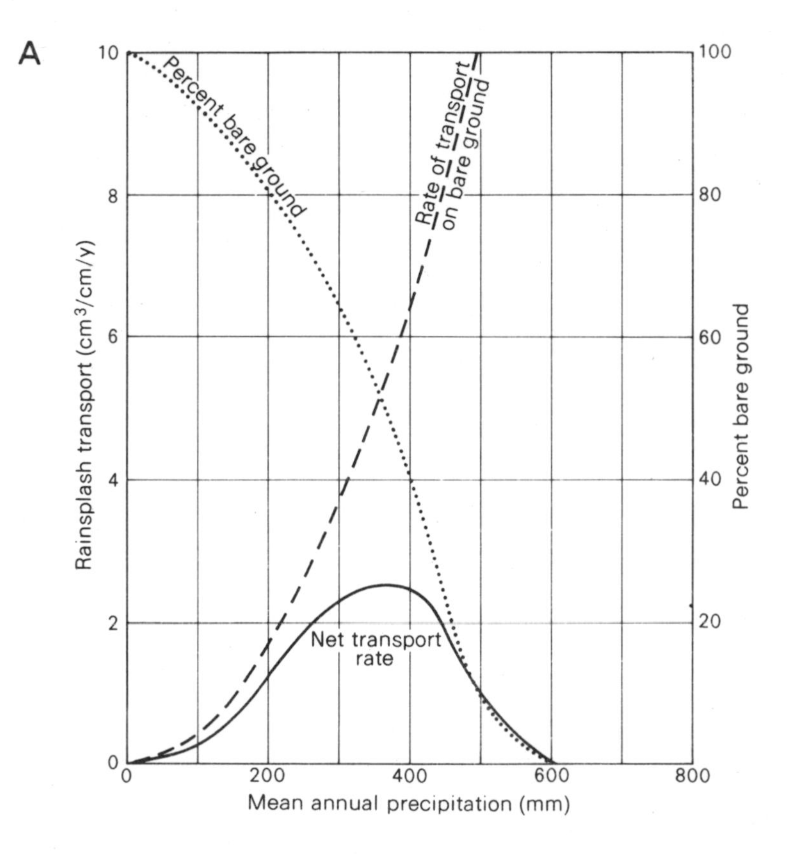

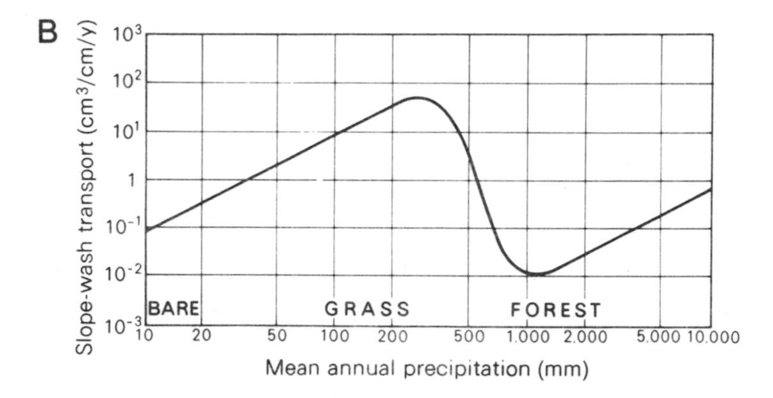

13. A. Generalised variation of rainsplash erosion with percentage bare ground, expressed in terms of mean annual precipitation in southern United States. After Kirkby, 1969.
B. Relationship between slope-wash transport and annual rainfall for a 10 m slope of 10° in southern United States. From Carson and Kirkby, 1972.

TABLE II **Erosion in Coyote Arroyo, New Mexico***

Rock type: Poorly consolidated calcareous silt, sand and gravel — Annual rainfall: approx. 30 cm

	Annual sediment yield (tonnes/km²)	% total yield
Surface erosion (wash)	8640	97.8
Gully erosion	125	1.4
Mass movement	63	0.7

*From Leopold, Emmett and Myrick, 1966.

Yair and Klein (1973) measured sediment concentrations of between 7500 and 15,000 ppm in runoff from the hillslopes of Nahal Yael, equivalent to an annual removal of 7 g/m². They found the amount of sediment transported, rather than the concentration

TABLE III **Runoff and sediment yield on selected hillslopes in Nahal Yael, Israel***

Plot	Length of plot	Area of plot	Size grading of slope debris						Mean infiltration rate+	Average runoff coefficient	Sediment collected by slope collectors	Mean sediment concentration
			Upper layer	Lower layer								
			Mean lengths of clasts (b-axis)	Coarse	Fine	Silt	Clay					
	m	m^2	mm	%	%	%	%		mm/sec	%	g	ppm
100	15.5	60	61	25.0	50.6	10.7	13.5		0.05	1.55	187	7740
101	16.1	60	38	19.6	54.0	11.2	15.1		0.02	1.56	218	9030
102	12.3	40	28	15.6	55.0	12.2	17.1		0.01	4.54	738	15,680

* From Yair and Klein, 1973.

+ Mean infiltration rate for nearby channel 0.64 mm/sec.

of sediment, to increase with the volume of runoff, as shown in Fig. 12b and Table III. As with runoff therefore, erosion of desert hillslopes by overland flow is likely to be determined more by the smoothness of the slope than by its length or declivity, and to be patchy and discontinuous in accord with patterns of runoff. The values cited refer to suspended load, which accounts for the bulk of the transport, and the amounts were naturally greatest on the slopes with the finer debris (Table III). Yair (1974) also found in the Negev that disturbance of the slope by burrowing animals was a major factor in sediment yield.

The sediment trapped on the Nahal Yael hillslopes was equivalent to an overall annual erosion rate of 7 tonnes/km^2 for the head catchment of 0.05 km^2. This is small compared with erosion rates for overland flow in Fig. 13 and for catchments in Fig. 14; clearly, with an annual rainfall of 30 mm, Nahal Yael is far too arid to yield extremely high overall rates of erosion.

In desert streamfloods

The sediment concentrations of 7–15,000 ppm in overland flow are approximately matched by suspended loads of 5–10,000 ppm in low-order channels of the same head catchment (Schick, 1970). Since the runoff coefficients of overland flow and streamflow have been shown to be comparable for larger rainfall events, it is apparent that the head-channels are competent to transport away most of the sediment carried to the slope foot. The prevalent conditions of non-accumulation at the slope foot have already been noted in relation to desert hillslope regimes, and they are also reflected in the typically rocky first-order channels with little alluvial fill.

Sediment concentrations in streamfloods tend to increase steadily downstream due to the increasing incorporation of loose channel sediment into the turbulent flood, to values of 20,000-30,000 ppm, but from the point where there is a significant depth of alluvial fill they increase more rapidly due to abstraction of discharge, sometimes doubling over a few hundred metres, and concentrations in excess of 150,000 ppm have been recorded on the exit fan of Nahal Yael. These are more than an order of magnitude greater than concentrations in comparable flood discharges in upland channels in humid regions, where sediment concentration is further diluted with down-channel increase in discharge.

Down-valley diminution of discharge and consequent increase in

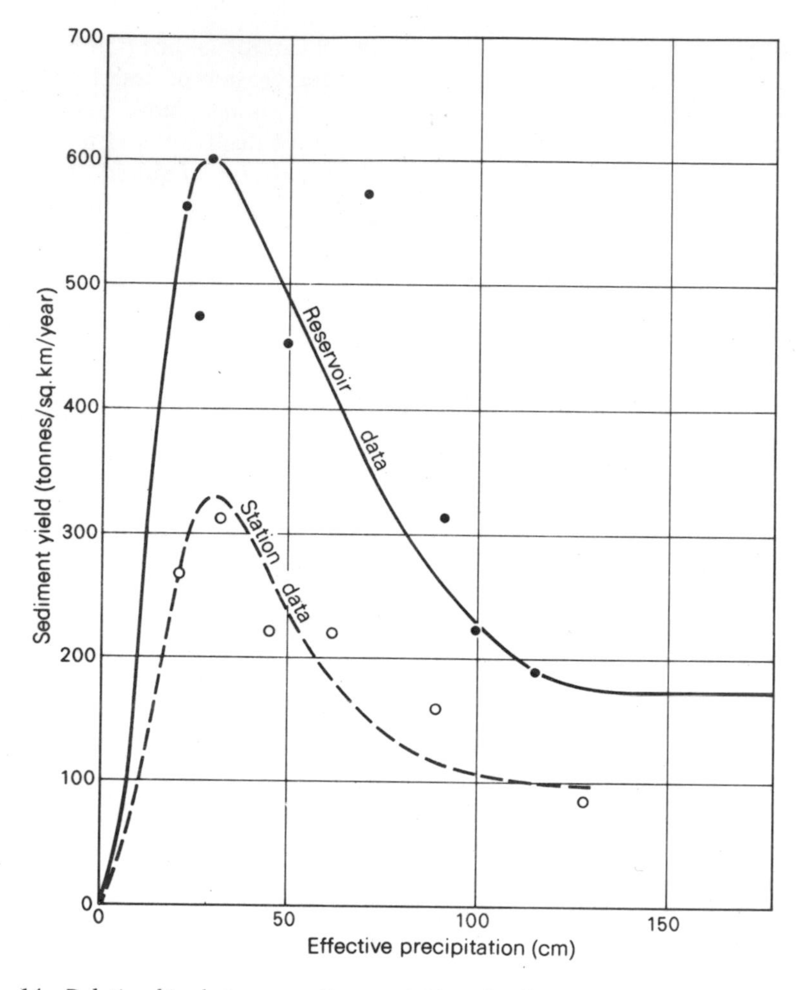

14. Relationship between sediment yield and effective precipitation from reservoir and gauging station data in small hilly catchments in midwestern United States. After Langbein and Schumm, 1958.

sediment concentration, with the complete cessation of smaller flows, result in a progressive down-valley increase in deposition in the lower sectors of desert upland drainage systems by floods of moderate dimensions.

With the passage of each flood the coarse bedload is mobilised and scoured to a depth of a metre or so, but these larger particles are generally redeposited a short distance downstream; for example

Schick (1970) found the median travel of gravel particles in Nahal Yael to be between 50 and 80 m per flood. On the assumption that the bedload transported was approximated by a downstream translation of a scoured layer measured by buried chains, the volume of bedload mobilised in Nahal Yael was at least twice and probably more than ten times the suspended load. These larger volumes may more than compensate for the shorter trajectories of individual particles, and it is likely in fact that bedload transport may equal or exceed suspended load transport, a markedly converse relationship to that found in upland floods in humid regions.

Schick (1970) estimated that about 1 per cent of the bedload in Nahal Yael was lost by comminution with each flow, indicating that the bulk of the gravel could be reduced to sand grades in traversing an upland channel a few kilometres long.

The bulk of the suspended load is deposited towards the lower limit reached by the flood, and in the Nahal Yael Schick (1970) found that virtually all the suspended load carried by recorded flows in the lowest channel sector could be accounted for by deposition on the exit fan.

Such self-contained systems of erosion and deposition allow an estimate of erosion rates in upland catchments where the rate of accumulation of alluvium on the terminal fan is known. Schick (1971) has in this way estimated an annual regional denudation rate of 0.3 mm for the southern Arava watershed, on the basis of an accretion of 1 mm annually on the piedmont fans. This is equivalent to a sediment yield of 160 tonnes/km^2, and for an effective precipitation of 50 mm this is not inconsistent with the rates for the western United States shown in Fig. 14. Denudation in head catchments of Nahal Yael (30 mm precipitation) may be at only half this rate, but Gerson and Inbar (1974) suggest an annual denudation of 1.0 mm for the friable marls of the nearby Mt Sdom, whilst Lustig (1965) gives figures suggesting 1–2 mm per annum for parts of the Great Basin, with precipitation ranging between 75 and 300 mm.

Debris flows

Individual small flows are by no means uncommon where heavy rain falls on steep unvegetated debris slopes, particularly where they include rocks such as shale that contribute a viscous muddy matrix. They commonly head in gullies or in short screes below cliffs and leave shallow channels flanked by bouldery levees (Sharp, 1942). Where they terminate on the slope they build lobate or cone-shaped accumulations of colluvium, and their contribution to boulder trains on debris slopes has been discussed in Chapter III.

The largest and most devastating debris flows are those which occupy desert mountain valleys, through increasing sediment concentration in streamfloods, often as a result of landsliding at the valley head and by the incorporation of bedload in a turbulent streamflow. A study of debris flows in the White Mountains of the western United States (Kesseli and Beaty, 1959) suggested that ideal conditions were provided by narrow steep-sided valleys with a depth of fill of between 1.5 and 5 m (Pl. 17), and that the flows were associated with intense rainstorms in excess of 100 mm.

The Wrightwood debris flow in the San Gabriel Mountains of southern California (Sharp and Nobles, 1953) had a mountain valley length of 5 km, with a gradient of between 10 and 15 per cent before it emerged on the plains. Frequently, such a flow begins with

17. Debris flow in Montgomery Creek, White Mountains, July 1965. Photo C. B. Beaty.

a valley-head slump. The advance of the flow is accompanied by rumbling and roaring noises, often with a rolling dust cloud as material is incorporated into the viscous mass. The flow occurs in surges as a result of debris blockages, and velocities of 15 kph have been recorded. Solid matter may account for between 25 and 75 per cent of the flow and the viscosity is such that large boulders can be carried, but apart from the incorporation of loose material mudflows have little erosive power and are generally confined to pre-existing channels. The mudflow is usually followed by streamfloods which trench the unstratified, poorly-sorted mudflow deposits. Mudflows contribute in varying degree to the construction of alluvial fans and this depositional role will be discussed in Chapter V.

FORMS OF UPLAND DRAINAGE

Drainage density

The high runoff and sediment yield from desert hillslopes are associated with dense channel networks. Melton (1957) has demonstrated how, within a number of small upland catchments in the southwest United States, with a range of relief and rock types, drainage density and channel frequency are positively correlated with aridity and with the consequent extent of bare ground, as shown in Fig. 15A and 15B. On soft rocks, ultrafine stream networks are developed in badlands, with drainage densities typically between 125 and 350 km per km^2, that is between 10 and 20 times the average range under humid climates.

Slope gullies

The concepts of critical erosion distance and belt of no erosion (Horton, 1933) have little meaning where wash is so effective. The initiation of channels is strongly influenced by the surface roughness and cohesion of the slope, particularly as determined by the amount and coarseness of slope debris. On soft-weathering and impervious rock, rills head at the hill crests (Pl. 18), but on hard-rock slopes with abundant coarse debris and structural irregularities the extension of connected channels is hampered. The patchiness and variability of runoff-yielding surfaces, the looseness and thinness of the mantles, and the diminished hydraulic advantage of rills over unconcentrated flow render the headmost

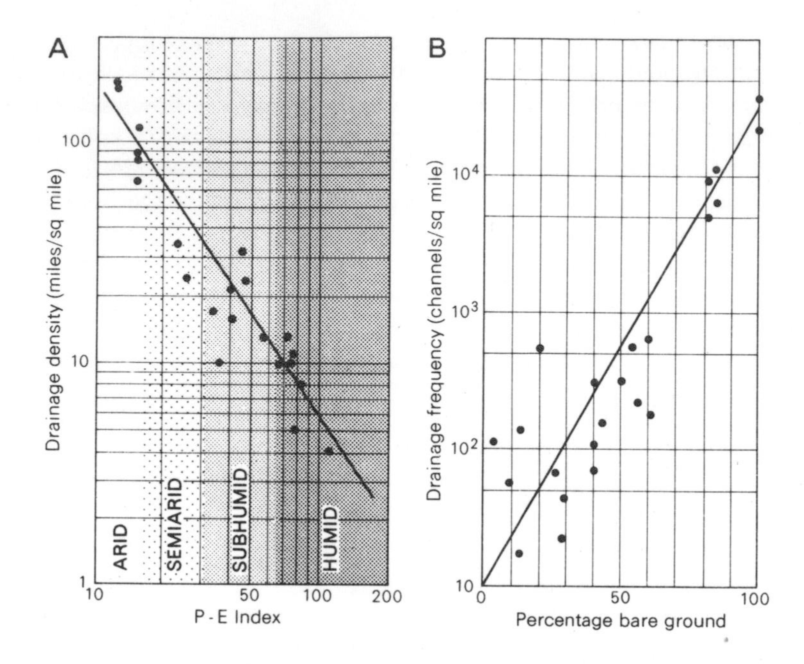

15. A. Variations of drainage density with aridity as measured by P-E index in southwestern United States. Each point represents one basin. From Melton, 1957.
B. Variation of channel frequency with percentage of bare ground in southwestern United States. From Melton, 1957.

lines of drainage on desert hillslopes shallow and impermanent. Fingertip channels may be temporarily obliterated where creep or soil-heaving becomes important (Schumm and Lusby, 1963.)

Where a debris slope is thickly mantled with large boulders, as discussed in Chapter III, linear flow may be so hampered and incipient depressions become so choked with waste that gullies cannot form. More commonly, however, there is a varying interaction between gullying and the ready transport of fine waste by wash on little-vegetated desert hillslopes. Surface debris tends to be coarsest in the heads and along the margins of gullies, indicating that finer components have been carried into and along the channels. Beaty (1959) has suggested that a balanced alternation of localised movement of waste from the interfluves into the gullies during light storms, together with a flushing of the gullies during occasional heavy rains, could result in an overall denudation of a hill-

18. *Close texture of first-order channels and rills on shale and weak sandstone, Black Mountains, Death Valley. Virtual absence of 'belt of no erosion'. Photo J. Stacy.*

slope without change in the position or average depth of incision of the gullies. Occasionally, however, gullies are displaced or choked by debris, although a regular replacement of interfluve by gully and vice versa, in the process described by Bryan (1940) as a gully gravure, appears improbable.

Gullies on steep desert hillslopes show a characteristic lack of branching due to an absence of cross-grading. This feature has not been well explained and may result from various causes. On hard rocks where erosion is 'weathering-limited' there can be little development of cross slopes, for here wash vies closely with the shallow channels in its effectiveness in slope erosion; in any case gully development is strongly hampered by rocky ledges and salients. Slopes in soft rocks may be severely eroded into deep parallel flutings, but the gullies are then so closely spaced that they interact by lateral abstraction; the interfluve crests are knife-edged and there is no scope for the development of tributaries. Whatever the reason, many gullies fail to link and die out singly on the footslopes, so contributing to the diminution of drainage density from upland to plain.

Head channels

Because of the predominance of wash transport over mass movement on the hillslopes, channels of the lowest orders in desert uplands receive most of their load in dimensions which they are competent to transport. This is particularly the case in uplands of crystalline rocks or of sandstone subject to granular weathering. In consequence the stream beds are commonly in or close to bedrock, the debris slopes normally lead smoothly to the channel, and undercut cliffs and blockages by boulder debris are exceptional. Longitudinal profiles are steep and irregular, with ledge falls and dry plunge-pools in strongly bedded and jointed rocks, for example the *sidds* of the upland wadis of eastern Egypt (Hume, 1925). Gradients commonly steepen below tributary junctions and minor tributaries may be left hanging because of more frequent flood scour in the major channels. However, within gross limits set by lithology, the rock channels of desert uplands show a fairly regular down-valley increase in depth and width with increasing catchment area and bankfull discharge, as with alluvial channels, and in that sense cannot be termed ungraded (Miller, 1958).

Larger upland valleys

The larger valley or canyon in a desert upland of strong relief typically consists of a narrow gravelly or sandy floor from which the hillslopes rise directly, locally as cliffs and more commonly as debris slopes, with no intervening footslope. Basal trimming of the slope by the channel is also uncommon. The floor is characteristically a complex of intermeshing longitudinal bars and washes (Pl.19), the latter including one or more active channels with abandoned courses at higher levels. In sandier channels deep scour pools may occur at bends or constrictions of the valley, together with transverse bars and other bed forms as evidence of the turbulence of the latest flood. Gravel bars and sandy depressions may alternate regularly along channels with mixed bedloads. Perched boulders and terrace remnants along the margins may attest to previous floods of great magnitude (Schick, 1974). Such valleys often have steep longitudinal gradients, as much as 10 per cent or greater.

The processes leading to these forms have been described in an

19. Valley floor in upland desert, with shallow braiding channel between gravel and sand bars. Henkries Valley, Namaqualand.

account of the 1952 flood in Milner Canyon, White Mountains, U.S.A. (Kesseli and Beaty, 1959):

> Observations in the lower canyon ... indicated that constant change in channel shape may take place without appreciable deepening of the channel itself. The main strand of current swings back and forth across the width of the channel within remarkably short periods, and gravel and cobble bars are built and just as quickly washed away. Relatively fixed obstacles in the channel, such as large boulders and partially buried logs, control the course of the main strand of current to a certain extent, but temporary masses of cobbles and gravel, built and torn apart by the stream itself, seem to be more important in bringing about shifts in the position of the main strand ...
>
> These course changes of the main strand of current were observed to take place so rapidly that any given part of the channel in the lower canyon would hardly be recognizable 5 or 10 minutes after first sight of it. Yet throughout the process of ever-changing channel shape, channel depth remained essentially constant. Material moved downstream discontinuously, perhaps 10 or 20 feet at a spurt. It would be incorporated into temporary mid-

stream or side-channel bars which might persist for 10 or 15 minutes. As they were washed away, cobbles or gravel would be shifted downstream another few feet, again to be lodged in a temporary resting place (p. 63).

Entrenched canyons

Where large desert rivers, including through-going allocthonous channels, have been rejuvenated by uplift, spectacular canyons may result from limited back-weathering of valley sides under an arid regime during channel incision. The interplay of rock structure and fluvial processes may be most striking, as in the canyons of the Colorado plateaux. Joint control of the trend and outline of canyons is strongest in headward-extending sectors and in tributary canyons. This progressively lessens as the canyon is opened by meandering, but the mechanisms are complex and include secondary processes activated by the down-cutting as well as fluvial erosion (Campbell, 1973). Seepage along major joints, working to watertables determined by the incising channel, leads to the formation and eventual collapse of alcoves in the canyon walls. In this way, natural arches are formed and small meander necks become detached. Patterns of alcove collapse at the heads of meanders commonly follow systems of exfoliation joints developed in conformity with the initial arcuate trace of channel incision, whereby original meanders become magnified. Such *ingrown* forms are a case of geomorphological convergence, in that they come to resemble smoothly sinuous true incised meanders such as the classic 'Goosenecks' of the San Juan River, Utah.

V

DESERT PIEDMONTS

Holmes (1955) has contrasted the steep detritus-yielding slope of the desert hillside with the piedmont slope of graded wash transport. Waste supply from the hillslope is limited by slow weathering under aridity; the finer debris tends to be readily washed away and accordingly the hillslope, whether determined by rock strength or by the angle of rest of larger debris as discussed in Chapter III, remains steep to its foot. On the other hand, wash transport can proceed at exceptionally low angles across desert piedmonts with little obstructing vegetation. These contrasted slope elements tend to meet in a sharp piedmont junction.

The desert piedmont links an upland of predominant erosion with a lowland in which surfaces are transportational or depositional in the main. It intervenes between the close, connected systems of upland drainage channels and the plains sectors with diminished numbers of channels, where local runoff is increasingly dispersed. The piedmont landforms express this transition. They may be erosional or depositional and the two types are considered separately before their relationship is discussed.

EROSIONAL PIEDMONTS: PEDIMENTS

The abrupt rise of desert hillslopes from the plains encouraged earlier views of the uplands as rock islands protruding from a sea of detritus, but piedmont plains cut in rock were noted by Gilbert around the Henry Mountains of Utah in 1877. The term *pediment* was first applied by McGee (1897) to similar surfaces in the Sonoran Desert in Arizona. It denoted the triangular facade above the portico of a Greek building, and fringing pediments can present such a profile at a distance, giving the upland a high-set appearance. The name was adopted by Bryan (1922) for granitic hill-foot plains in southern Arizona, and in the next twenty years most writing on pediments continued to emanate from the southwestern United States.

A review by Tator (1952–3) based primarily on that literature reveals a wide application of the term pediment. It had been used of interfluves, valley straths, and stepped surfaces, with thicknesses of covering sediment ranging from nil to more than 6 m. Settings comprised granitic terrain in southern Arizona and California, faulted basin-and-range topography of the Great Basin, and footslopes of the southern Rockies and Colorado plateaux.

The range of pediment forms broadened further with wider geographical application. It embraced the French *glacis d'érosion,* which described gravel-mantled terraces on soft rocks below the Saharan Atlas (Dresch, 1949; Joly, 1950). Pediments in South Africa were part of high plains eroded in weak subhorizontal strata beneath steep *kopjes* crowned with hard dolerite or sandstone (Fair, 1948). Australian pediments included gibber-strewn footslopes below silcrete-capped tablelands. Surfaces described as pediments range in extent from hundreds of square kilometres to miniature footslopes only a few metres wide in badlands.

A recent bibliography of pediments, with many definitions, is that by Whitaker (1973).

The definition of pediment adopted here is *a piedmont plane cut in bedrock and separated from the backing hillslope by an abrupt change of gradient*. It will be seen that the form of the pediment is strongly influenced by the discharge of water and sediment from

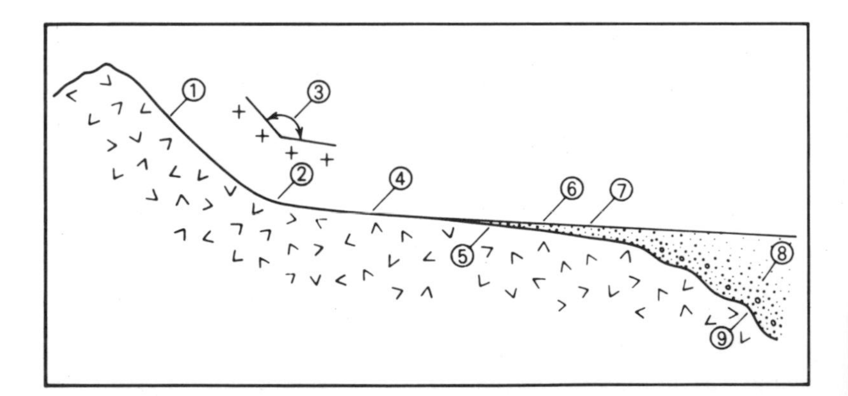

16. Terminology of pediments and related features of the desert piedmont zone, as employed in this volume: 1. backing hillslope; 2. piedmont junction; 3. piedmont angle; 4. pediment; 5. mantled pediment; 6. mantle; 7. alluvial plain; 8. alluvial fill; 9. suballuvial floor.

the backing hillslope, and the term should not be extended to include those more distant erosional plains which have been shaped independently. It may include plains where an original upland has been eroded away, as long as the position of the former piedmont junction is reflected in the overall profile. The pediment may be mantled with detritus but its surface form should be determined by the erosional base. *Pediment* and *mantled pediment* are used of subaerial and covered rock planes respectively, the latter where the cover is clearly not in immediate transit. Associated features are illustrated and defined in Fig. 16.

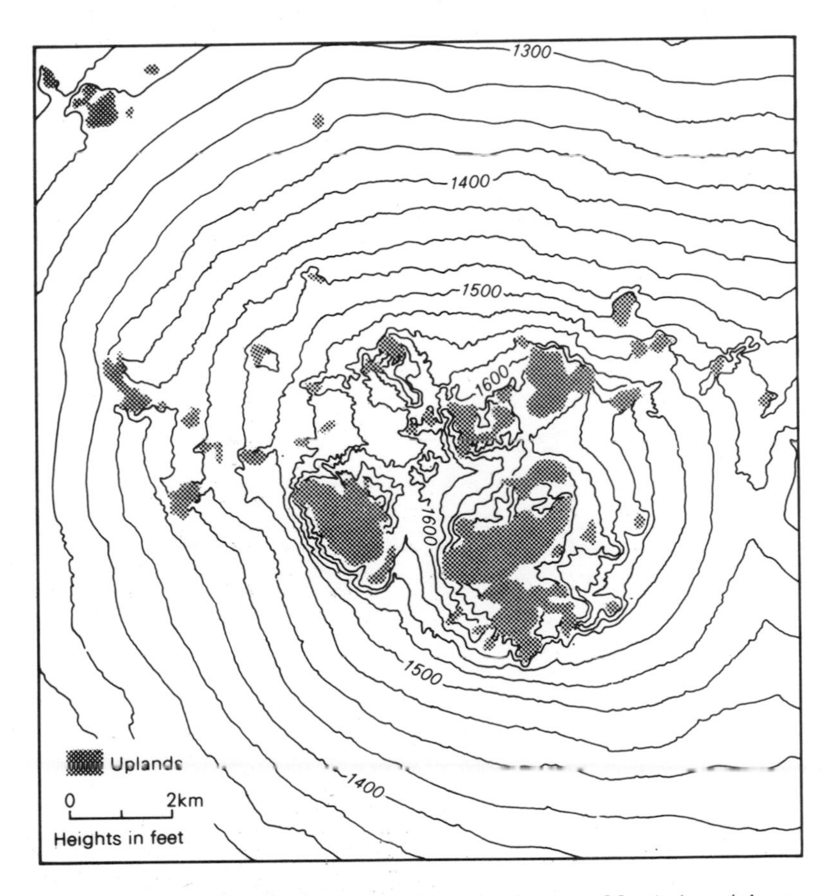

17 Contours on a pediment surrounding the Sacaton Mountains, Arizona. From Yi-Fu Tuan, 1959.

Forms of pediments

*Longitudinal profiles.** Pediments commonly extend from a hill foot to a trunk stream channel or the upslope margin of an aggradational plain. The profiles of pediment or pediment-mantle surfaces are typically concave to straight, the gradient decreasing with distance downslope to approximate to a logarithmic curve. Slopes diminish from as much as 10° at the head, where the smoothness of the slope may disguise its steepness, to as little as 0°15' at the foot. A convex lower sector may result from incision of a trunk stream channel. Pediments in homogeneous rock surrounding isolated uplands tend to have smooth radial profiles with downslope increase in contour spacing as in Fig. 17. Rock outcrops and outlying hills may protrude above the pediment in its upper parts.

As a graded slope of wash transport the overall gradient of the pediment is influenced by factors that affect hydraulic roughness, such as the coarseness of the surface debris and the density of perennial vegetation, by the magnitude and frequency of runoff across it and the amount and calibre of the sediment load, and by the rate of lowering of a trunk stream channel relative to its distance from the mountain front.

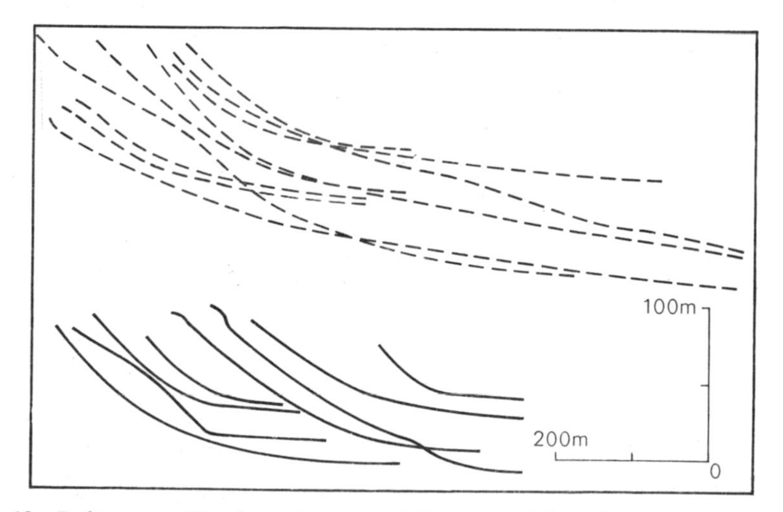

18. Pediment profiles from the semiarid Karroo (solid) and sub-humid Natal (broken), South Africa. From Fair, 1948.

*'Longitudinal' is used of the downslope profile normal to the hill foot and in the direction of water flow, by analogy with stream profiles. 'Transverse' is used of profiles along the main contour.

For example, pediment gradients are lessened and the piedmont junction is accentuated where aridity is greater, probably because of the smaller yield and finer dimension of debris from the hillslope and possibly also because of the diminished vegetation. This is shown by comparison of piedmont profiles from subhumid Natal with the more angular junctions on similar lithology in the semiarid Karroo in Fig. 18 (Fair, 1947, 1948). A maximum contrast is provided in the extreme-arid Sahara, for instance in western Tibesti where towers of flat-bedded sandstone stand sheer above sand-veneered pediments of gentlest gradient (Pl. 20; Peel, 1966).

An overriding control is bedrock lithology, particularly on the hill but also beneath the pediment, as a determinant of the calibre of the detritus which veneers or mantles the pediment. Granite and similar massive rocks, including some sandstones, which weather directly from large boulders to sand and grit with few particles of intermediate size yield well-defined pediments (Fig. 19), for the sharp contrast between boulders or rock on the hill face and sand or grit on the pediment is matched in the distinct separation of the slope elements by a *piedmont angle* (Figs. 16 and 20 Aa; Pl. 21). Even greater contrasts may result where fine-grained limestone is

20. Flat sandstone pediments in the extreme-arid Tibesti, eastern Sahara. Photo R. F. Peel.

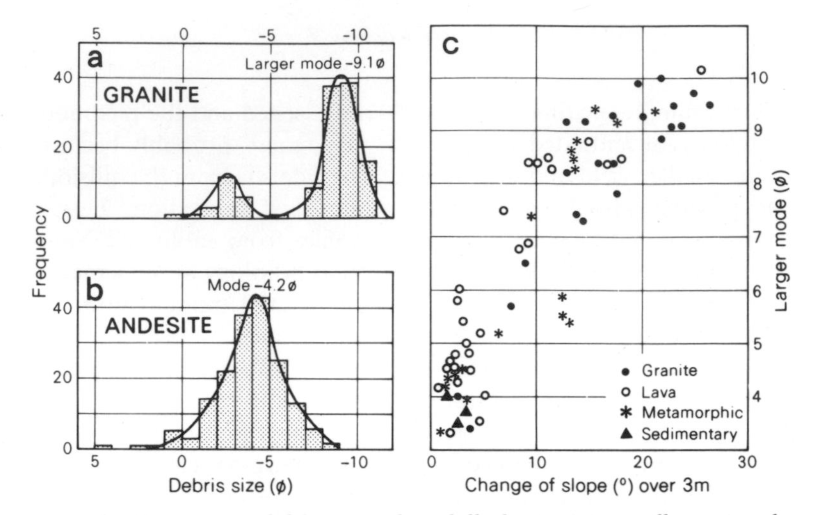

19. a, b. Histograms of debris size from hillsides in Arizona illustrating the bimodal size-distribution of granitic debris, as boulders and grus, and the unimodal size-distribution of andesitic debris, around the cobble range; c. rate of change of slope at the piedmont junction in relation to the modal diameter of largest debris. The extremes are dominated by igneous intrusive (granitic) and volcanic rocks. All from Rahn, 1966.

21. Piedmont angle on a granitic pediment, central Australia. The slope contrast is matched in the size-distribution of debris, with joint blocks on the hillslope and sandy grus on the pediment. Photo by the author, CSIRO, Division of Land Use Research.

subject mainly to solution weathering, giving an undercut angular junction (Fig. 20 Ab). Pediments of this general type, which may be classed as *stone-free* and which are predominantly granitic, typically have low gradients, rarely exceeding 5°, and rather straight profiles, expressive of the fine texture of the detritus and the lack of change in grain-size downslope.

Other rock settings, including mixed hard and soft rocks, may give a more complete size-range of debris, which then generally extends from the hill across the pediment with gradual downslope decrease in the calibre of the coarse fraction. In such cases the change of gradient is equally gradual. The piedmont junction is typically a concave transitional segment (Pl. 22) and over a distance of 50 m it may be impossible to define where the hillslope ends and the pediment begins (Fig. 20Ac). These forms may be classed as *stony pediments* and they are generally steeper and more concave than stone-free pediments. Gradients are steeper on rocks yielding much coarse debris and, for the same reason, where upland relief is strong.

22. Concave piedmont junction on a stony pediment, central Australia. A silcrete capping provides the gibber pavement on a pediment cut in shale and thin sandstone.

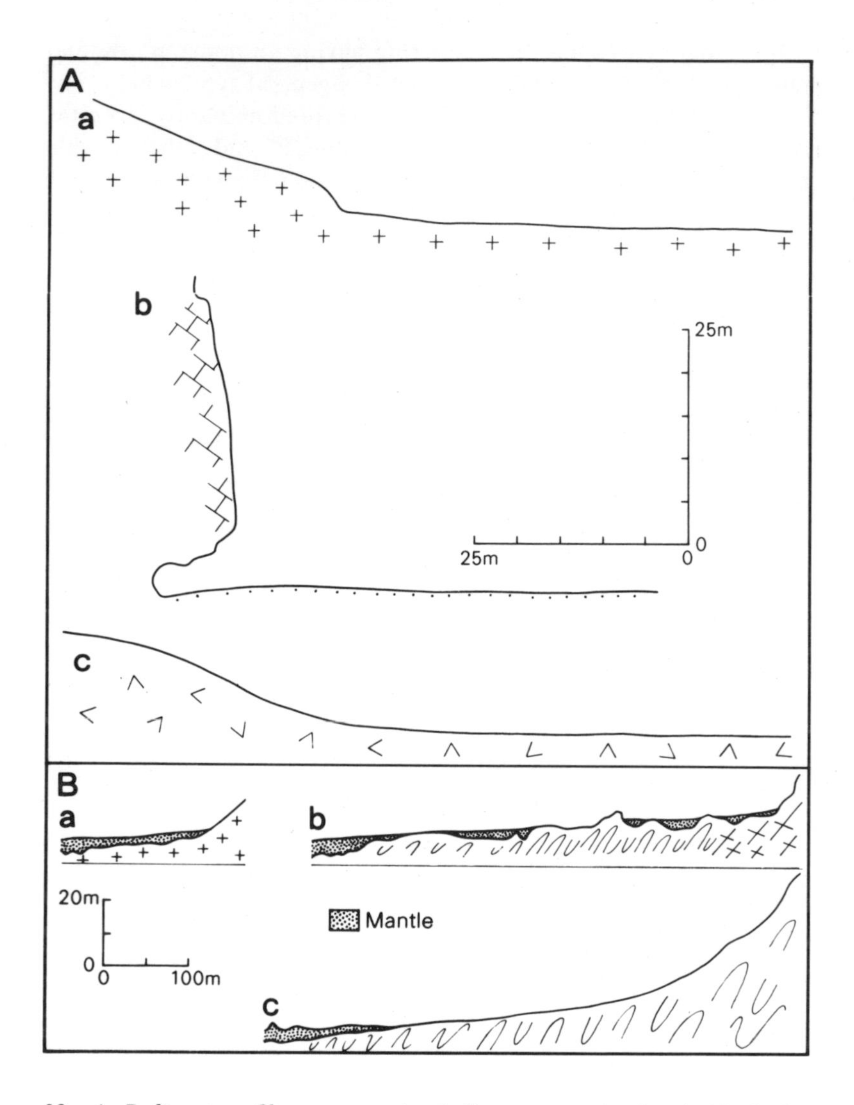

20. A. Pediment profiles on a. granite, b. limestone, and c. basalt, Kimberley District, Western Australia. (a and c from unpublished material supplied by C. R. Whitaker; b from Jennings and Sweeting, 1963).
B. Pediment profiles on a. mantled granitic pediment, central Australia; b. mantle smoothing an irregular pediment on gneiss and schist, central Australia; c. stony pediment on metamorphic rocks, mantled in lower sector only, western New South Wales. After Langford-Smith and Dury, 1964.

Drainage on pediments. At the heads of granitic stone-free pediments are distributary networks of tiny rills representing patterns of dying runoff that may at its maximum have approached sheetflow. They are shallowly cut into a superficial apron of loose grus. The rills head at the piedmont angle or stem from runnels on the hillslope but do not indent the hill base; the head of the pediment commonly consists of straight joint-controlled sectors, and extensions into the upland are typically broad blunt-headed embayments (Pl. 23). The rills continue a hundred metres or so down the pediment depending on the extent of the apron, which in turn is related to the dimensions of the upland, and then give place to small gullies incised a few decimetres into mantle or bedrock (Mabbutt, 1955). Few larger channels continue from the upland across the pediment, a lack which results in minimal transverse relief.

23. *Pediment embayment on granite, Namaqualand, South Africa. Note the piedmont angle, the light sandy wash adjacent, and the absence of channels from the upper pediment. Aerial photograph reproduced under Copyright Authority 5659 of 6.5. 1976 of the Government Printer of the Republic of South Africa.*

On stony pediments, wash processes doubtless operate on the interfluves but rills tend to be obstructed in their development or obscured by the stone pavements. The main drainage features on these pediments are stream channels which continue from upland valleys and which enter the pediment at re-entrants in the piedmont junction. Since slopes determined by flow in channels are gentler than those formed by wash, interfluvial sectors of stony pediments are steeper than and remain above channel tracts in the upper pediment sectors, giving an appreciable transverse relief.

Stream channels of all orders on pediments are shallow and sandy.

Mantles. Early descriptions of pediments allowed a metre or so of overburden as being in transit downslope; some later workers, particularly those favouring a fluvial origin for pediments, have accepted mantles tens of metres thick, claiming this to be no more than the depth of flood scour in large channels. The distinction between pediment and alluvial apron becomes blurred if it cannot be judged whether the rock base determines the surface form.

On granitic stone-free pediments the mantle commonly extends to the head without exceeding a few metres in thickness (Fig. 20Ba); in contrast, the mantle on many stony pediments is confined to the lower sector as a feathering wedge and encroaches upslope only along stream courses (Fig. 20Bc). Particularly in enclosed basins, the mantle may pass outwards into a thick fill above a non-planate *suballuvial floor* (Fig. 16). *Peripediment* has been applied to an intervening sector bevelling an older basin fill (Howard, 1942).

In interstream sectors where it has been washed from the hillslope, the mantle is poorly stratified and sorted, particularly near the piedmont junction, and has been termed *pedisediment*. In character it is intermediate between colluvium and alluvium, and it commonly grades into layered alluvium downslope and near stream courses. On granitic stone-free pediments the mantle consists mainly of sand and grit, often with a clay subsoil which may incorporate weathered bedrock, indicating erosional stability. Limestone pediments in the Kimberley District, Australia, have almost complete clay mantles, partly transported and partly residual (Jennings and Sweeting, 1963). Mantles on stony pediments may contain stone with finer materials, and the former may also be concentrated in a pavement. Older pediment mantles may be cemented by calcrete or gypcrete.

Hypotheses of pediment formation

Some explanations of pediments emphasise the retreat of the backing hillslope and with it the *extension* of the pediment. Since the pediment normally grades smoothly upwards to the piedmont junction, such explanations must tend to assume that hillslope retreat and pediment planation are the joint expression of a common group of processes. Some explanations of this type emphasise the work of minor rills and sheetflow following preparatory weathering, whilst others stress planation by stream channels.

Explanations invoking weathering, rill work and sheetflow. The scheme of Lawson (1915) is fundamental. He attributed retreat of the hillslope to weathering, wash and mass movement and claimed that the hillslope angle would remain constant, as set by structure and the angle of repose of large boulders. A rock bench would extend at its foot, buried and protected from further degradation by the growing apron of detritus yielded from the hill (Fig. 21a). The piedmont junction would thus be determined by the leading edge of the apron. The yield of detritus would decrease as the hillslope diminished through the rising encroachment of the apron, and this would be spread across an ever-increasing apron surface; consequently the apron would advance progressively more slowly and the rock floor under its protection would be increasingly flattened by weathering and erosion at its leading edge, to give a convex suballuvial bench.

This scheme was doubtless the result of acquaintance with the basin-and-range topography of the southwestern United States, where prominent fault scarps suggest an initiating tectonic form and where, in the enclosed basins, a detrital apron commonly extends to the hill foot.

Lawson held that at a late stage, with decrease in upland relief and sediment supply, the alluvial apron might be stripped near the hill foot and the underlying bench exposed as a rock floor. He considered the likely agent to be sheetfloods following cloudburst rains, probably because the original mention of pediments by McGee (1897) was linked with an account of a sheetflood that stressed its erosive capacity. Davis (1938) also considered that underloaded sheetfloods might at a late stage remove the alluvial cover and bevel the exposed and weathering pediment by 'rock floor robbing'. Overland flow as a continuous sheet on the pediment is probably localised and restricted to phases of maximum discharge;

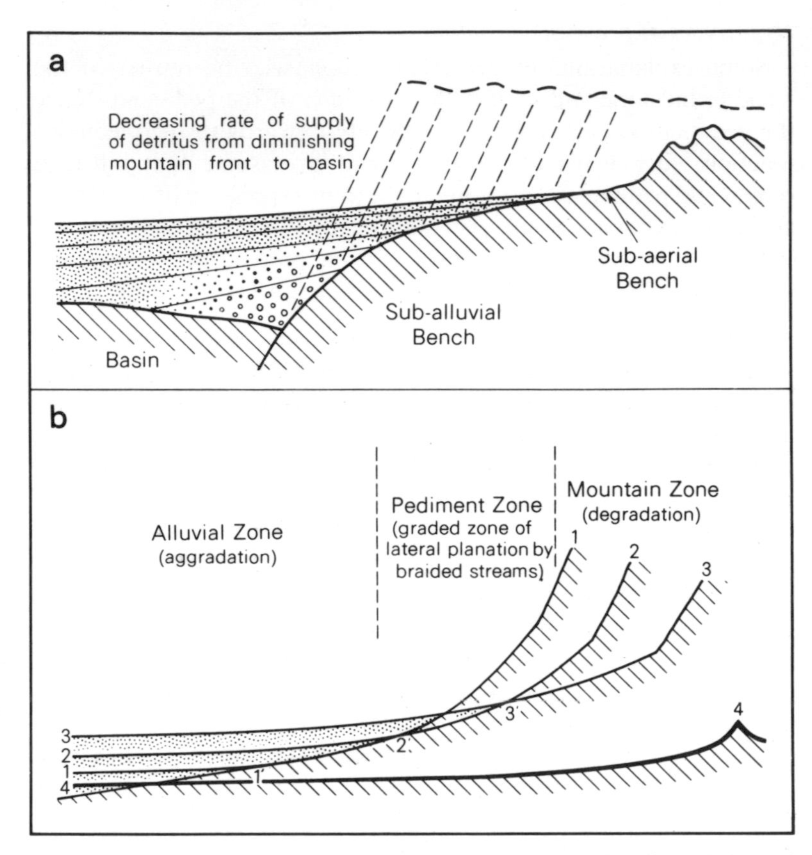

21. *Schemes of pediment formation: a. exposure of a suballuvial bench formed by weathering-retreat of mountain front (from Lawson, 1915); b. lateral stream planation (from Johnson, 1932b).*

at other times it takes the form of intermeshed tiny rills, which Davis included under 'sheetflood' and which Bryan (1922, 1925a) termed 'rill work'. Granted the transportive and erosive power of overland flow, it seemed feasible to these writers that it might also contribute to headward extension of the pediment by wearing back the hillslope.

Explanations invoking lateral planation by streams. Lateral erosion by migrating stream channels was described by Gilbert (1877), accepted as a contributory process by Bryan (1922), but particularly emphasised by Johnson (1932a, 1932b). He saw the pediment as a graded slope of transport between an upland subject to vertical erosion and an aggrading alluvial lowland; stream erosion in this intermediate zone would be limited to lateral cutting by shifting braided channels. In this way the pediment was smoothed and also extended headwards through undercutting by stream trimming at the hill foot (Fig. 21b). With diminishing upland area and sediment supply, the pediment would advance by undercutting of the hill base more rapidly than the aggrading plain, and as it lengthened it would be progressively regraded by streams. Undercutting of the hillslope would naturally be greatest at stream outlets from the upland, where triangular re-entrants would be opened, with rock fans.

Features postulated by Johnson have been identified, but evidence does not support his hypothesis as a general explanation. Even on pediments with large channels the lateral range of stream swinging is restricted; for example Sharp (1940) concluded from the present angles of stream outlets that not more than 40 per cent of the front of the East Humboldt Range in Nevada could have been impinged upon by channels. Yet pediment slope and piedmont junction are well-developed in the intervening parts inaccessible to stream channels. The evidence of stream action is least impressive on granitic pediments, including those in Arizona cited in support of lateral planation (Howard, 1942). Basal steepening of granitic hills is rare; embayments are blunt-headed rather than apical, their floors have rills rather than true channels, and they are concave in cross-profile rather than fan-shaped; the few small channels leaving the upland generally run normal to its front, yet the pediment is smoothest and the piedmont angle sharpest on the interfluves.

Retreat of the hillslope and the associated extension of the piedmont need not have been important during the fashioning of existing pediments, however, and accordingly some explanations of pediments do not stress this evolution.

Mantle-controlled planation. In many stable shield deserts, for example, the occurrence of relict weathered layers and duricrusts in the lowlands indicates that the main outlines of hill and plain were determined in the geologic past by compartmentation of deep weathering, that the lowlands have formed as partial etchplains during one or more cycles of stripping, and that the desert pediments of today have resulted from modest lowering and trimming of exposed former weathering fronts or other boundaries of differential weathering.

Since granitic rocks in deserts are physically resistant above-ground but susceptible to chemical weathering below the surface where moisture is retained, a mantle may contribute to planation of the pediment in various ways (Mabbutt, 1966a). Subsurface weathering of bedrock beneath the mantle is marked at the piedmont junction, which receives runoff from the hillslope; it attacks and may notch the hill base (Pl. 24), and so may contribute to oversteepening (Twidale, 1962) and certainly to the maintenance of the piedmont angle. Protrusions of an uneven bedrock surface into the mantle are similarly subject to enhanced weathering. These weathered layers are then removed during episodes of stripping of the mantle, whilst in some areas of fine weathering products lateral eluviation through the mantle is indicated (Ruxton, 1958).

24. Notching of granitic piedmont junction by subsurface weathering, central Australia. Photo by the author, CSIRO, Division of Land Use Research.

Where bedrock protrudes above it, the mantle surface constitutes a plane of concentrated physico-chemical weathering, as shown by ground-level tafoni, and also offers a baselevel for erosion by rills which remove and trim weathering rock.

In these ways the levelness of a blanketing depositional mantle may be transferred to the bedrock surface. (Fig. 20Bb).

Composite explanations. The relative importance of the above processes varies locally; for instance back-weathering, rill erosion and mantle-controlled planation may dominate on interfluves and on granitic pediments generally, whereas lateral stream planation is predictably more effective along drainage tracts, particularly on stony pediments. There is in any case no clear separation of the roles of changing interlacing rills and migrating stream channels.

Control of pediment form by wash load and pavement roughness

That various processes can produce a common pediment form suggests the operation of general controls, conceivably through the conditions of transport of waste across the piedmont. The planatory action of running water on the pediment results from an abundant though not excessive wash load which imposes a braiding habit on channels and rills alike, hampers the concentration of local runoff, and chokes incipient incisions. Suppression of dissection in the wash apron on a granitic pediment is shown in Pl. 23, where gullies head at the lower fringe. To the question whether unconcentrated runoff is a *cause* or an *effect* of smooth pediment surfaces, the answer is that process and form develop together under the control of the balance of runoff and sediment supply.

Another hindrance to linear water action on stony pediments is the relatively immobile pavement of coarser debris, which in the absence of close vegetation determines the hydraulic roughness, and becomes a dispersive impediment to channelled runoff. As a determinant of surface roughness, it also partly controls the gradient of the pediment. The link between pediment gradient and the calibre of the coarser pavement stone is illustrated in Fig. 22, on metamorphic rocks in the Mojave Desert (Cooke, 1970a; Cooke and Reeves, 1972). This explains the contrast between the straighter and overall gentler slopes of stone-free pediments and the more concave profiles of stony pediments with a cobble pavement.

Dury (1970) noted that the downslope decrease in the size of the larger gibbers on stony pediments in western New South Wales was

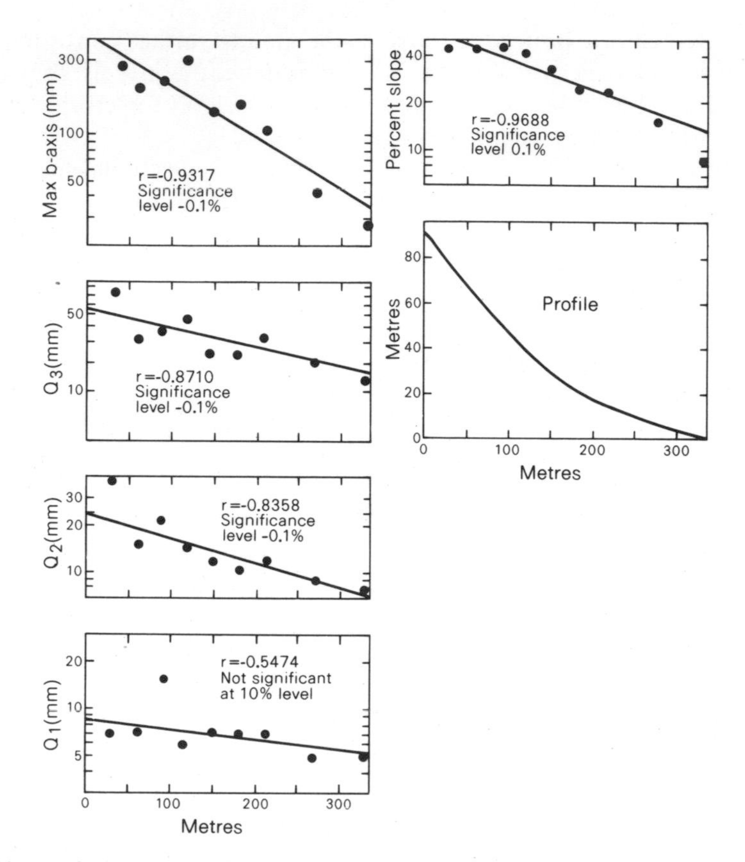

22. *Relationships between debris size and distance down the pediment, and hence with pediment slope, Apple Valley, Mojave Desert, California. Q_1, Q_2, Q_3 represent the three quartiles of pavement stone size. From Cooke and Reeves,* 1972.

not accompanied by improved sorting. This suggests that, following an initial gravity-assisted descent on the hillslope, there is little further selective transport of gibbers, as is also indicated by their inclusion in patterned ground, and that their reduced size downslope may be the result of weathering in place and reflect greater age as set by the period elapsed since the passage of a receding hillslope. With time, and concurrent decrease in pavement roughness, the lower pediment sectors have been regraded to gentler slopes.

In badlands without coarse debris, Schumm (1962) has shown

that the difference between the hydraulic roughness of a fissured clay hillside and a silt-veneered pediment allows the juxtaposition of slopes of 40° and 6° at the piedmont junction without effect on the depth or velocity of overland flow across it.

Baselevel control of pediment form

It has been claimed that ideal conditions for the extension and regrading of concave pediments are provided where alluvium is steadily removed by external drainage to provide a slowly subsiding baselevel (Bryan, 1922, 1925a); that a stable or slowly aggrading basin fill could result in a partly mantled, straight or convex pediment; and that rapid aggradation or removal of fill could give rise to buried or dissected pediments respectively. However, in the basin of the exoreic Gila River of southwest Arizona, cited by Bryan as an ideal case, the tributary streams have tended to form pediment terraces rather than extending undissected erosional planes, since stream activity has consisted of rapid erosion or aggradation in response to climatic and other variations, rather than of steady lowering. All that can be stated under this heading is that the effects of incision of a trunk channel should not regress sufficiently rapidly to cause incision of its tributaries in the piedmont sector.

Pediments, pediplains and slope recession

Similarities between desert pediments and concave valley-side floors in humid regions were recognised early (Davis, 1930), and the pediment was also seen as a special case of the concave basal segment of a standard faceted hillslope (Wood, 1942). The extension and coalescence of pediments through slope recession and elimination of uplands, or *pediplanation*, was invoked on a continental scale to explain widespread multi-concave erosion surfaces or *pediplains* with steep-sided hills in arid regions such as southern Africa (King, 1947). These concepts were subsequently extended and the pediment was claimed to be 'the fundamental landform to which epigene (subaerial) landscapes tend to be reduced the world over' (King, 1953, 1957, 1962).

This controversy cannot be entered into fully here, where the object is to explain features characteristic of desert pediments, for example the long smooth concave profiles and abrupt piedmont junctions that are unmatched elsewhere. Attention is, however, drawn to the fact that hill slope recession is not a prerequisite for the existence of a pediment, and to the danger of overstressing the genetic link between the pediment — as usefully defined — and broader patterns of desert upland and plain.

As already stated, pediment-fringed granitic uplands in many shield deserts were already isolated during more humid Tertiary periods of planation and deep weathering. In central Australia (Mabbutt, 1965a) and in the Mojave Desert (Oberlander, 1972) such hills stand above nearby fragments of weathered plains marked by laterite or by leached palaeosols. On those plains the piedmont profiles were more broadly concave than those of today, and subsequent evolution has merely involved the excavation of a piedmont angle in a former transitional slope, mainly by stripping the fossil weathered profile. In some cases the present piedmont angle was largely predetermined by a Tertiary weathering front sensitive to structural contrasts in the granite, although the angularity of the exposed surface has been further emphasised under the present arid climate by processes such as mantle-controlled planation. The extent of such an etched pediment bears little relation to contemporary subaerial back-wearing of the hillslope, which may have been negligible.

Strike lowlands backed by ranges of steeply dipping hard rocks, as in the Flinders Ranges of South Australia (Twidale, 1967), offer further examples of pediments below non-retreating hillsides. Hillslope recession is here opposed by fixed outcrop, and the close correspondence between stratigraphy and present-day landforms demonstrates that it can only have been minor; lithological boundaries have been so exploited during episodes of vertical erosion at the piedmont that the structural control of relief has constantly been reasserted. In these settings, sequences of pediments may be formed at successively lower levels without significant shift in the piedmont junction. Pediments of this type commonly bear protective gravel mantles, favouring the preservation of staged piedmont surfaces, where episodes of stream incision have intervened (Pl. 50).

In these quite common cases where the retreat of a frontal escarpment does not dominate in the reduction of an upland mass, denudation within upland catchments becomes the major factor in any extension of piedmont lowlands and eventually of pediments, by whatever processes. This pattern of evolution, as noted by Lustig (1969), has been sadly neglected in accounts of pediment formation.

Where extension of a pediment *is* linked with scarp retreat, modifications of form must follow from that recession. With increasing distance from the upland, load and pavement conditions may no longer maintain diffuse runoff on the lowest sector of the pediment, which then becomes subject to incision. In the part-duricrusted Interior Lowlands of arid Australia, for example, there is a downslope passage from concave pediments to multi-convex rolling lowlands with diminished control by the escarpment. It is for such reasons that the lowermost parts of pediments show considerable range of form, including convexity (Mammerickx, 1964). Because of the likelihood of such changes occurring as the landscape is opened out, extrapolation of the conditions of pediment planation on the scale of continent-wide erosion cycles appears unjustified.

Regrading of pediments

The maintenance of concavity on an extending pediment implies regrading. This may be achieved continuously through erosion by diffuse runoff, but many pediments show evidence of periodic dissection followed by renewed planation at lower levels; for example Davis (1938) postulated an alternation of sheetfloods and incising streamfloods on the extensive pediments of the Sonoran Desert. The effects vary with the setting.

Extensive shallow mantles on granitic pediments are subject to periodic dissection and stripping, with some removal of the weathered bedrock. Such forms have been described by Yi-Fu Tuan (1959) in southeast Arizona, where they are linked with the incision of arroyos into alluvial flats. Repeated mantling and shallow stripping of this kind, linked with the weathering and erosion described earlier, will result in a general lowering and levelling of the pediment over time.

Pediments at the foot of fixed strike ranges of steeply-dipping strata are subject to more spectacular incision, resulting in staged or nested pediments (*glacis emboîtés*) such as front the Saharan Atlas (Joly, 1950). These multiple surfaces may owe their preservation to gravel mantles or to reinforcement by duricrust. Trenching at the hill foot and the isolation of pediment spurs by headward cutting of strike channels may be facilitated by structurally-determined exceptional weathering of soft rock through long-continued infiltration at the piedmont junction (Twidale, 1967).

Dissection tends to be greatest at the hill foot, with revival of the piedmont angle, whence the staged pediments converge downslope. Staged surfaces are best developed at the outlets of major channels and tend to be replaced by a single polygenetic pediment in intervening sectors (Mensching and Raynal, 1954). Periodic general dissection of pediments in basins of interior drainage raises the question of climatic change, to be discussed in Chapter X.

On pediments below rapidly retreating scarps with thin caprock the stone mantle is generally too thin to preserve former profiles and the complex interplay of planation and slight incision results in undulating surfaces.

AGGRADATIONAL PIEDMONTS: FANS AND BAJADAS

The individual piedmont depositional landform is the *alluvial fan,* a body of detritus formed at the outlet of a mountain valley. It has the shape of a segment of a cone with its apex at the channel exit, and is concave in radial profile and convex transversely. *Alluvial cone* is sometimes used synonymously with alluvial fan, but may be reserved for smaller, steeper hill-foot forms which grade into talus cones of more angular gravity-fed debris. An alluvial fan typically has a radial distributary drainage, including active and fossil feeder channels as well as gullies heading on the fan itself. Fans range from less than a few hundred metres to tens of kilometres in radial extent. Slopes at the head may attain between 5° and 10°, whilst at the toe they are generally less than 1°, depending on the texture of the alluvium. An alluvial fan surface may be prolonged up-valley as an alluvial floodplain or terrace.

These features are expressed in the concentric arcuate contours of fans depicted in Fig. 23. The similarity with the pediment in Fig. 17 reflects the convergence of depositional and erosional forms through their common functional role as graded piedmont slopes of transport.

Distribution of arid-region alluvial fans

Alluvial fans are not restricted to deserts, but a combination of plentiful coarse debris on steep catchment slopes, occasional intense rains and sediment-charged flash floods, sharp topographic breaks at the upland front, and interior drainage with sediment accumulation favours their development in arid regions. They are generally

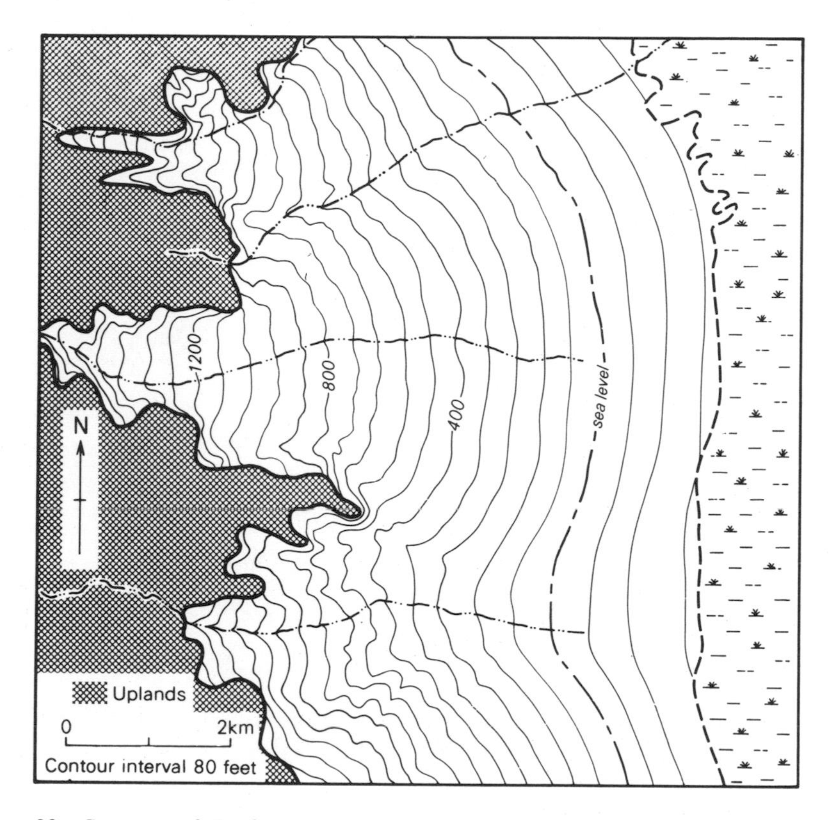

23. Contours of Trail Canyon Fan in the west of Death Valley, California (from Denny, 1965) for comparison with the pediment contours of Fig. 17. The similarity is enhanced because of the recent extension of deposition at the foot of this segmented fan; in other circumstances the concavity of the lowest sector would have been less marked.

associated with the vigorous tectonic relief of mountain-and-basin deserts; for example they flank young fold mountains in the dry Andes and central Asia, and reach their maximum development in enclosed tectonic basins or *bolsons* limited by prominent fault scarps, as in the southwestern United States. The floor of Death Valley in California, at 85 m below sealevel, is overlooked by summits above 2000 m. Such relief brings climatic gradients which further reinforce the tendency for deposition in the basin, resulting in the impressive fans in Pl. 25.

25. Coalescing alluvial fans, Death Valley, with contrast between dark pavements and lighter washes. These fans are extending due to uplift of the upland. Photo H. E. Malde.

In shield and platform deserts, fans occur along the margins of rift valleys, as along the Gulf of Agaba on the east of Sinai and its extension in the Jordan Rift. Otherwise they are poorly developed in these stable tectonic settings. For this reason fans are generally not well developed in arid Australia, an important exception being those along the faulted western front of the Flinders Ranges, extending into the Lake Torrens sunkland (Williams, 1973).

Fan-building processes and associated deposits

Streamfloods, debris flows and various intermediate forms of heavily charged flash runoff from steep catchments contribute to the growth of alluvial fans (Bull, 1972).

Characteristics of streamfloods in desert uplands have been

considered in the previous chapter, and the fan-building floods of desert piedmonts are similarly short-lived, highly sediment-charged and often violent. The active drainage tracts or *washes* of desert alluvial fans continue those of the upland drainage, with the same gravel or sand bars and interlacing shallow channels, the whole being subject to episodic overbank flows with alternate scouring and deposition. Near the apex of the fan these washes tend to be entrenched and the floods confined, but they become progressively less incised downslope and commonly intersect the general fan surface in its middle or lower sector. Below this *intersection point* is a flood-out zone of unconfined deposition marked by sandy or gravelly banks and lobes and terminal distributary channels. On fans with steep catchments yielding predominantly coarse debris, transverse bars and overbank splays of loose gravel may be left stranded at intermediate points along the washes due to excessive infiltration. These accumulations have been termed *sieve deposits* (Hooke, 1967) because the infiltrating waters remove the matrix of finer sediment downwards.

We are here only concerned with the terminal depositional stage of the debris flows or mudflows discussed in Chapter IV. These tend to issue on to the fan as elongate lobes, mainly confined within the defined wash tracts in upper fan sectors, although locally spilling over the banks in subsidiary lobes, and to spread out and terminate in multiple distal lobes at the intersection point as they become grounded by infiltration (Blackwelder, 1928). The greater viscosity shows in the steep margins of the lobes, which form prominent levees and bars where later surges have breached and forced aside earlier deposits. On these rims are perched the very large boulders that can be carried by viscous flows. Beaty (1963,1970) has described a debris flow of 1952 on a fan below the White Mountains of California and Nevada, which formed a lobe extending 3 km down the fan, attained several hundred metres width near its terminus, and contained more than 1 million m^3 of sediment. Debris flows normally last only a short time but may be followed by streamfloods which can scour out the channel fill and redistribute it down the fan.

Streamflood deposits on alluvial fans differ from those of debris flows in form and internal structure. The former show the stratified graded successions, lensed layering and current bedding characteristic of deposits from braiding streams. Deposits from debris flows on the other hand consist of overlapping irregular layers in which poorly sorted gravel 'floats' within a matrix of vesicular silty sand. The matrix may subsequently be washed out by rain from thin debris lobes leaving haphazardly strewn boulders. Attempts to distinguish systematically between streamflood and debris flow deposits on the basis of size-grading and sorting (cf. Bull, 1972) have proved unsatisfactory because of local effects of catchment lithology and topography, whilst the problem is further confused by the many intergrades between the two types. Fan sediments of whatever origin occur as lenticular beds less than a metre thick and a few tens of metres wide, in individual bodies of limited extent. This reflects the restricted area of active deposition on the fan at any time, and its shifting character.

Some alluvial fans, such as those of the White Mountains (Beaty, 1974), consist predominantly of debris flows, others almost entirely of waterlaid deposits, and others again of some mixture of the two. Blissenbach (1954) suggested that debris flows, as the result of localised cloudburst storms, might be expected to predominate in very dry areas, whereas under higher rainfalls there would be a greater likelihood of fluvial re-sorting. Debris flows might also be predicted to be important in cold deserts with seasonal frost and thaw on upland slopes, although this has yet to be determined. Geologic and topographic conditions in the catchment are likely to be the paramount controls, however. For instance debris flows are favoured by steep slopes of tectonically shattered or weathered rock, particularly where these include shale as a source of muddy matrix, as on the fans of the San Joaquin Valley adjoining the youthful Coast Range of California (Bull, 1964). Where both types of deposit occur, debris flows tend to be more important in the upper sectors and stream-laid sediments to predominate in the lower parts of the fan.

Surface features of alluvial fans

Since only part of a large alluvial fan is receiving sediment at any time, its surface is made up of sectors of differing age and dynamic status as revealed by contrasts in microrelief and degree of weathering. This is exemplified by Shadow Rock Fan in Deep Springs Valley, California, shown in Fig. 24 (Hooke, 1967). The youngest surfaces are active washes, typically a metre or so deep, with one or more braiding channels. They mainly lead from the head of the fan, but can also originate locally from the dissection

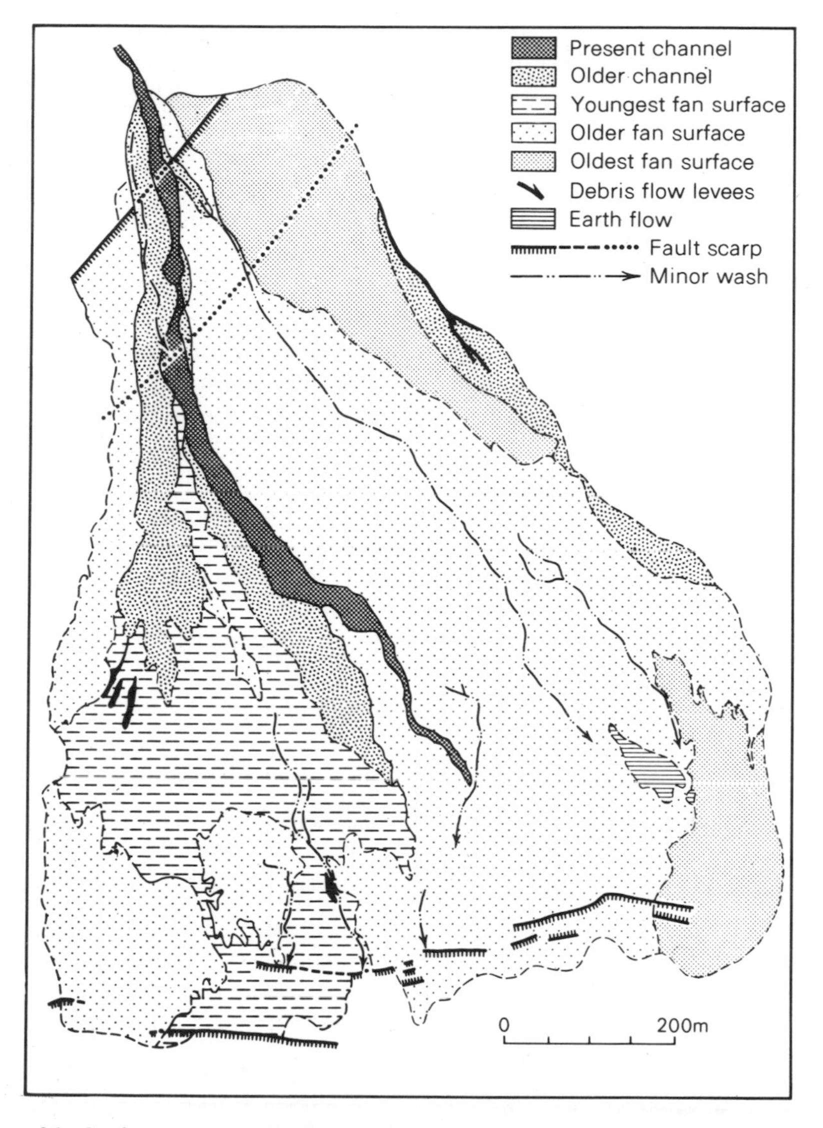

24. Surface types on Shadow Rock Fan, California. After Hooke, 1967.

of older fan surfaces. They are occupied during all floods and the freshly abraded gravel gives them a lighter tone, as shown in Pl. 25. They interbranch with older washes that are now followed only by the larger floods, with more subdued channel forms and lightly-varnished gravels. Washes are abandoned through braiding or crevassing, through choking by back-filling with alluvium, or through piracy by actively incising channels.

The washes constitute the zones of active or recent aggradation and can be markedly uneven in detail, particularly where the deposits are coarse. Fan surfaces in Sinai and Death Valley may, for example, have a transverse relief of a couple of metres, formed by the margins of depositional lobes, banks of scoured washes, or by the trimmed walls of older fan sectors.

Between these channel tracts are more stable sectors in which the relief and patterns of washes have been obliterated with time, leaving smooth or at most gently corrugated uniform stony pavements. The compaction and smoothing of fan pavements is assisted by gravitational settling of gravels when the finer matrix becomes saturated, helped by surface expansion and contraction, possibly enhanced by impregnating salts. Another contributory factor may be surface creep on appreciable upper fan slopes, as in Death Valley, where it has resulted in low boulder-fronted terracettes (Hunt *et al.*, 1966; Denny, 1965). The greater ages of the pavements are also shown by a fuller development of dark desert varnish on the gravel and by weathering, providing fine interstitial material and forming even surfaces. Fan pavements resemble stony pavements in other desert settings, which will be discussed in Chapter VI. The oldest pavements, becoming relatively impervious and subject to runoff, may generate dissecting gullies.

Causes of alluvial-fan formation

It has been traditional to stress the decrease of slope between upland and lowland as a major determinant of fan-building at the margin, and to regard the desert piedmont as a particularly favourable setting on this account. Enclosed desert basins in which a youthful, possibly active fault scarp sharply separates an upland catchment of bold, rejuvenated relief from an encroaching aggradational surface of moderate gradient are under this view regarded as ideal environments for the growth of alluvial fans. The coarse detritus yielded from steep desert mountain slopes and the characteristic high sediment concentrations of desert streamfloods should favour deposition at the valley outlet and the development of prominent constructional slopes. A straight mountain front with minor valley re-entrants, as in a fault-bounded desert basin, allows free shifting of aggradational washes below the outlet, resulting in well-formed and clearly demarcated fans.

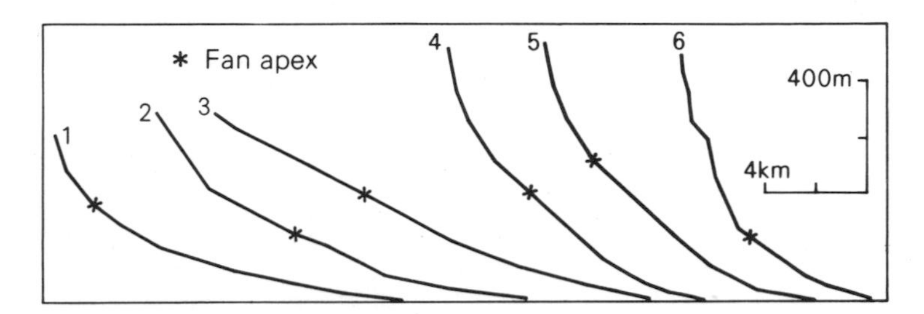

25. Radial profiles of main washes on fans in the Amargosa Valley, Death Valley region, California. From Denny, 1965.

These general tectonic, physiographic and climatic factors certainly determine the location and relative development of alluvial fans; for instance we see them combined in the excellent development of the Death Valley fans. However, they do not explain the mechanisms involved and are inadequate to account for the continuation of alluviation immediately below the outlet—essential for active fan formation—where a fan-head slope has already been built up to prolong that of an erosional valley floor, as it does for example in the fans of the Death Valley region shown in Fig. 25 (Denny, 1965). A reason must be sought in the changes in channel form and habit and in load-discharge relationships which occur at the outlet from the upland valley, particularly that from the single braiding channel of the confined valley floor to the bifurcating multiple washes of the open fan, with diminished depth and width of flood flow in the several channels. This constitutes a decrease in transportive power which is reinforced by marked loss of discharge through infiltration into pervious fan gravels, and consequently causes deposition.

Controls of fan form

In consequence of deposition along them, the active washes tend to become higher parts on the fan surface in its middle and upper sectors, as shown in Fig. 26A. This eventually results in the lateral shift of the wash to an adjacent lower-lying tract or to its abandonment, either through choking with alluvium or through

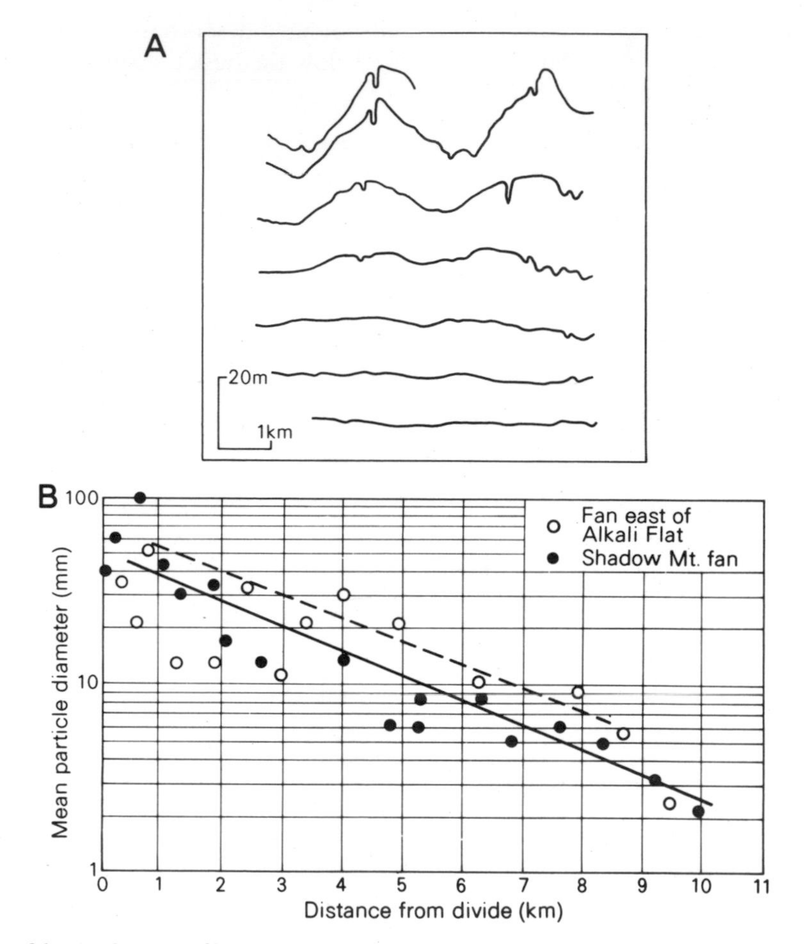

26. A. Cross-profiles of fans in the San Joaquin Valley, California, showing tendency for washes to form higher tracts in middle and upper sectors of fan. From Bull, 1964.
B. Down-fan decrease in mean particle size on two fans in the Death Valley region. From Denny, 1965.

piracy by a more advantageously placed channel. Even more important are periodic shifts in the loci of deposition along the lengths of the washes. As a result, the whole fan surface becomes subject to aggradation from migrating or changing feeder washes; however, deposition will tend to occur most frequently towards the head of the fan and along its median axis, producing the semi-

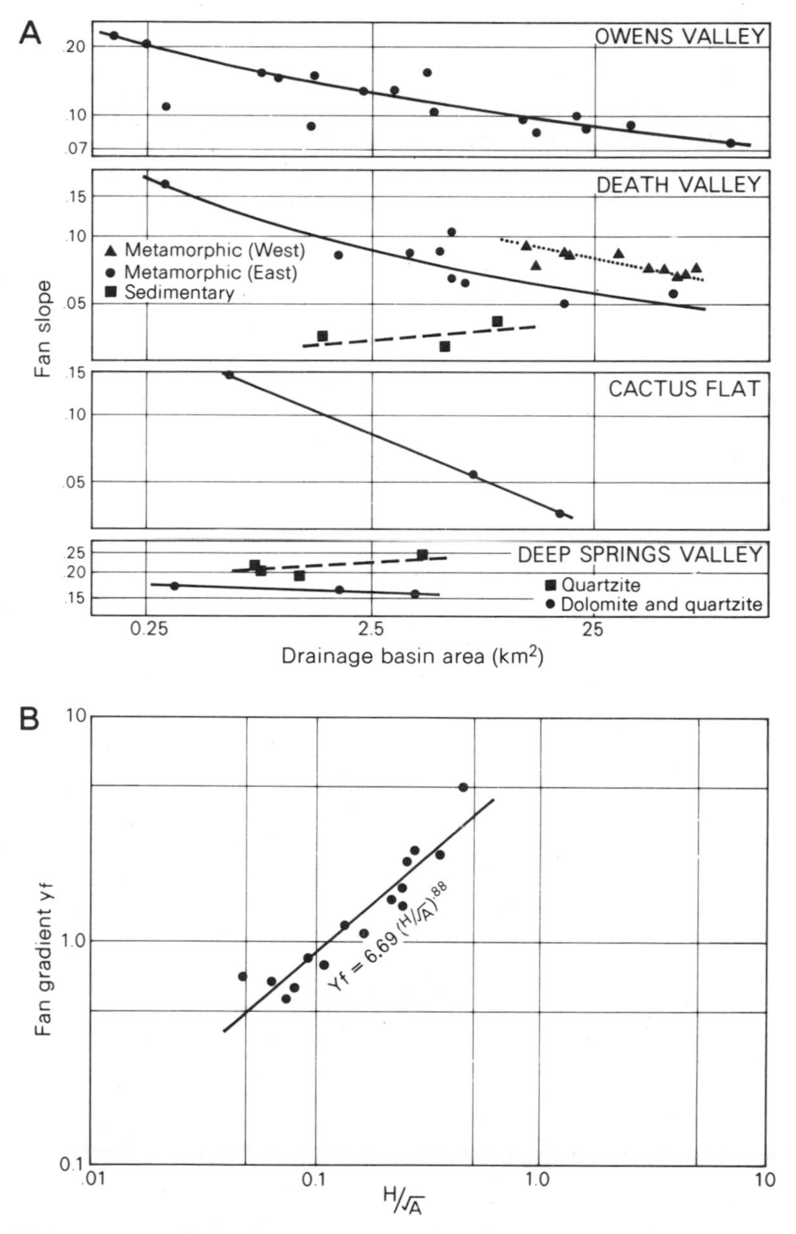

27. A. Inverse relationship between overall fan slope and drainage-basin area, southwestern United States. From Hooke, 1968.
B. Increase of gradient of fan-head washes with catchment relief in southern Arizona. From Melton, 1965b.

conical form. Radial profiles tend to be concave, as shown in Fig. 25, associated with the general decrease in mean grain size of the fan materials with distance from the apex (Fig. 26B).

In a given setting, fan gradients are determined by catchment conditions. A main control is exercised through the calibre of debris supplied to the fan, with which its slope is directly related. This is indicated incidentally in Fig. 27A, in which Death Valley fans with catchments in metamorphic rocks yielding coarser debris are seen to have significantly steeper gradients than those to which finer alluvium is supplied from younger sedimentary rocks. Linked with this is the control exercised by the predominant depositional process; for example, fans in Deep Springs Valley with catchments in quartzite are formed mainly of debris flows and sieve deposits and have steeper gradients than those derived from dolomite and quartzite, which are fed by debris flows and streamfloods (Fig. 27A), whilst these in turn are steeper than fans on which streamfloods predominate.

Overriding and encompassing such influences, however, is the control exercised by the vigour of erosion and deposition, generally determined by the tectonic movements which have fixed the vertical and horizontal dimensions of local relief (Bull, 1964). Despite their generally fine-textured deposits, for example, fans along the west side of the San Joaquin Valley of California are surprisingly steep. This is a function of the marked and active tectonic relief, and the resulting high rate of aggradation on the fans. In this area, fans derived from shale are steeper than those fed from sandstone because of the greater erodibility, more frequent debris flows and more rapid aggradation.

For this general reason, fan-head slope correlates well with basin relief as depicted in Fig. 27B (Melton, 1965b), probably through its influence over the gradient of the upland feeder channel by way of the supply and coarseness of debris. Conversely, fan slope generally decreases with increase in catchment area as shown in Fig. 27A, presumably as an association of diminished catchment slope and finer detritus but also because the greater discharge is enabled to transport the finer-textured alluvium further and across gentler gradients.

Segmented fans

Since fans are well developed in tectonically unstable areas, they are prone to modification as a result of earth movements. Furthermore, as dynamic landforms they can be expected to reflect morphologically any changes in regime introduced by climatic oscillations. Rejuvenation of a fault scarp or increased transportive power of feeder channels in consequence of a change of climate can result in dissection of the fan head and a shift downfan in the locus of deposition as shown in Fig. 28a. In this way a *segmented fan* develops by

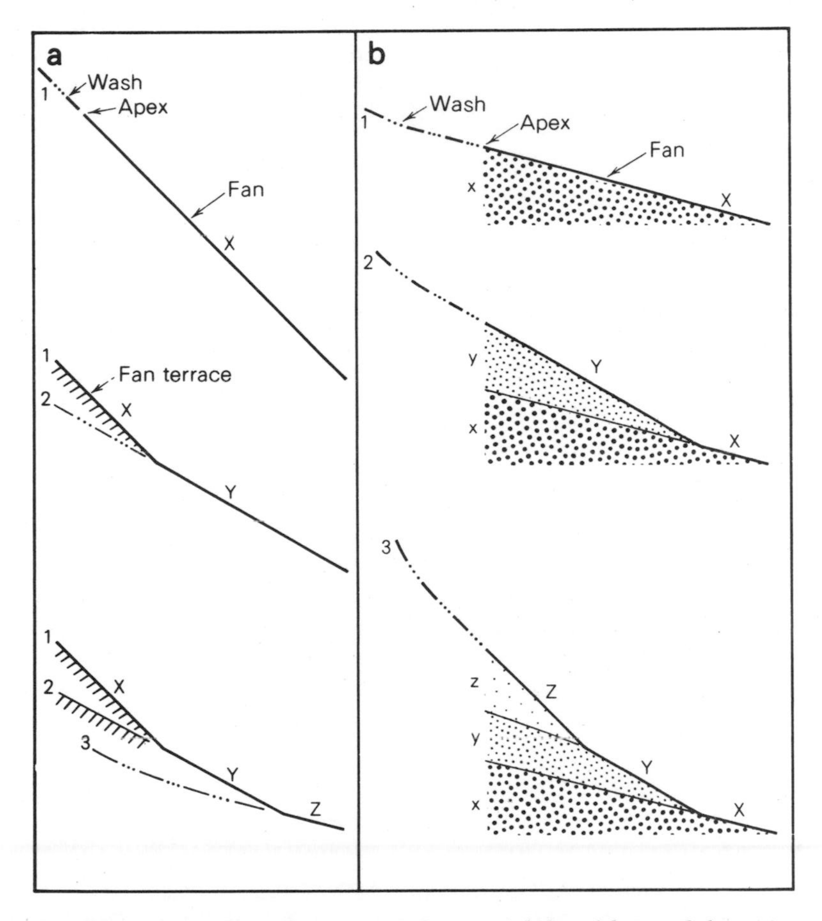

28. Schematic profiles of segmented fans: a. shifts of locus of deposition down-fan due to down-cutting and lowering of gradient of feeder channel; b. shifts of locus of deposition up-fan where gradient of feeder channel is steepened by successive uplifts. From Bull, 1964.

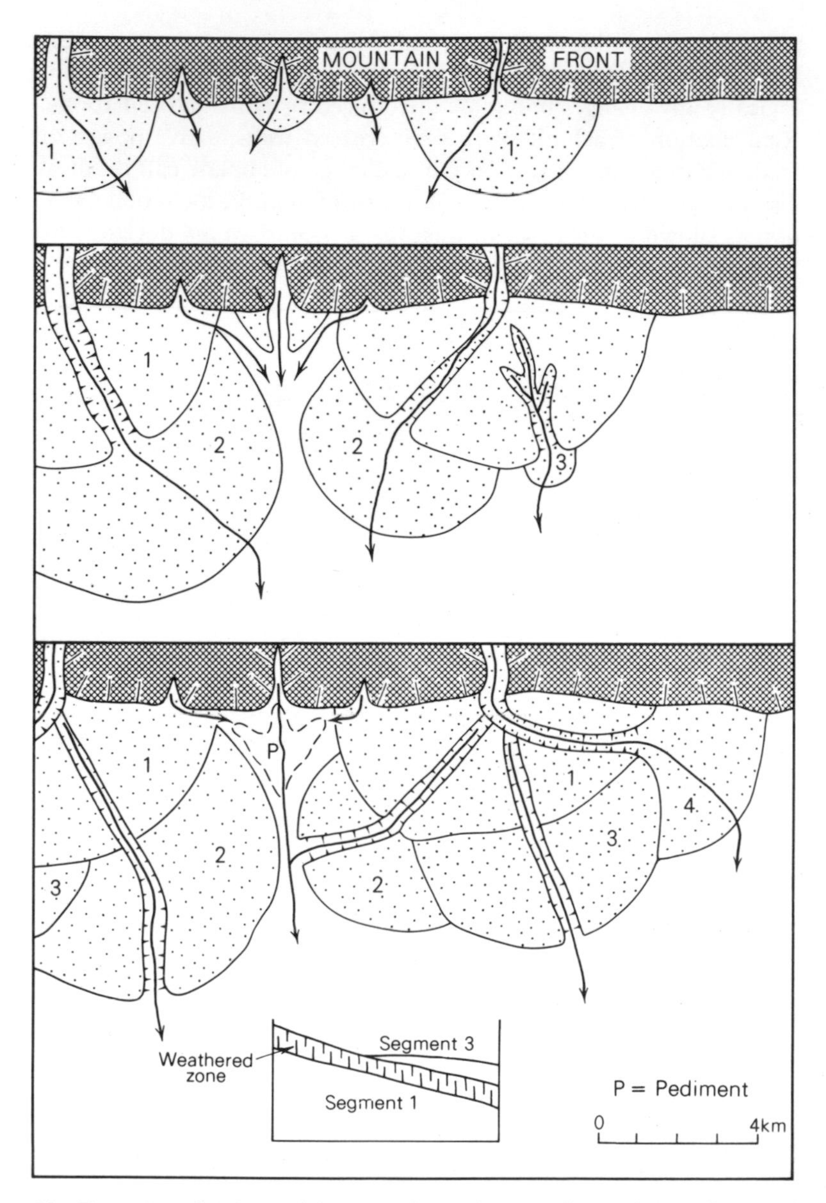

29. Extension of segmented fans into the piedmont under conditions illustrated in Fig. 28a. Successive stages of accretion are linked with shifts of entrenched washes. Stratigraphic relationship of segments shown in inset. After Denny, 1967.

the accretion of new sectors (Pl. 26), with a radial profile consisting of a number of segments of fairly uniform gradient rather than a continuous concave curve. Alternatively, segmentation can result where differential uplift beyond the rate of stream-cutting results in a steepening of the gradient of the upland stream, leading to renewed aggradation and the formation of a younger depositional segment at the fan head (Fig. 28b). Segment boundaries may be regular arcs extending from side to side of the fan, but the accretion of new segments may be irregular as in Fig. 29, because rejuvenation of a feeder channel is commonly accompanied by a shift of course. Segment boundaries form limits of overlap of the various alluvial formations making up a fan.

Fans fronting youthful fault scarps are commonly segmented. For example the fans along the west side of Death Valley have become incised at their heads and extended downslope due to down-tilting of the valley floor to the east, and for the same reason the fans on the east side are undergoing renewed alluviation at their heads where the slope has been rendered inadequate (Hunt *et al* .,

26. Segmented fan and playa margin, Death Valley. As a consequence of faulting, the feeder wash has become entrenched and a younger fan segment formed. Photo C. B. Beaty.

1966; Hooke, 1972). Segmentation is expressed in many of the fan profiles, as can be seen in Fig. 25, and this is confirmed by the stratigraphy of the fans. Two episodes of tilting are indicated. From the San Joaquin Valley also, Bull (1964) has cited examples of segmentation through renewed aggradation at the fan head as a consequence of repeated up-tilting. Each stage of the progressively steepening stream profile is prolonged in a related segment of the aggrading fan, which in turn prolongs an alluvial terrace.

Dissection of a fan head associated with a shift in the locus of deposition must be distinguished from the entrenchment of feeder washes arising merely from changes in the patterns of drainage or the scouring out of channel alluvium. Confirmation of a change in fan regime may not be simple however, and there are important differences of interpretation of the forms of alluvial fans in the southwestern United States, for example.

Bajadas

Compound or *coalesced fans* link to form aggradational piedmont plains, particularly in enclosed basins. These are termed *bajadas* from the southwestern United States, where they occupy bolsons. Their slopes are those of the confluent fans and the profile parallel with the upland front is undulating due to the convexities of fan elements (Blackwelder, 1931). The bajada typically leads to flats of finer-textured alluvium with playas.

Dynamics of alluvial fans

Rates of fan-formation. Fans are among the most rapidly developing desert landforms. For example Beaty (1970) considers that the Milner Creek Fan west of the White Mountains postdates the Bishop Tuff and is therefore younger than 700,000 y. This gives an average annual accumulation of 3000 m^3, or an upward growth of between 7.5 and 15 cm per thousand years. Fans on the west side of the San Joaquin Valley are of comparable age (cited in Hooke, 1968), with an indicated upward growth of 20 cm per thousand years. Hooke (1972) has estimated the average rate of accumulation on the Death Valley fans as 1 m per thousand years, on the assumption that it is equal to that on the adjacent playas.

Fans as evidence of dynamic equilibrium. Such an interpretation is suggested by the empirical relationship between *fan area* (A_f) and drainage catchment area (A_d) of the form

$$A_f = cA_d^{\ n}$$

established by several investigators in the western United States (Bull, 1964; Denny, 1965, 1967; Hooke, 1968). In Fig. 30 the mean value of the exponent lies close to 0.9, similar to that linking catch-

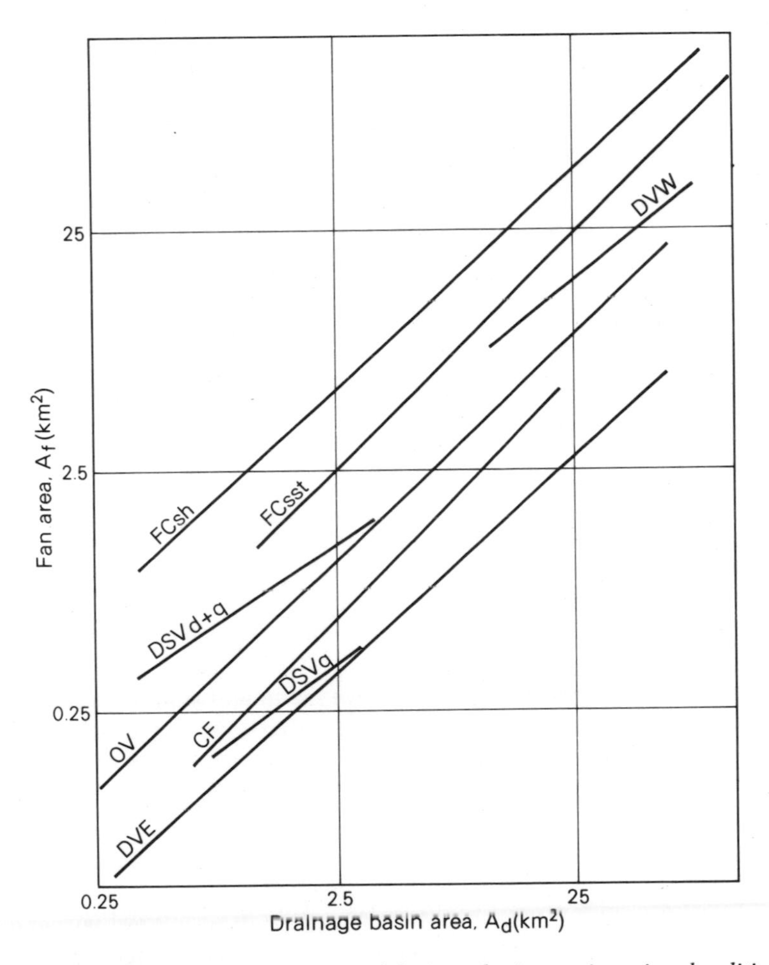

30. Relationship between fan area and drainage basin area in various localities in the southwestern United States. From Hooke, 1968. FC-Fresno County, DVW-Death Valley-west, DSV-Deep Springs Valley, CF-Cactus Flats, OV-Owens Valley, DVE-Death Valley east; sh-shale, sst-sandstone, d-dolomite, q-quartzite.

ment area and stream-sediment yield. This underlines the effectiveness of an alluvial fan as a trap for most of the sediment yielded from its drainage basin, as mentioned for example in Chapter IV in connection with the sediment budget of the Nahal Yael. That the exponent is less than unity probably reflects the lower average slopes of larger catchments, an excess of catchment area over that of the average runoff-yielding storm, and increased opportunity for storage of alluvium.

Denny (1967) held that this equilibrium was attained through a balance between aggradation and denudation on the fan itself, as exemplified in the opposed action of washes heading in the upland and those originating on the fan, the former building up the fan surface and the latter removing finer secondary products to the basin flats. As a fan grows by the accretion of new sectors, the proportion aggrading at any time must decrease if discharge and sediment supply remain constant, and the area subject to erosion will accordingly enlarge. When a balance is eventually attained, the fan may be regarded as a store of debris awaiting further comminution before undergoing renewed transport beyond the piedmont. This is achieved through the weathering and erosion of fan pavements, which may constitute more than half the surface of a mature fan.

On the other hand, Hooke (1968) has stressed the importance of rising baselevels in the bolsons in which these relationships have been established. If deposition is taking place uniformly over its surface, the extent of a fan should reflect a competitive balance with the aggrading fans on its margins and the playa flats below. The fan will accordingly extend or diminish until the sediment supplied achieves a common rate of aggradation. The value c in the relationship between fan area and catchment area, which determines the intercepts of the regression lines in Fig. 30, will vary *between* bolsons in accordance with aggradation rates set by the relative areas of basin and upland; for example it is lower in steeply enclosed basins such as Owens Valley, and higher in the more open settings of the Mojave Desert. *Within* bolsons, c will vary with tectonic, lithologic and relief controls of catchment dynamics; this is why catchments in erodible shale in the San Joaquin Valley (Fresno County) have produced fans about twice as extensive as catchments of comparable size underlain by sandstone. This concept of an equilibrium state gives greater attention to the shape of the basin of aggradation and consequent variation in the thickness of the alluvial deposits.

Those who support this interpretation would cite in further support the relationships between fan gradient and catchment characteristics discussed earlier.

On the other hand, those who view the control of tectonics over the course of fan formation as primary present an alternative view (Bull, pers. comm.). With tectonic activation of erosion and aggradation, fans might be expected to extend in proportion with catchment area as the determinant of sediment supply. At this stage the exponential relationship between fan area and catchment area is one of allometric growth rather than of steady-state conditions. For example, it holds for Death Valley fans which are actively growing and which cannot be in equilibrium. In due course the fan may achieve a size that is optimal in terms of its competition for space with adjacent landforms, as proposed by Hooke (1968), but with cessation of tectonic activity and consequent fall-off in sediment supply, erosion of the fan deposits will predominate. In any setting, therefore, dynamic equilibrium will occupy only a moment of geologic time as allowed by tectonic controls.

Fans as cyclic landforms. Davis (1905) considered that the growth of alluvial fans belonged to a youthful stage of the cycle of desert landforms, when strong relief yielded abundant waste to the piedmont. He envisaged that fans would encroach in *fan-bays* as the upland drowned in its own detritus (Davis, 1938). Eventually, fans and bajadas would drown the retreating uplands and link across basin divides. In the long-continuing later stages of the cycle, with diminished relief and sediment yield and under the influence of lowering by deflation, aggradation would be overtaken by erosion, sheetfloods and streamfloods would strip the alluvial formations, and erosional surfaces or pediments would predominate.

RELATIONSHIP BETWEEN DEPOSITIONAL AND EROSIONAL PIEDMONTS

Under the cyclic scheme of the evolution of desert landforms as propounded by Davis (1905), aggradational forms were regarded as characteristic of the early stages of strong relief, with streamfloods in the uplands and an abundant supply of waste to the adjacent lowland. Erosional piedmonts were seen as developing in a slowly evolving late stage of the cycle, under the influence of slow lowering of baselevel through deflation. At this stage, underloaded sheetfloods, under conditions of diminished relief and waste production, were held to be capable of exposing the rock floors of the piedmont zone and of regrading pediment slopes.

At a time-independent scale, Denny (1967) has stressed the dynamic balance of present-day processes as a determinant of the relative importance of aggradational and erosional piedmonts in deserts. A major control is held to be the relative extent of upland and plain as determined by structure. The argument runs as follows. The shale piedmont lowlands about the Henry Mountains of Utah are large and the uplands are small islands of resistant porphyry, and accordingly pediments are extensive; in contrast, Death Valley is a small basin between large and vigorously uplifted blocks, and here fans and bajadas dominate. Where fans and pediments occur together, Denny has claimed, as evidence of dynamic equilibrium, that the areas of the pediments, like those of the fans, are commonly proportional to their catchment areas. Pediments are likely to predominate where sediment is removed by out-going drainage, as by the Rio Grande and the Colorado River systems in much of the arid United States.

However, differences of structural setting invalidate presentation of the landscapes of the Mojave Desert and of the Great Basin as contrasted evolutionary stages as under the Davisian scheme, just as they oppose an explanation of the pediments of the Henry Mountains and the bajadas of Death Valley as contrasting steady-state expressions of the differing relative extent of upland and plain in the two areas. Shadow Mountain Fan, cited by Denny (1967) as a combined fan and pediment landscape, appears more properly a partly mantled pediment, fed by multiple rills from the backing hillslope and with no master stream.

There is no *general* tendency for piedmont aggradation under aridity. In the stable settings of shield and platform deserts, upland valleys and piedmonts have become graded over long periods, the predominantly fine waste from extensive granitic rocks is swept clear and spread thinly over extensive lowlands, often with out-going drainage, and pediments are widespread. For converse reasons, piedmonts in the tectonically youthful, vigorous relief of mountain-and-basin deserts are dominated by aggradation, with extensive alluvial fans and bajadas.

VI

STONY DESERTS

DESERT PAVEMENTS

Stone mantles are extensive in deserts, not only on hillslopes but on a range of lowland surfaces including alluvial fans and pediments, and are termed *pavements* where the stones are closely packed on flat or moderately inclined plane surfaces. These are rendered prominent by lack of vegetation, but their formation and persistence are also favoured by processes typical of if not exclusive to arid regions. They achieve their maximal effect in areas of low relief, where it is the surface itself that impresses rather than the contours of the land. Hume (1925:62–3) wrote of their overwhelming monotony in the Egyptian Desert:

> Far as the eye can reach there is nothing but plain or at most gentle undulation, unbroken by any conspicuous elevation; a plain of bare rock fragments shimmering in the sunlight, a vast expanse from whence life has fled, where mirage distorts the lowest scarp or long upward slope into a high cliff, and the cairns which mark the main camel tracks are weirdly tremulous as seen through the medium of the heated air.

Hamada and reg

Desert stone pavements range from rocky or boulder-strewn surfaces to smooth plains of finest gravel, and in the Old World deserts these differences are recognised in the vernacular. *Hamada* (Arabic: 'unfruitful') describes difficult bouldery terrain and *reg* (Arabic: 'becoming smaller') trafficable finer pavements, with *serir* replacing reg in the central Sahara. Most hamada pavements are residual or consist of boulders transported only short distances, whereas a larger proportion of regs consist mainly of transported stones. However, it will be seen that a distinction based on the size of the pavement material accords only in some degree with that by origin. Inadequate genetic distinctions become still more blurred as older pavements are compacted and made smooth by weathering to smaller fragments and interstitial soil, as their components are redistributed by wash, or as they are levelled by trapping waterborne or aeolian sediment.

Another claimed distinction is that reg pavements are commonly underlain by a soil profile, more or less developed depending on the age and stability of the surface, whereas hamada boulders rest on bedrock. This contrast may follow from the greater resistance of the rocks yielding residual hamada boulders, or from closeness to the source area where the hamada boulders have been transported.

Hamada also has the connotation of a structural tableland in much of the Sahara, generally in association with flat-bedded rocks or near-horizonal weathered crusts. In contrast, regs are generally plains. The Tibesti Serir and Gattusa Serir of the eastern Sahara for instance are alluvial plains with gradients of between 0.1 and 0.2 per cent and with a local relief of less than 50 cm (Fürst, 1966). Many such alluvial pavements now appear as relict surfaces, beyond the limits of active drainage or left standing above it through dissection, for they are durable features of the desert landscape which require a period of stability for their full development.

Hamadas. There are two types of surface to which the name is applied, namely outcrop, or *rock hamada* (Pl. 27), and *boulder*

27. Rock hamada, Rajasthan Desert, India. Photo C. S. Christian, CSIRO, Division of Land Use Research.

hamada. These may occur together in a complex, but as Fürst (1965) has noted on hamada tablelands in the Fezzan, there is commonly a downslope sequence from rock hamada behind the scarp, through boulder hamada to the more compact even surfaces in the lower parts and shallow depressions for which the term reg is appropriate. This follows from differences of surface stability, with stripping in exposed sites and re-deposition of finer materials downslope.

A residual origin is shown in the angularity and lack of sorting of the rock fragments in many hamada pavements, and in an affinity with the rock below. Nevertheless boulders such as basalt may be rounded *in situ* by weathering, and residual hamada pavements may also include boulders rounded by transport where these are yielded from the local rock, as in the 'conglomeratic hamada' of the Negev (Evenari *et al.,* 1971). Hamada pavements that are not residual include the bouldery upper sectors of many alluvial fans.

Hamada surface characteristics are largely determined by the mode of breakdown of the underlying rock. In the Libyan Sahara, for example, compacted bedded limestones yield stepped rock pavements with loose slabs, whereas softer limestones form smoother surfaces with smaller fragments of bedrock and weathered-out flints. Hamada on Nubian Sandstone is dark-toned

28. Basalt boulder hamada, Libyan Sahara. Photo W. Meckelein.

and uneven, with large blocks and much sand. Most daunting are the black basaltic hamadas as in the Fezzan, with packed spherical boulders up to a metre across (Meckelein, 1959; Pl. 28). Stony tableland surfaces with silcrete duricrust in Australia range from bossy rock pavements to expanses of brown boulders or 'gibbers' with varying soil cover.

Regs. The alluvial character of a reg pavement may be indicated by sorted rounded gravel of mixed composition and distant origin (Pl. 29), and also by its occurrence near to or in extension of a stream channel. As in the Tibesti Serir, airphotos may reveal traces of former distributary drainage systems where such a pavement appears to lie beyond present flood limits (Fürst, 1966). Residual regs, which are usually closely associated with hamadas, generally consist of angular flakes of local bedrock or of a less weatherable residue, as in flint pavements on soft limestone in the Negev. However, they may contain rounded stones from a source rock, as in parts of the Gattusa Serir derived from conglomeratic Nubian Sandstone (Meckelein, 1959). Problems are posed by *mixed reg* pavements where gravels have been fractured in weathering or where residual material has been redistributed by surface wash

29. Reg with rounded gravel, Rajasthan Desert. Photo C. S. Christian, CSIRO, Division of Land Use Research.

30. Compact stone pavement with polygonal arrangement of larger silcrete gibbers, central Australia. Photo R. W. Millington, CSIRO, Division of Land Use Research.

(Alimen, 1958).

Reg pavements comprise two elements in varying proportion: a compact mosaic of small stones embedded in soil and a more uneven component of larger fragments lying loose on that surface or protruding through it (Pl. 30). Sharon (1962)* designated these the 'basal' and 'loose' horizons. The latter naturally predominates in boulder hamada.

Formation of stone pavements

The pavements generally consist of a concentration of stone in a surface layer a few centimetres thick, above soil which may contain stones dispersed through it or which may be relatively stone-free.

Concentration by deflation. A traditional explanation is that the pavement is a lag or veneer left by the deflation of fines, which continued until the stone was sufficiently closely spaced and protuberant to act as a protective *desert armour*. Symmons and Hemming (1968) estimated that a reg pavement in the southern Sahara was equivalent to the stone contained in the soil over a depth

* Despite the title of his paper, the surfaces described are generally accepted as regs.

of 4.3 cm, and hence represented a lowering of the surface by that amount.

The effectiveness of deflation on desert pavements is less than is commonly asserted, however. Sharon (1962) recorded initial deflation on newly cleared and disturbed plots on limestone reg pavements in the Negev, but noted that under natural conditions the soils resisted wind erosion because they were silty and cohesive and tended to form a thin crust on repeated moistening and drying. This situation applies to many desert pavements. In any case, the amount of lowering and stone-concentration by deflation is limited because the efficacy of wind erosion diminishes markedly as the protective stone cover increases and is virtually ineffective when it reaches 50 per cent. Undisturbed stone pavements are among the most wind-stable of desert surfaces.

Concentration by wash erosion. Rainbeat and scour by overland flow appear to be more effective than wind in eroding fine-textured soil on sloping desert pavements. The plots cleared of stone by Sharon in the Negev had slopes of 5°, and wash accounted for most of a lowering of the surface by 5–50 cm over a period of five years, during which time the stone pavement was renewed from below by differential erosion.

Concentration by upward displacement of stone. Pavements on stony tablelands in arid Australia commonly occur on and in the topsoils of weakly solonetzic profiles in which the subsoils are almost stone-free, as shown in Fig. 31 (Mabbutt, 1965b). The pavement stones are silcrete gibbers which recur in quantity beneath the subsoil and which have originated from an underlying duricrust. For the pavements illustrated to have formed as residues would have required a stripping of more than 1.5 m of the solum, and it is much more probable that the stone has been brought to the surface by processes acting within the soil. Comparable profiles are known in soils formed from basalt in the Nevadan Desert (Springer, 1958), and in alluvial soils in the Atacama and Californian deserts (Cooke, 1970b). The soils which exhibit this phenomenon contain expansive clays in base-saturated complexes and are subject to swelling and heaving on wetting and to shrinkage and deep cracking on drying.

Where the soil is residual it appears probable that the stones have moved upwards through the solum. Such movements have been demonstrated experimentally in similar soil material, although the

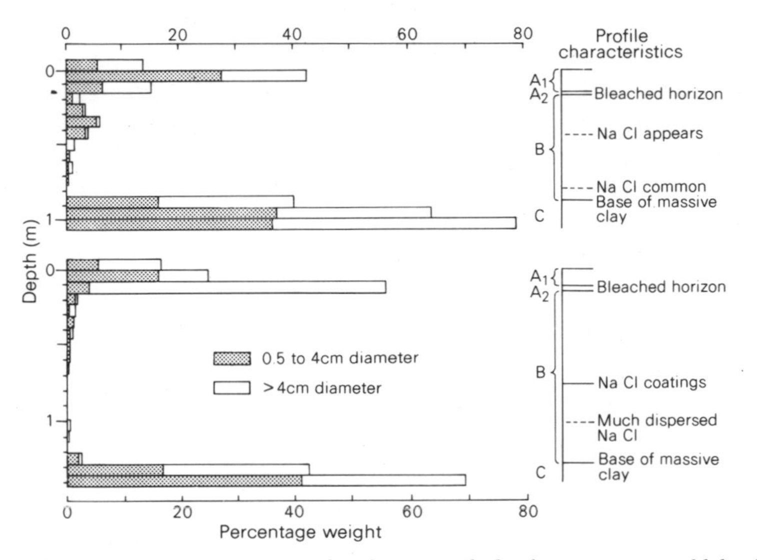

31. Variations in percentage weight of stone with depth in two stony tableland soil profiles in central Australia. From Mabbutt, 1965b.

mechanism in a soil profile is not exactly known. It is possible that the stones shift upwards as the underlying soil swells and that fine particles falling down the cracks that develop as the soil dries prevent their complete return. The few stones remaining in the subsoils in Fig. 31 are small grades that could be accommodated in cracks. It may be that the stones induce differential swelling in the soil by speeding the downward advance of the wetting front locally, and are displaced upwards in consequence (Yaalon, pers. comm.), or that the mere existence of an inert stone within a swelling subsoil produces those effects. The processes suggested resemble those postulated for types of sorted patterned ground in deserts with which upward-sorted stone pavements are associated.

An alternative form of displacement is possible where the soil material has been transported, particularly by wind. For example Jessup (1960) has described from South Australia a layer of gypseous clay considered to be aeolian, with a stone pavement similar to that on a buried pediment beneath. It is possible that, as the original pavement trapped wind-borne dust, some of the stone was displaced upwards *pari passu* through swelling of the aeolian clays on wetting during the course of their accumulation, and that the stones were never deeply buried.

Relative concentration by surface weathering. Stone contained in the relatively moister environment of a desert soil is more prone to weathering than that on the surface, particularly where the soil is impregnated with salt or gypsum. Consequently a stone pavement may survive above a soil with little stone. The phenomenon is marked in granitic gravels; for example some fan terraces in eastern Sinai bear stone pavements above horizons of granitic grus formed by breakdown of boulders beneath the surface. This subsurface horizon is deepest and most free of stone on the highest, oldest terraces, where it may exceed 50 cm thick.

Salt-impregnated horizons beneath stable pavements may undergo extreme comminution through hydration and other forms of salt-weathering, and pulvurent saline soils (*fech fech*) below reg plains in the central Libyan Saharas may have originated in this way. Meckelein (1959) found solonetzic soils with crack-fillings of fine dust in this area which he represented as a stage in this extreme reduction. Residual dust soils (*Staubboden*) in the extreme desert of northern Chile, with splinters of bedrock in shallow saline profiles, seem to be a more extreme development (Mortensen, 1927).

Evolution of desert pavements

Concentration of a stone pavement through winnowing by wind or wash can be relatively rapid, although the contribution by movement through a soil may take longer. Once formed, the pavement is relatively stable. The closely-spaced stones act as a drag on the surface wind, restrict the entrainment of finer intervening materials and so limit deflation. On moderate slopes runoff is maintained in a dispersed form by the stone mantle, as discussed in relation to pediments in Chapter V, and drainage incision is checked in consequence. Selective erosion is also countered by resulting increase in stone density. Whether residual or transported, pavements naturally consist of materials resistant to weathering. Hence the pavement exercises a protective role on the surface which is reflected in the name *desert armour*.

Nevertheless the pavement is subject to further slow evolution over thousands of years, involving weathering of the stone and redistribution of the secondary products (Cooke, 1970b). This evolution is generally towards an increasingly even surface of finer calibre and greater compaction, with progressive darkening by superficial weathering. For instance these characteristics identify older pavement surfaces on alluvial fans in Death Valley (Hunt *et al.*, 1966).

The pavement stones are subject to the various forms of desert weathering reviewed in Chapter II, but their occurrence as detached fragments on or in topsoils on stable, evenly sloping surfaces influences the relative effectiveness and expression of the weathering processes. For example, pavement stones may be wetted frequently by dew, which contributes salts as well as moisture to weathering (Yaalon and Ganor, 1968). Being generally unchannelled and shedding little runoff, the pavements are particularly subject to episodic or seasonal cycles of shallow wetting by rainfall and evaporative drying, through which weathering is activated. Bare and generally dark-coloured pavements are among the most strongly heated surfaces of the desert and are subject to strong evaporation fed by capillarity from soil solutions that are commonly saline; consequently salt efflorescences are widespread, the soil itself is commonly impregnated with chlorides and sulphates of sodium and calcium, and salt-weathering is accordingly important. Pavements trap wind-borne dust and salts which lodge between the stones and inside fissures and foster salt-weathering and dirt-cracking. Lichens and algae exploiting the shaded and somewhat moister environments under the stones contribute an organic element to chemical weathering.

In the breakdown of pavement stone there tends to be a further selective concentration of resistant fine-grained material, for example flints on soft limestone surfaces. Such stones eventually break down by spalling, blocky fracturing ('crazing') or by complete cleavage or radial splitting, whereas coarser-grained stones undergo granular disintegration and pitting. Fracturing of pavement stone has traditionally been attributed to insolation weathering, but thermal expansion may be no more than contributory; most of the broken stone is superficially chemically altered, and developing cracks are commonly filled with fine dirt, salts or oxidised coatings that are subject to expansion on hydration or further crystallisation (Yaalon, 1970). There are many signs of stones being prised apart; for example the perfectly fitting parallel flakes of vertically cleaved pavement boulders are generally held a centimetre or so apart by earth fillings. That fracturing is *not* dependent on surface heating is shown conclusively by the occurrence of cleaved boulders in underlying soils, particularly in gypseous and calcareous horizons, where the impregnating substances extend into the cracks as wedge-shaped fillings (Young, 1964). By these various means a matrix of increasingly fine particles is supplied to the pavement. The proportion of such secondary material measures the maturity of development of a desert stone pavement.

Redistribution of the fine particles among the stones by wash and rainbeat contributes to the smoothing and compaction of the pavement. Bare soil is generally crusted above a vesicular horizon 1–3 cm thick which also extends around and beneath the surface stones. These features result from puddling and sealing of the exposed soil by heavy rainfall or wash run-on, above a saturated layer that can flow into hollows between the stones and then set on drying, the vesicles resulting from the escape of entrapped air (Evenari *et al.*,1974). Equally important in compaction is the gravitational settling of stone during expansion and contraction of the surface, on heating or cooling and more notably on wetting and drying, particularly in saline or expansive clay soils. Pavements of this type are generally soft and puffy after rain, but on drying the stones become firmly embedded in a tight mosaic.

31. Dark brown desert varnish on upper surfaces of boulders removed from stone pavement, Biskra region, Algeria. Photo G. E. Williams.

Creep of saturated surface materials may assist the smoothing of a sloping pavement, particularly where dispersal of the matrix is accentuated by salinity. Microrelief is reduced by differential flowage and the pavement stones become more evenly distributed and further embedded. Contour-aligned steps attributed to surface creep on alluvial fans in Death Valley have been described in Chapter V.

Desert varnish

Dark staining is typical of weathered rock surfaces in deserts (see Chapter II) but develops into a lustrous dark patina on stones in pavements and is then termed *desert varnish* (Pl. 31). It is an amorphous coating 0.02–0.1 mm thick above a weathered rind up to 3 mm thick (Engel and Sharp, 1958), due to enrichment in hydrous oxides of iron and manganese, the former producing red-brown and the latter dark brown to black hues. The underlying rind is partly depleted of mobile elements such as calcium and magnesium and is stained brown by impregnation with oxides of iron and manganese.

The coating consists of an inner layer enriched in alumina and silica and a thicker outer layer enriched in iron and manganese (Hooke *et al.*, 1969). The concentration of the additives in the inner layer normally decreases inwards, suggesting an origin in the stone, whereas that in the outer layer normally increases outwards, indicating an external source. The proportion of manganese relative to iron in the outer layer *always* increases outwards, perhaps the most striking aspect of the chemistry of desert varnish.

Desert varnish forms best on materials resistant to weathering, with slightly rough and porous surfaces. Stones rich in iron and manganese take on a dark varnish more rapidly, but composition of the stone is not an essential factor. Fine-grained basic lavas and some metamorphic rocks develop a thick varnish, unlike granite and friable sedimentary rocks subject to granular disintegration. The stones should not be subject to movement, and degree of darkening by varnish can indicate the stability and relative age of a desert stone pavement.

Small loose stones tend to be varnished dark and shiny all over. On larger stones embedded in the soil the underside is generally orange to red-brown and there is commonly a dark shiny *ground-*

line band of thicker varnish extending from as much as 4 cm below the surface to about 0.5 cm above. These variations are due to the movements of evaporating films of moisture from which the external components in the varnish are deposited, and to an accompanying fractionation of iron and manganese due to the greater solubility of the latter in the non-acid and oxidising environment of a desert soil. The ground-line band marks the capillary fringe of evaporating soil moisture and an end-stage of preferred deposition of manganese after the deposition of iron oxides on the underside of the stone.

Possible sources of iron and manganese are the rock itself, solutions in the soil, and atmospheric dust. The fact that stone on the ground is generally more varnished than rock outcrop indicates the importance of the soil as a source or as a medium of enrichment. Lichens, algae and soil microflora may contribute by acting as reducing agents and mobilising iron in solution in the form of hydroxysols (Laudermilk, 1931; Scheffer *et al.*, 1963; Krumbein, 1971).

It seems likely that rates of formation of desert varnish vary widely with the setting. It is questionably claimed that stones moved only twenty-five years previously in the Mojave Desert have become varnished in conformity with their present positions (Engel and Sharp, 1958), whereas artefacts dated as 2000 y old on fan pavements in Death Valley are only very lightly varnished (Hunt and Mabey, 1966). The process appears to be very slow on surfaces above the ground; for example some Egyptian monuments have gained a barely perceptible staining in 2000 y and others 5000 y old bear only a slight, light brown varnish (Blackwelder, 1948). Comparative dating by degree of varnishing is probably risky in view of variations between sites.

Varnish may be destroyed by granular breakdown or spalling, or by sand blast. What has been interpreted by some as evidence of periodicity or stratigraphy in varnish (cf. Hunt, 1961) is seen by others to reflect a balanced process of weathering and renewal, or merely differences in the dates of emplacement of differently varnished stones.

PATTERNED GROUND IN STONY DESERTS

Stones on desert pavements are sometimes arranged in patterns, commonly in association with systems of microrelief. Stony deserts rival the periglacial zones as areas of patterned ground, with the role of frost action largely assumed by the expansion and contraction of clay soils on wetting and drying. Comparable effects in highly saline soils on or near playas are dealt with in Chapter VIII.

Stony gilgai

The Australian Aboriginal word *gilgai,* for a small waterhole, has been applied to various forms of hummocky microrelief on clay-rich soils (Hallsworth *et al.*, 1955; Harris, 1959; Verger, 1964). Those that occur on stone pavements and involve stone sorting have been grouped as *stony gilgai*. They are widespread on stony table-lands and plains underlain by claystone or shale in the south and east of the Australian arid zone, in association with soils having weakly developed solodized solonetzic profiles and known as 'desert loams' (Stace *et al.*, 1968).

Gilgai microrelief appears to result from the forcing upwards of large blocks of soil detached by horizontal fracture in the solum, probably caused by the swelling of wetted subsoils, including soil material fallen down cracks when the soil was previously dry. Edaphic requirements, either singly or in combination, are a clay content that increases down the profile, from 10–30 per cent in the topsoil of a desert loam to 40–50 per cent in the massive B horizon, a significant proportion of expansive clay minerals such as montmorillonite, and a high degree of base-saturation, normally with calcium as the dominant cation of the exchange complex but with increasing sodium at depth. These characteristics determine the large swelling capacity of gilgai soils on wetting, particularly in the subsoil, and the formation on drying of systems of cracks extending from the surface and widening below. Regular cracking through the profile is essential for the return of fine topsoil to lower levels, for rapid but uneven wetting of subsoils, including the fillings of the cracks, and for the detachment and upward displacement of blocks of subsoil. The main climatic requirement is the occurrence of hot dry periods to dry and crack the soil in depth, and gilgai are not confined to arid regions, but restricted leaching under low rain-fall favours base-saturation and the incidence of montmorillonite.

Two forms of stony gilgai are widespread in arid Australia. The *round gilgai* found on flattish ground has a slightly higher stony outer rim or 'mound' up to 8 m in diameter and a less stony flat inner part or depression, up to 3 m diameter, which may have a small central 'crabhole'; (Fig. 32a). The stony pavement between gilgai is termed the 'shelf'. Larger surface stones embedded in the mound may show signs of heaving, but the ground between is generally sealed with a silt crust. The soil immediately beneath is a strongly pedal reddish clay consisting of subsoil that has been displaced to the surface in the mound, and it is typically almost stone-free, for the soil movements forming gilgai seem to resemble those responsible for upward stone sorting. In contrast, the soil below the depression is sandier, looser-structured and lighter in colour to a depth of 30–50 cm, where it begins to resemble the mound soil at the surface, and it is stony throughout. Round stony gilgai may occur randomly or in networks.

These forms occur towards the dry limits of gilgai in Australia, and rainfalls as low as 150 mm may account for the typically small relief of between 5 and 15 cm between mound and depression. It is possible that some are fossil from previous wetter conditions, but comparable forms have been illustrated from the much more arid Libyan Sahara, in basalt hamada of the Djebel es-Soda and in sandstone hamada of the Djebel en-Namus (Meckelein, 1959).

On a range of slopes between 1 and 10 per cent, stony gilgai become distorted and may link laterally in contour-aligned *lattice systems* in which the downslope mound becomes prominent as a steeper stony riser and the shelf forms a stone-free flatter tread (Fig. 32b). Shallow depressions a metre or two across, each with one or more crabholes, occur towards the inner side of the treads and introduce a lobate element into the lattice, for the riser below tends to swing concentrically about them. These systems are related to sorted stone steps on adjacent hillslopes, described in Chapter III. They cover thousands of square kilometres of undulating lowlands with silcrete gibbers on red clay soils known as 'stony downs' in south-eastern arid Australia.

Each lattice constitutes a series of repetitive hydrologic units, with risers and treads forming zones of runoff and run-on respectively, for the stony risers, like the mounds of round gilgai, have strongly pedal soils which resist infiltration, whereas the soils of the depressions are lighter-structured and allow deeper wetting. Presumably the microrelief is regenerated through the resultant inequalities in wetting and drying, and so resists the processes tending to smooth the slope.

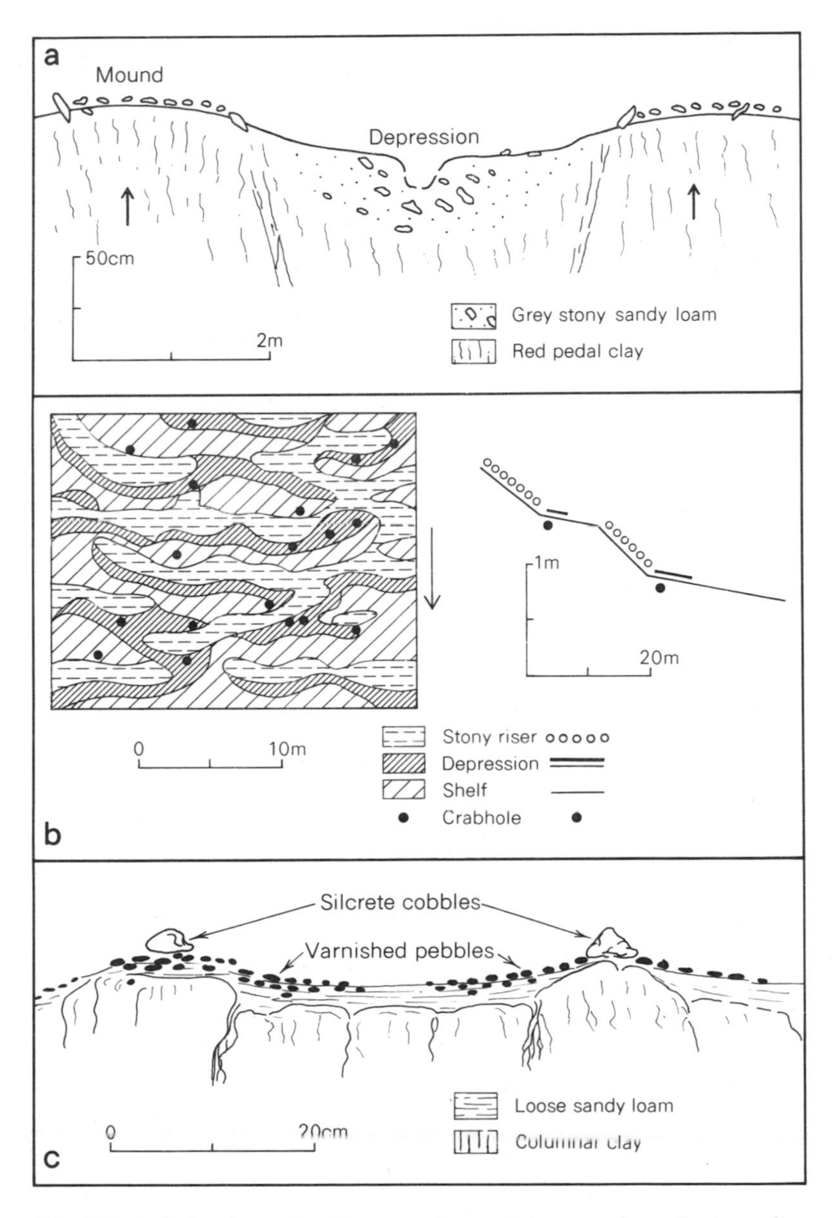

32. Microrelief and associated patterned ground in stony desert in Australia. a. Section across round stony gilgai; b. plan and section of lattice gilgai on slope of 1–2°, western New South Wales (from Mabbutt et al., *1973); c. Section across stone polygon on reg pavement, central Australia.*

Sorted stone polygons

These are found with structured subsoils subject to minor heaving, as on stony plains with silty desert loam soils in the southeast of the Northern Territory of Australia. The polygons are between 40 and 80 cm in diameter and are outlined by silcrete cobbles mainly lying loose on a close pavement of small pebbles of siltstone (Pl. 30; Fig. 32c). The boundaries of the polygons coincide with columnar partings in the subsoil, and movement of the soil is indicated by vertical fractures beneath the borders of the polygons where the subsoil is within a few centimetres of the surface; the inner cells have thicker topsoils. Meckelein (1959) has illustrated somewhat larger polygons in fine reg pavements in the terminal basin of Wadi el-Faregh in the Libyan Sahara. These appear to have been determined by sand-filled polygonal cracks in an underlying gypsum horizon.

STONY DESERT TABLELANDS

Extent and control

Hamada also signifies tableland in the Old World deserts, and similar forms such as the *stony tablelands* of arid Australia occur in

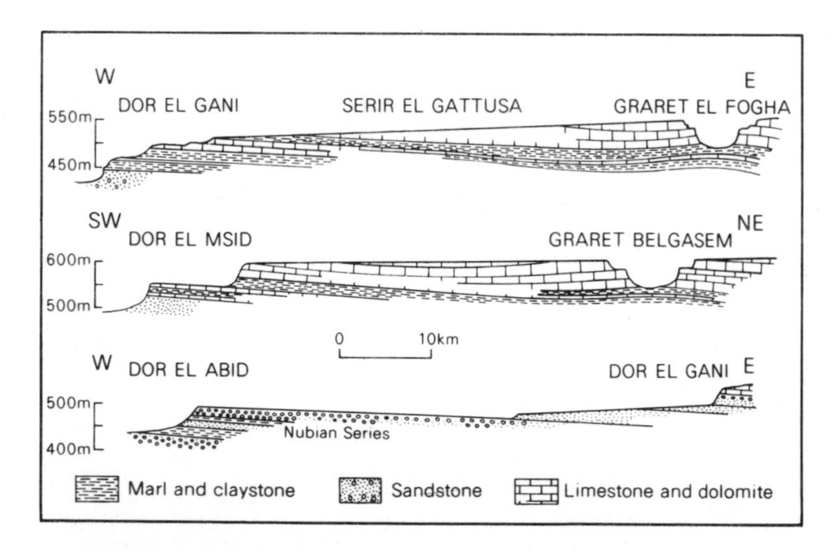

33. Regional profiles and stratigraphy of hamada tablelands in the northeast Fezzan, Libyan Sahara. From Fürst, 1965.

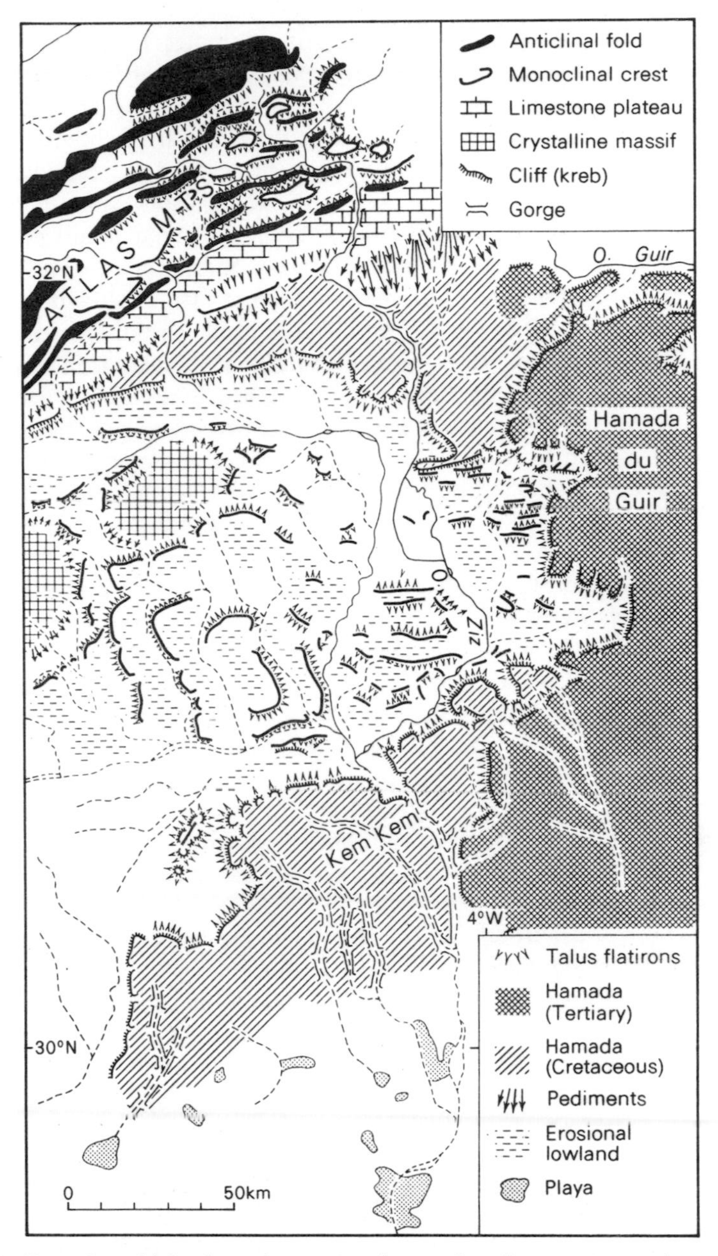

34. Hamada tablelands and associated staged pediments in southeastern Morocco and northwestern Algeria. From Joly, 1950.

comparable settings elsewhere. In part this reflects the prevalence of flat-lying rocks and the corresponding extent of level surfaces formed by resistant strata in the platform deserts, of which hamadas are typical. Structural tablelands dominate much of the northern Sahara; for example the gentle northward descent of the Tademaït Plateau in central Algeria expresses the slight dip of a capping of Cretaceous limestone (Mensching *et al.*, 1970), and limestones of similar or younger age and the underlying Nubian Sandstone yield a stepped tabular relief in the Fezzan further east, as shown in Fig. 33 (Meckelein, 1959; Fürst, 1966). But factors other than structure contribute to the formation and persistence of tablelands in deserts.

One is the common occurrence of constructional surfaces in desert basins of interior drainage. This is exemplified in the hamadas of southeastern Morocco (Fig. 34) where silicified flaggy freshwater limestones cap the highest level in a series of depositional surfaces ending in the mantled pediments of the Saharan Atlas (Joly *et al.*, 1954; Conrad, 1969).

A second factor is the tendency for protective *duricrusts* to form owing to restricted leaching in deserts, and for older crusts of non-desertic origin to survive for similar resons, favouring the preservation of associated land surfaces (Pl. 15). In the Lake Eyre basin of Australia, for example, desertic crusts include *gypcrete* on inner plains close to a former baselevel and *calcrete* nearer the outer margin. Of less certain origin is the *silcrete* on older, higher surfaces in the basin, whilst fossil *ferricrete* survives extensively on the crystalline rocks of the basin rim. Many structural hamada surfaces have been reinforced by calcification or silicification under arid weathering.

Lastly, as discussed in Chapter V, stone-mantled surfaces of moderate gradient tend to survive undissected because sediment-laden runoff is readily dispersed and channelling suppressed by the pavement cover. Under these conditions, and with the slow rates of arid erosion, planate land surfaces can survive for long periods. Many structurally conditioned hamada surfaces in fact represent ancient cyclic plains protected from dissection under subsequent aridity; for example the hamada surfaces of the Fezzan in Fig. 33 truncate the limestone caprock sufficiently to suggest such an origin.

Desert duricrusts

Calcrete. This term and its equivalent *caliche* are used of cemented calcareous horizons ranging from soil hardpans to massive caprocks, which may originate as sedimentary limestones in ephemeral desert lakes, as horizons of lime-enrichment in soils and sediments, or from solution and re-deposition on limestone bedrock.

Lacustrine and fluvial limestones are extensive in basin depressions and in shallow valleys in shield and platform deserts, and much of the Kalahari Limestone (Pl. 32) is of this type (Boocock and van Straten, 1962), as are the calcreted valley trains of central and Western Australia (Mabbutt, 1963a, 1967). They have formed in ponded systems of interior drainage.

32. Urinanib Escarpment forming the western rim of the Kalahari Desert in South West Africa, with massive calcrete capping (Kalahari Limestone).

Calcretes formed as horizons of lime-enrichment and cementation, or K-layers, develop towards the lower limit of leaching in soils or sediments, and their illuvial character is commonly indicated by a downward decrease in compactness and lime content and by the occurrence of more soluble substances such as gypsum at greater depth. Early stages of growth as shown in Table IV are influenced by the texture of the matrix, and development may be more rapid in permeable coarse sediments such as gravels, but in all cases a continuous hardpan can result. Crusting then follows through periodic solution and re-deposition of lime at the top of the 'plugged' and relatively impervious hardpan, and at this stage the

TABLE IV **Evolutionary scheme of formation and breakdown of calcretes***

<table>
<tr><td>BREAKDOWN</td><td>Gravel pavement</td><td colspan="2">Calcrete boulders</td></tr>
<tr><td>CRUSTING</td><td colspan="3">Flaggy caprock
Laminar crust
Crusted layer</td></tr>
<tr><td>CEMENTATION</td><td>Cemented gravel or sand hardpan</td><td>Cemented nodular ('honeycomb') hardpan</td><td>Cemented chalky hardpan</td></tr>
<tr><td rowspan="3">ACCUMULATION</td><td>Granular interfill</td><td>Concretionary horizon</td><td>Soft chalky horizon</td></tr>
<tr><td rowspan="2">Pebble or grain coatings</td><td>Concretions</td><td>Powdery masses</td></tr>
<tr><td colspan="2">Filaments</td></tr>
<tr><td>MATRIX</td><td>GRAVEL OR COARSE SAND</td><td colspan="2">SAND TO CLAY</td></tr>
</table>

* After Beaudet *et al.*, 1967; Gile *et al.*, 1966; Goudie, 1973.

calcrete can become physiographically significant as a protective residual horizon. Evolved profiles incorporate the various stages in upward succession (Fig. 35; Pl. 33). and radiocarbon dating has confirmed an upward growth. Massive laminar crusts may reflect several cycles of calcrete formation, and the hardest flaggy caprocks commonly exhibit silicification. These pedogenic calcretes are typical of moderately arid to semiarid climates with between 100 and 400 mm rainfall, particularly on the temperate margins of zonal deserts and extending into the midlatitude deserts (Goudie, 1973). Depth of calcification and development of calcrete diminish with rainfall to a thin near-surface rubble at the dry limit. In hamada settings any original overlying soil will generally have been stripped and the upper calcrete broken into a bouldery surface.

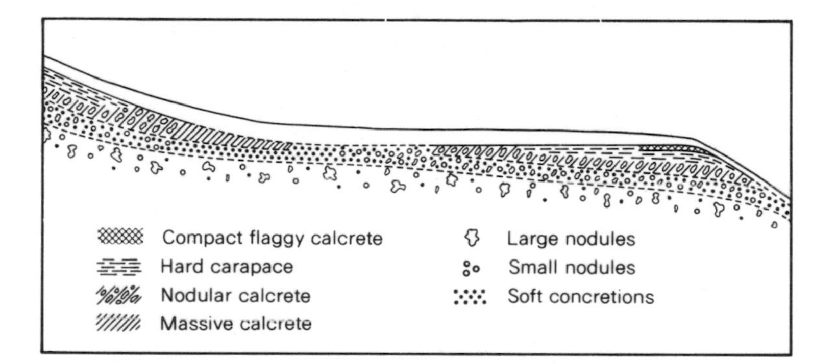

35. Lateral variations in a calcrete profile about 1.5 to 2 m thick on a dissected piedmont terrace of the Middle Atlas, Morocco. From Ruellan, 1967.

Calcretes of this type may attain several metres thick and may take from tens to hundreds of thousands of years to develop fully. In Morocco only the oldest surfaces, of Lower Pleistocene age, have massive flaggy calcretes, and younger surfaces show progressively lesser development, with those later than Pleistocene showing little calcification as in Fig. 88A (Ruellan, 1967).

The main source of lime for these pedogenic crusts is from solution after extensive shallow flooding or from laterally migrating soil water. This source is particularly important on slopes below limestone outcrop, as on the crusted mantled pediments south of the Saharan Atlas. An alternative source is wind-blown calcareous

33. Detail of calcrete capping, Libyan Sahara, laminar above and nodular below. Photo M. Fürst.

dust which lies long enough for rain to leach lime downwards into the soil (Gile *et al.*, 1966). Such a process must be invoked to explain the continuing development of calcrete on surfaces isolated by dissection, or wherever there is no local source of lime.

Examples of hamada calcretes formed in flat-lying limestone bedrock are the caliche of the southern High Plains of the United States, developed in the Pliocene Ogalalla Formation, and those of the Hamada du Guir in the Moroccan Sahara formed in Neogene freshwater limestone (Conrad, 1969). In Israel, the *nari* calcrete formed on folded Eocene chalk follows the rolling contours of the Judaean Hills, thicker on summits and footslopes and thinnest on the steep midslopes.

Gypcrete. Crusts cemented by gypsum tend to form in the lowest parts of desert basins where sulphates have crystallised with the evaporation of saline groundwater near the surface. Topography and the water table are primary controls therefore, but such a soluble cement can only survive under aridity. The gypcrete-capped tablelands which stand between 12 and 45 m above the western margin of Lake Eyre in central Australia appear to represent such a surface which was later faulted and dissected (Wopfner and Twidale, 1967). The crust here is between 0.5 and 1.5 m thick and cements gravel, sand and lacustrine clay. An alternative or additional source of gypsum is dust blown from dried salt lakes (Jessup and Norris, 1971), particularly where crusts extend up hillslopes of non-gypseous rock and across uneven surfaces above the water table. Chlorides blown with the gypsum are presumed to have been removed by leaching.

Silcrete. Siliceous crusts range from silica-cemented sands and gravels to an amorphous matrix of microcrystalline silica with floating quartz grains. They are not restricted to arid areas but are known from many deserts, including the Namib, the Kalahari, and the Sahara (Goudie, 1973). Nowhere are they more extensive and geomorphologically prominent than in arid Australia, particularly in the Lake Eyre basin where silcretes of Cainozoic age cap stony tablelands above desert plains mantled with derived gibbers.

The Australian silcretes are typically between 1 and 5 m thick and in their upper part tend to be massive or laminar with columnar jointing, becoming nodular towards the base. They are light grey to brownish-red when fresh, but weathered surfaces take on the red-brown patina which imparts a sombre hue to the Australian stony deserts. The duricrust typically overlies a weathered profile which may attain 50 m thick, consisting of an indurated and disturbed 'fragmental' zone up to 10 m thick above a kaolinitic pallid zone with localised ferruginisation near the base (cf. Wopfner and Twidale, 1967). There is a growing body of opinion that the deep profile represents an older episode of weathering and that it has subsequently shared the silicification of the surface horizon; whatever the genetic relationship, however, the juxtaposition of hard crust and weak underlayer gives a characteristic escarpment of cliff and gibber-mantled debris slope (Pl. 15).

No comparable crust is known to be forming today, and consequently the mode and environment of formation of the Australian silcretes remain uncertain. That silcrete indeed formed under dry conditions is suggested by the occurrence of gypsum and other soluble salts in the underlying profile, by known high pH requirements for the mobilisation of silica, by the superficial and hence possibly evaporative character of the crust, and by the occurrence of a younger siliceous hardpan in parts of Western Australia receiving less than 200 mm rainfall (Litchfield and Mabbutt, 1962). The occurrence of the silcretes suggests that they mainly formed as surface-related layers by deposition of silica, from solution as monosilicic acid or as a colloid, in pervious sediments or in weathered rock. This picture is locally complicated where silicification has been guided by stratigraphic and structural traps.

In Australia, silcrete is mainly associated with arenaceous and argillaceous rocks in lowlands formed in sedimentary basins (Fig. 36), and the most massive crusts occur in pervious quartzose

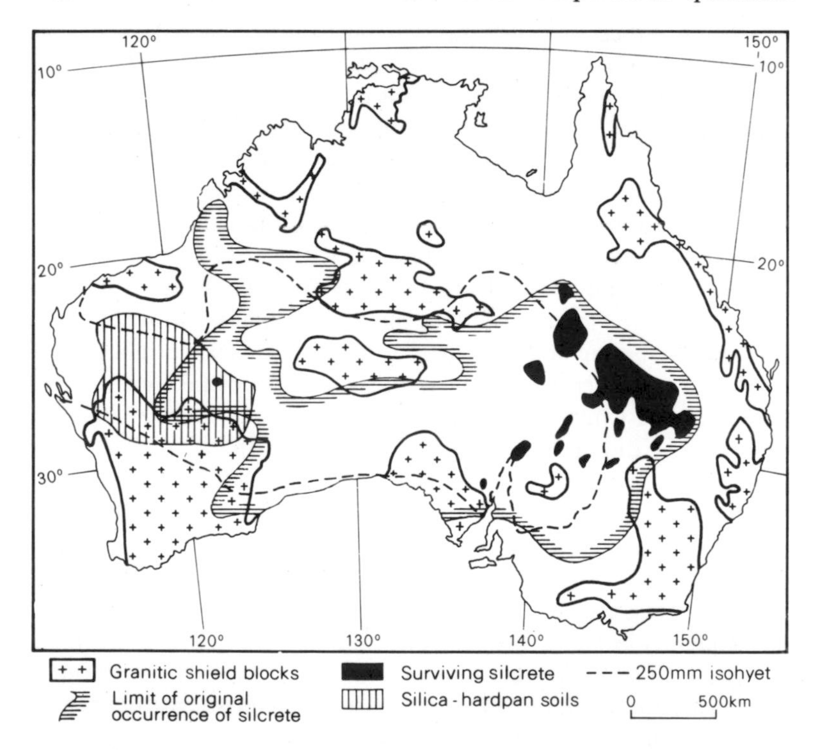

36. Distribution of silcrete in Australia, modified after Stephens, 1971, in relation to rainfall and the extent of sedimentary basins as delimited by granitic shield blocks.

sediments. It extensively preserves a planate land surface, locally depositional, known as the Australian pediplain, and it is this surface that survives on the stony tablelands of the Australian deserts. The intact silcrete forms a bossy rock pavement which breaks down into blocks and spherical boulders, and these subsequently spall into angular fragments; it is extremely durable and may continue to maintain a level surface although no longer strictly in place.

Forms of desert tablelands

The association of a subhorizontal hard capping with a weaker lower layer yields a steep escarpment with a sheer upper face–the *kreb* of the Sahara or *breakaway* of the Australian stony desert–and a debris-mantled or structurally-benched slope below (Pls. 15, 32). The face is subject to honeycombing behind a hardened crust and particularly to undermining by cavernous weakening at the base. Overhangs are particularly common with limestone cappings.

Valleys incised into the tablelands are also rimmed by cliffs (Fig. 37). The upper sectors are generally V-shaped or box-shaped in section; they have steep or amphitheatral heads and their profiles are broken by rocky steps. The lower sectors tend to be wider sandy trenches seamed with minor channels. In the driest areas such as the eastern Sahara the tableland catchments yield only rare floods.

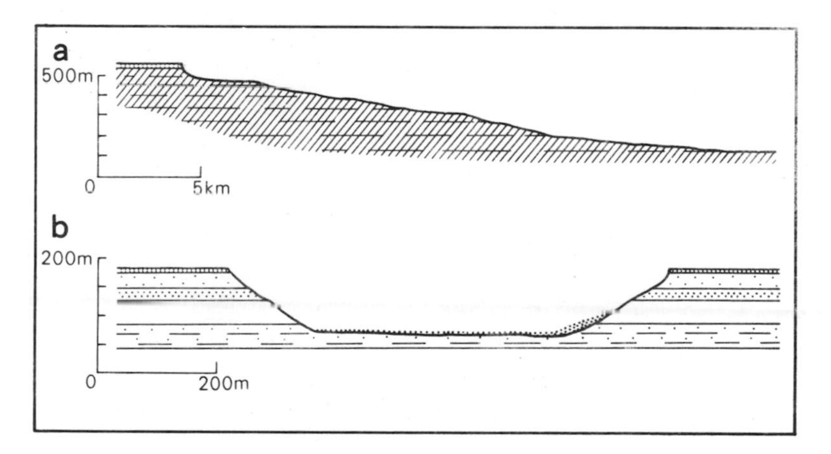

37. a. Longitudinal profile, and b. transverse section of a typical entrenched wadi in the Gilf Kebir, central Libyan Desert. From Peel, 1941.

Peel (1941) attributed the widening and extension of wadi canyons in the Gilf Kebir to sapping at the base of the cliffs, as evidenced by lines of caves in seepage zones fed by occasional rains infiltrating into the porous Nubian Sandstone of the plateau surface. This was considered to have operated only during wetter periods in the past.

Whether duricrusted or formed by an exposed resistant bed, desert tablelands provide excellent examples of structural and tectonic relief. In the Lake Eyre basin of Australia, silcrete duricrust has been deformed in elongate domes which have determined the patterns of its dissection and survival; generally the stony tablelands and cuestas occupy the flanks of anticlines, with younger stony lowlands eroded in the axial tracts and depositional surfaces in the synclines (Fig. 38). Structural control of hamada forms in the Fezzan by slightly truncated resistant limestone is demonstrated by the profiles in Fig. 33. These also include anticlinal depressions or *garaet*, akin to the more spectacular *makhtesh* of the Negev.

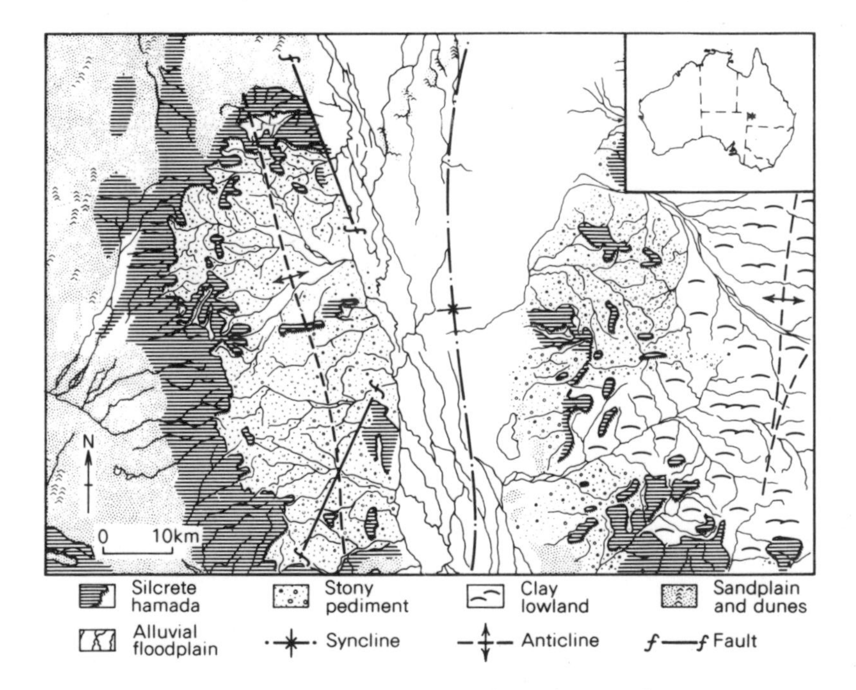

38. Dissection of silcrete-capped hamada tablelands in southwestern Queensland, in relation to fold axes.

Features of tablelands with calcrete cappings, and to a smaller extent of those with gypcrete, include closed depressions such as the *dayas* of Algeria and Morocco, considered to be in part of solutional origin. Since these involve modification and partial destruction of the calcrete, possibly under changed climatic conditions, they are discussed in Chapter X.

WIND EROSION IN STONY DESERTS

Stable open surfaces such as stone pavements or rock plains favour the rapid drift of sand and dust from local sources and commonly exhibit forms due to sand blast and dust abrasion.

Ventifacts

Pavement stones shaped by sand blast and dust abrasion are known as *ventifacts*. An upwind face tends to be faceted at right angles to the dominant wind, and since at this scale abrasion

34. On L *basalt* Dreikanter *32 cm long, with wind-polished original upper surface. Photo W. Czajka;* on R, *granite* Dreikanter *showing fluting and less marked faceting characteristic of crystalline ventifacts. Photo R. P. Sharp.*

increases with height above ground, the facet slopes forward (Sharp, 1949). In plan, the face is typically slightly convex upwind, and it tapers towards its ends. The abrasive effect decreases as the face is lowered to an angle of about 30°, by which stage the pebble is likely to have been undermined by wind-scour and rolled over about an axis at right angles to the wind to rest on the facet, when the process can be repeated. If the rock is homogeneous the faces will be smooth and will meet in sharp angles. The surfaces are commonly finely rippled and may have a dull glaze or wind polish due to dust abrasion (Pl. 34a).

The most evolved form of ventifact is the three-ridged *dreikanter,* shaped like a brazil nut; single-ridged stones or *einkanter* are an earlier stage, and pyramidal shapes are typical of smaller stones moved more frequently and in irregular fashion by the wind. The original shape of the stone and any structural planes can influence the form of a ventifact and some are no more than slightly faceted rounded pebbles. Faceted forms do not develop well in heterogeneous or structured materials, which tend rather to become pitted or grooved (Pl. 34b), and facets do not survive well on weatherable rocks such as soluble limestone.

Ventifacts are less common than the literature might suggest, and except in areas of exceptional sand drift such as the Coachella Valley of the Mojave Desert (Sharp, 1964) they characterise pavements of considerable age. They also occur in periglacial settings.

Wind-sculpted rocks

As shown in Fig. 39, the bulk of the curtain of drifting sand moves within 30 cm of the ground, and the combination of size and density of grains yields a maximum abrasive effect at a height of 20–25 cm (Sharp, 1964). Rock outcrops and large blocks are undercut at this level, above which the paths of the bombarding grains disperse, giving paraboloidal upwind faces (Pl. 35; Czajka, 1972). In heterogeneous rocks, faces at angles greater than 55° to the wind tend to be pitted by erosion of softer grains, whereas faces more gently inclined become fluted or grooved, particularly where bedding planes trend downwind.

Wind-sculpted terrain

South of the Tibesti in the southern Sahara are the wind-sculpted plains of the Borkou lowland (Hagedorn, 1968; Mainguet, 1968,

35. Keel-shaped sand-blasted blocks of andesite, Campo Negro, Argentina. Photo W. Czajka.

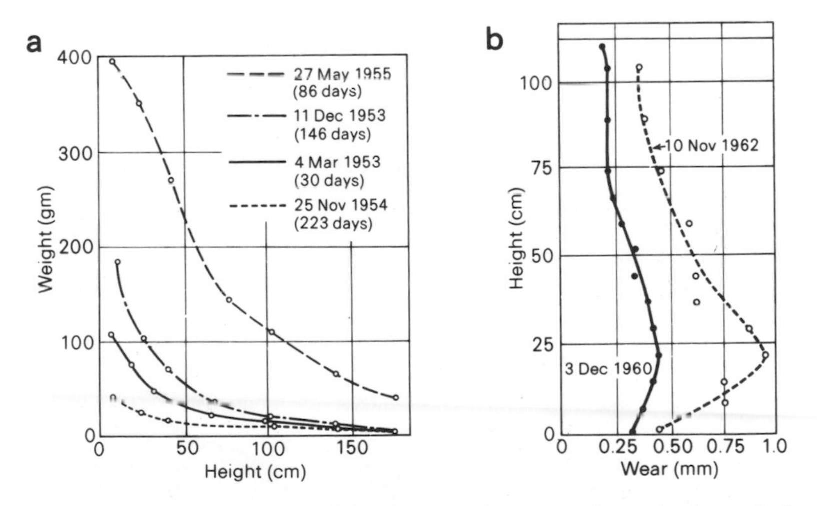

39. a. Distribution of mass with height in a saltating sand curtain; b. vertical distribution of wear by sand blast on a leucite rod, Coachella Valley, California. From Sharp, 1964.

1970). This is a terrain of parallel corridors and flat-crested ridges formed in gently dipping sandstone, as shown in Fig. 40a and Pl. 36. The corridors are up to a kilometre wide and several kilometres long; they are partly rock-floored, with occasional enclosed depressions containing remnants of lake beds, and partly sand-covered with areas of small barkhans. The ridges are of similar dimensions and up to 50 m high, and their steep flanks are freshly undercut, grooved, pitted and polished by sand blast to a height of 3 m (Fig. 40c). The erosion of the corridors has exploited a master joint direction in the sandstone which coincides with the NE trade winds, which here blow strongly for 8 months of the year, well-armed with sand. The forms are not developed in rocks other than sandstone, for a derived sand supply is essential, and then not in the most massive strata, nor where the joint trend departs by more than 15° from the dominant wind direction. Not all the sculpture has been by wind corrasion, however, for the history includes past episodes of weathering, stream erosion and deposition, and subsequent deflation.

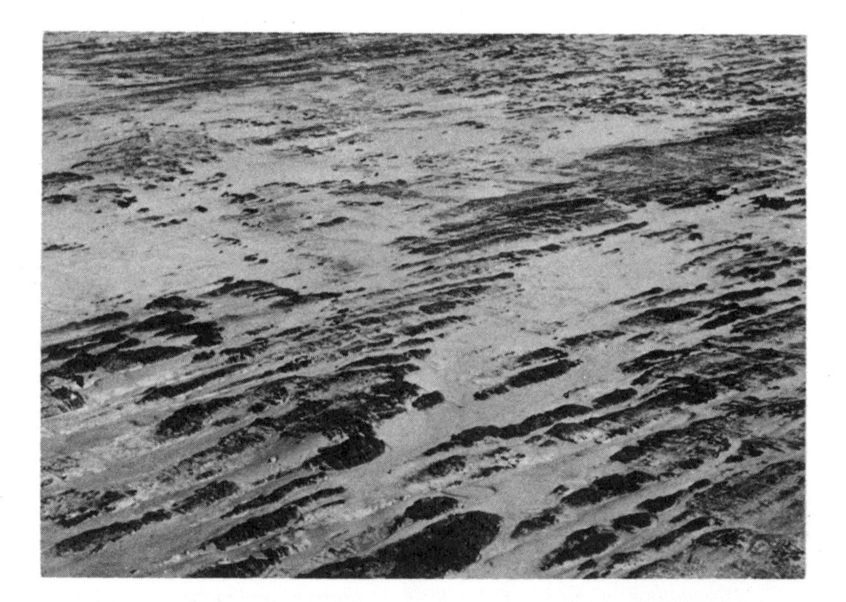

36. Rock yardangs in sandstone, Borkou, Chad Republic. The lightertoned rock faces indicate vertical range of sandblasting. Photo H. Hagedorn.

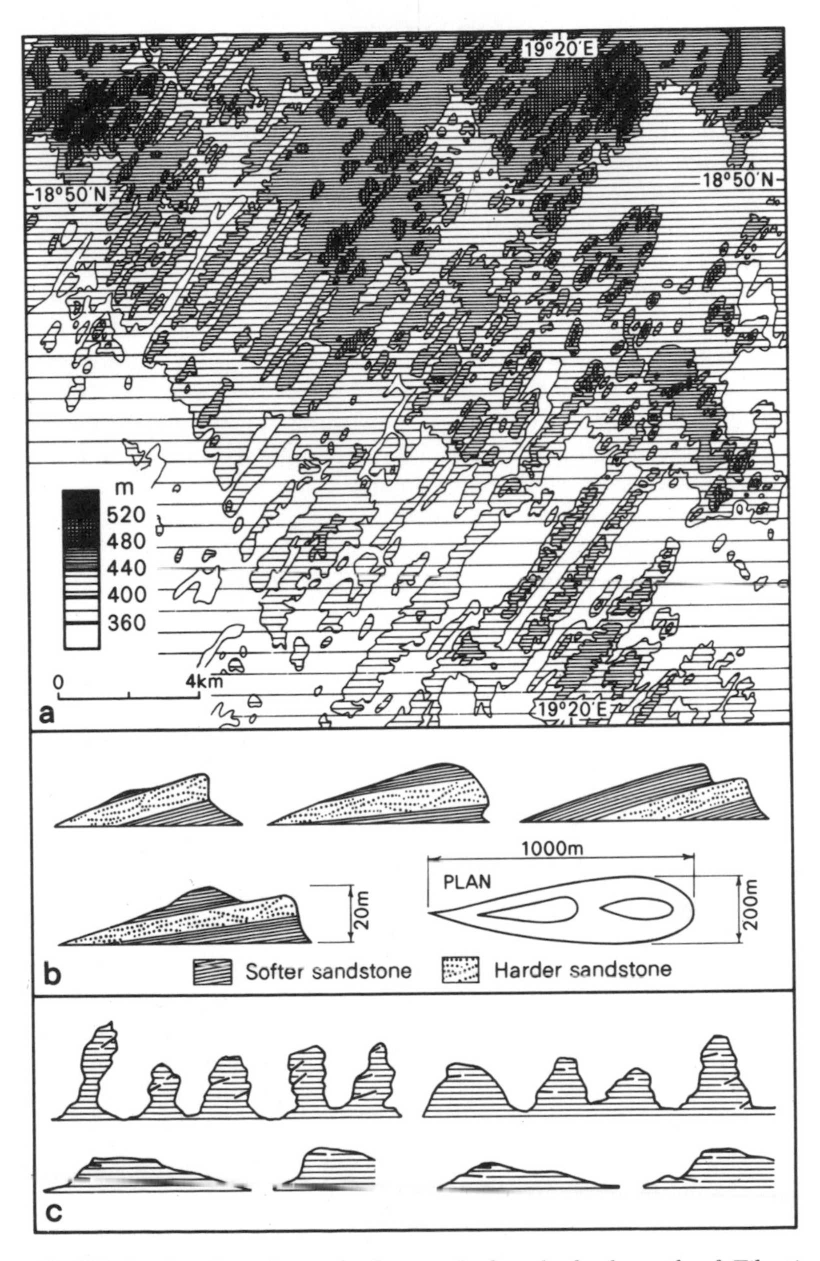

40. Wind-sculpted sandstone landscape, Borkou lowland, south of Tibesti Massif, Chad Republic. a. Contours of parallel ridges and corridors; b. longitudinal profiles and c. cross-sections of rock yardangs; a and c from Hagedorn, 1968; b from Mainguet, 1968.

On exposed low crests rising in the face of the dominant winds, and particularly on thinner-bedded sandstone, corrasion has progressed further and the ridges have been reduced to *rock yardangs*. These are dynamically stable, keel-shaped forms up to 20 m high and a kilometre long, aligned in the direction of the nearby ridges. They have characteristic convex longitudinal profiles, but the shapes vary with the arrangements of softer and harder sandstone beds (Fig. 40b). The isolation of the yardangs is aided by the occurrence of marked vertical jointing in the direction of the wind, whereby larger forms are cleaved and reduced to downwind strings of smaller yardangs. Their shapes are similar to those of yardangs eroded in soft lacustrine sediments, as described in Chapter VIII, although the latter are generally smaller.

Another aeolian landscape of ridges and troughs occurs in the coastal tract of the southern Namib, where northerly winds have etched out weathered arkose and sandstone in closely folded strata of similar trend, leaving interbedded dolomite in sharp relief (Kaiser, 1926).

Wind-formed depressions in stony deserts

Where a protective stone mantle is breached or locally inadequate, or where the surface is rendered less resistant by virtue of soil texture or salinity, the wind may scour hollows in soft underlying material. Probably the largest of this type are P'ang Kiang depressions cut in indurated basin fill below the stony plains or *gobis* of the Mongolian deserts (Berkey and Morris, 1927). These attain 100 m deep and 10 km diameter. Evidence of wind erosion is seen in sand blasting and undercutting of the cliff margins, particularly those parallel with the prevalent wind, and more generally in dust storms. Weathering and rilling of the marginal slopes contribute to the enlargement of the depressions, and sediment spread on the floor is in turn subject to deflation on drying. Groundwater can exercise an important control over the excavation of large depressions by wind, often in association with tectonic controls, and features of this type are discussed in Chapter VIII in connection with wind erosion of desert lake basins.

VII

DESERT RIVERS AND FLOODPLAINS

Desert rivers fall into three groups according to their sources. First are the exogenous rivers that enter and may also leave the desert. In the zonal shield and platform deserts most large rivers of this type have their sources in distant uplands in humid areas, for example the Nile and Orange rivers; in the mountain-and-basin deserts of midlatitudes they typically rise in the encircling high mountains, as do the Colorado River in the United States and the Tarim and similar rivers of central Asia. The exogenous rivers are commonly perennial, although like the Darling River in Australia they may vary greatly in seasonal discharge. Where they reach the sea they form an outlet for sediments and act as general baselevels, as the Rio Grande does for the deserts of New Mexico, but in the mountain-and-basin deserts they mainly form part of systems of interior drainage with aggrading lowlands, as exemplified by the rivers draining to the Aral Sea.

A second group of desert rivers rise in moister uplands *within* the deserts. Where these are snow-fed or spring-fed, like the Jordan, they may be perennial, but generally they are at most seasonal and become increasingly intermittent out into the plains.

The third and by far the commonest are the ephemeral rivers dependent on desert storms, and it is mainly these that are discussed in this chapter.

STREAMFLOODS IN DESERT LOWLANDS

Floods in central Australia

Most desert streamflows are entirely storm-fed and contribute to rather than draw from groundwater; consequently they reflect the character of much desert rainfall in their rapid onset and short duration, their infrequency, and in their localisation and variability within a large drainage system. This was shown by floods in the Finke River in central Australia in 1967, when an average of 250 mm of rain fell in the northwestern catchments of the Lake Eyre basin (Fig. 41) in five weeks in late summer. The rains included intense falls, many stations receiving more than 100 mm over-night, and they were regionally variable. The head catchments in the ranges of central Australia received their main rainfalls in late February, whereas further south the heavier rains fell in early March.

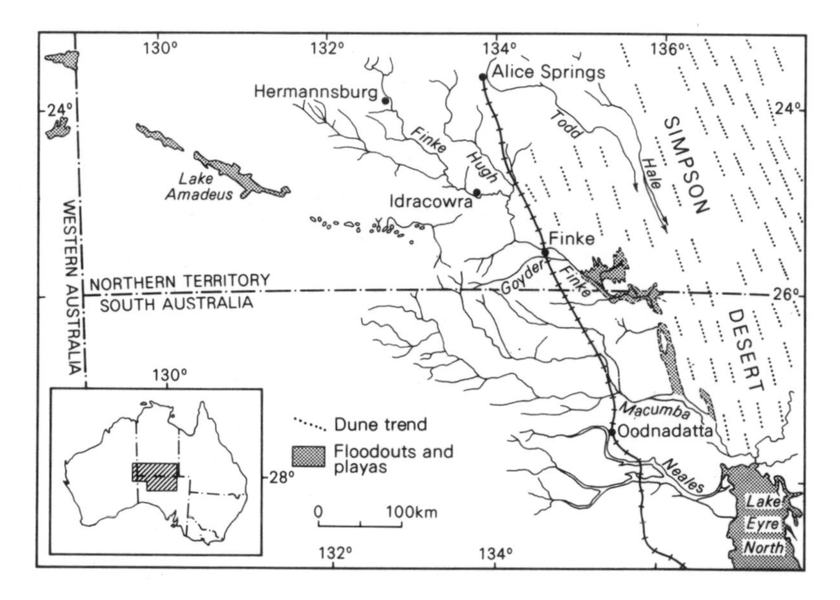

41. The Finke-Macumba catchment in the northwest of the Lake Eyre basin, central Australia.

Figure 42 depicts the durations of flows and heights of flood crests at three stations along the Finke River in relation to rainfalls in upper and lower sectors of the catchment, based on records collected by Williams (1970). It reveals a reduction in the duration and magnitude of the early floods in the upper plains sector between Hermannsburg in the ranges and Idracowra 250 km downstream, as a result of losses into the sandy channel. What at Hermannsburg was one flood with three peaks became three separate flows before reaching Idracowra. At Finke Township, however, these losses were offset by floods from a major tributary, the Hugh River. The records at Finke show the flood response to have been most rapid and abrupt at the onset of rains, with later responses slightly delayed as rain continued to fall on wet catchments. The flood waves steepened as the flood crest advanced, as shown by comparison of hydrographs at Hermannsburg with those at Idracowra. The time of travel of floodwaters from Idracowra to Finke was 28 hours, giving a flood advance of about 2 m/sec. Variability of the floods

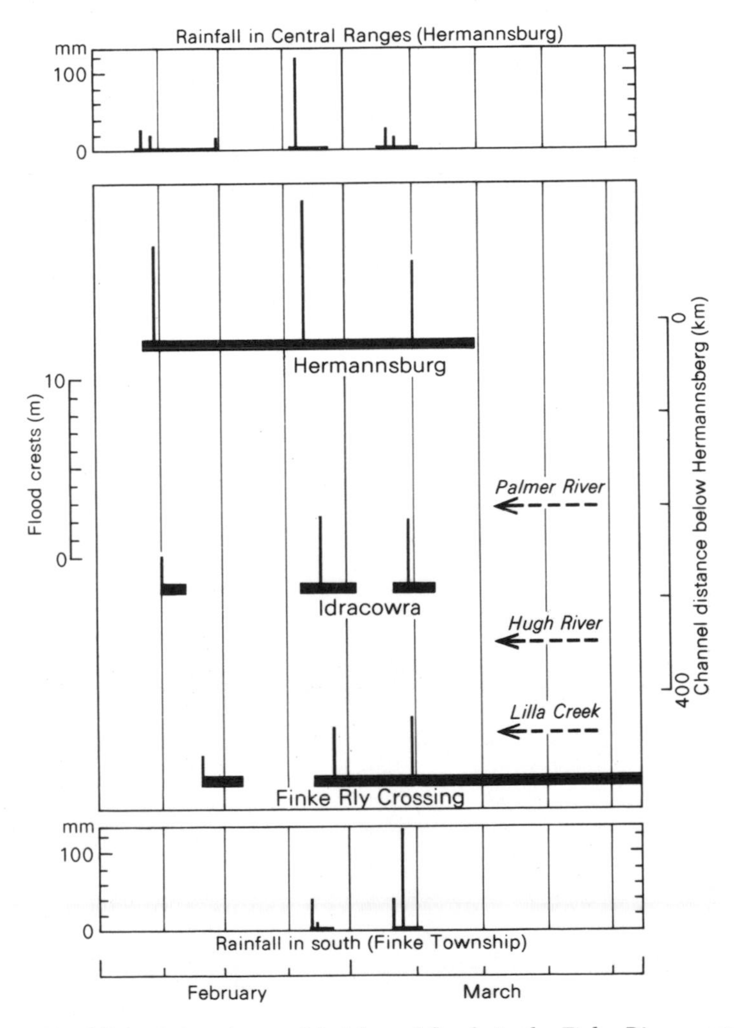

42. Rainfalls and durations and heights of floods in the Finke River, central Australia in February-March 1967 (localities in Fig. 41). Based on data in Williams, 1970.

along the river is clearly shown, particularly the continuation of flows in the lower Finke in response to local rains after the cessation of flooding up-valley. At Finke Township, flow continued until almost the end of April.

On the evidence of bedforms the mean velocity of the late February flood in the middle sector of the Finke was estimated at 3 m/s, giving a peak discharge of 1200 m^3/s (Williams, 1971). Despite the magnitude of the flood, however, the waters failed to reach topographic baselevel in Lake Eyre. In the flood-out shown in Fig. 41 they spread out over a width of 11 km with an average depth as little as 0.8 m, and were eventually absorbed into the sands of the Simpson Desert, where in some interdune corridors water remained standing for up to 17 months. A comparable flood in the Stevenson River, further south, is shown in Pl. 37.

Flood regimes in the northern Sahara

Among the most studied of desert river systems is that of the Guir-Saoura-Messaoud shown in Fig. 43, which drains from the

37. *The Stevenson River, 300 m wide, flooding into the Simpson Desert dunefield, March 1967. Floods of this magnitude occur at intervals of a decade or so. Photo Adelaide* Advertiser.

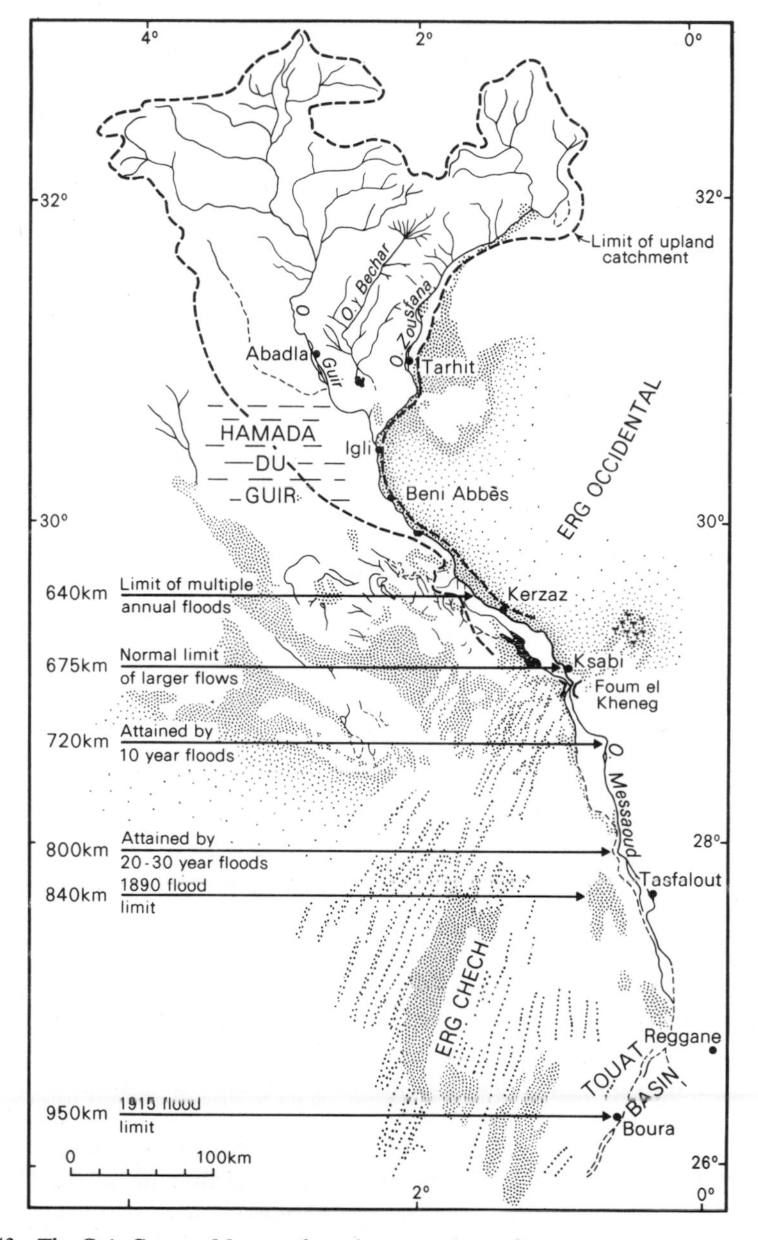

43. The Guir-Saoura-Messaoud catchment in the northwestern Sahara. Limits of flooding from Vanney, 1960.

High Atlas into northern Algeria (Dubief, 1953b; Vanney, 1960). Its catchment of 44,000 km^2 includes an appreciable montane sector receiving more than 150 mm annual rainfall, making this the most active of the large wadis entering the northern Sahara. There are two components in its flood regime: longer periods of flow, often with several peaks, following winter-season cyclonic rains in the uplands, and shorter floods with a characteristic abrupt rise caused by convective summer storms (Fig. 44a).

Flood records reveal a decrease in the frequency of flow and in discharge down-channel from the piedmont, characteristic of ephemeral desert rivers. At Abadla, 300 km from the source, floods occur in an average of 4.5 months per year; at Beni Abbès, a further 150 km and below the confluence of the Zousfana, the annual frequency has fallen to 2.5 months, whilst Kerzaz, 640 km from the source, is reached by floodwaters once each year on an average. Only exceptional floods at intervals of a few years traverse the defile of Foum el Kheneg, and these mainly branch back northwards, into the floodout or *sebkha* of el Melah. Over a period of 85 years up to 1960, only seven important flows had continued along the Messaoud towards topographic baselevel in the Touat basin. In 1915, the largest of these persisted into a lower arm of the Messaoud along the southeast margin of the Erg Chech, to a distance of 975 km from the source.

The accompanying down-valley diminution of discharge, both in peak flow and total volume, is shown by comparison of station hydrographs for the 1952 flood in Fig. 44b and is depicted diagrammatically for the major flood of March 1959 in Fig. 44c. The peak flows are considerable, that for Abadla in 1959 being calculated as 4000–5000 m^3/s and the corresponding discharge at Beni Abbès as 1000 m^3/s. Irregularities may occur in the down-valley decline of a particular flood due to local patterns of rainfall in the catchment and to the timing of flood waves in tributaries; for example local rains at Igli in 1952 caused flow there before flooding began in the Guir up-valley at Abadla (Fig. 44b). Nevertheless, transmission losses tend to be fairly regular in the middle sectors of large channels where they are mainly the result of infiltration into a sandy bed or of evaporation where floodwaters spread, as they do in the Guir below Abadla. In the lower sectors, however, major losses occur irregularly at points along the course due to lateral discharge into sump basins by distributaries. As shown in Fig. 44c, almost half

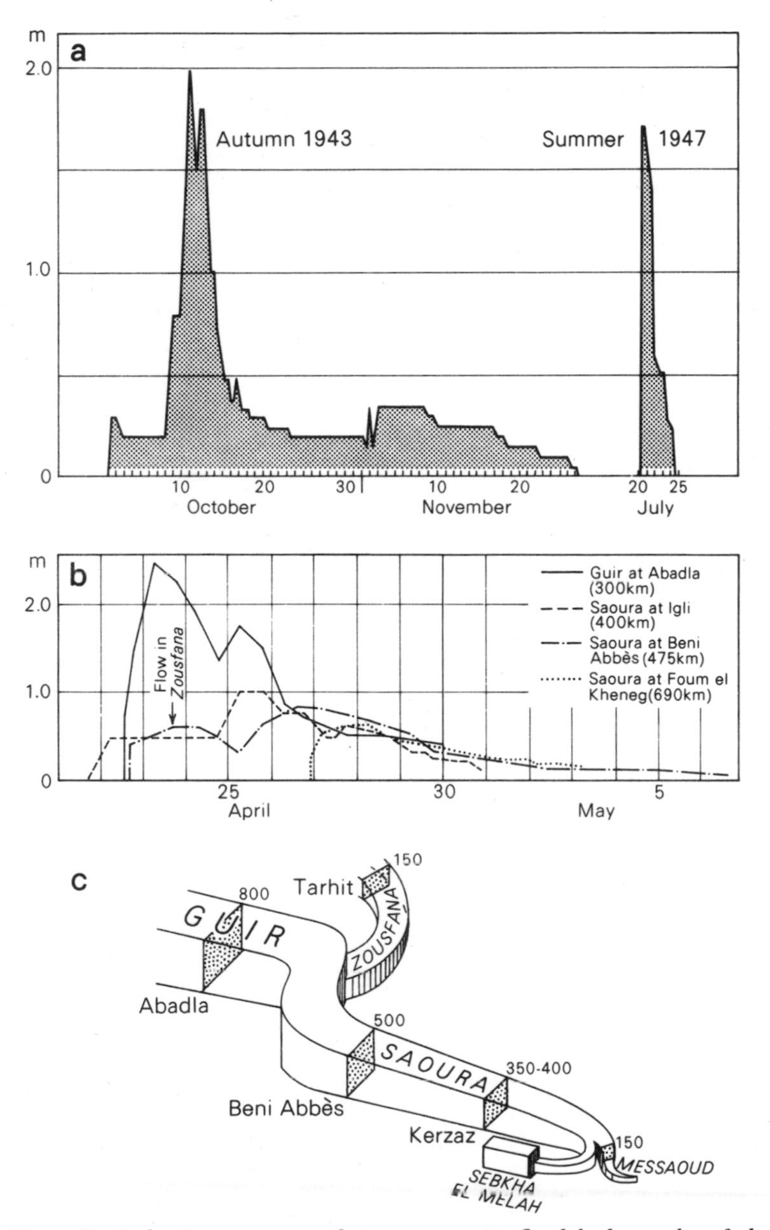

44. a. Typical winter-season and summer-season flood hydrographs of the Guir at Abadla. From Dubief, 1953b; b. hydrographs of a single flood in the Guir-Saoura system in May 1952 (localities in Fig. 43; from Dubief, 1953b); c. diagrammatic representation of down-channel diminution of flood discharges ($m^3 \times 10^6$) in the Guir-Saoura and Zousfana, March 1959. Peak discharges shown in m^3/s. From Vanney, 1960.

the 1959 flood through Foum el Kheneg was taken into the Sebkha el Melah, which has an extent of 250 km^2. Losses into dune sands also increase in importance in lower sectors.

The considerable flood discharges attained in desert channels reflect the rapid and high runoff response from heavy rainfall on sloping desert surfaces. Vanney (1960) estimated that the 1959 discharge at Abadla represented 80 per cent of the rain falling on the catchment upstream, a response which decreased down-valley to 20 per cent at Beni Abbès and to 15 per cent at Kerzaz. This decrease reflects the failure of the lower runoff yields from desert plains to counterbalance transmission losses, and demonstrates the inadequacy of figures for desert runoff averaged over large catchments.

The torrential nature of the flood response is also shown in high flow velocities. Peak velocities of 5.5–7 m/s were recorded in the 1959 flood at Abadla and the average velocity of the flood calculated from bedforms was 3 m/s, with rather higher values in the lowest reaches because of higher gradients and a restricted valley cross-section. The advance of the initial flow in the lower Saoura from Beni Abbès to Foum el Kheneg averaged 0.6 m/s, and the passage of flood crests between 1.2 and 1.6 m/s. These are in the upper range of flood velocities in the northern Sahara (Teissier, 1965).

The increasingly ephemeral regime down-valley is reflected in the attenuation of the drainage net of the Guir-Saoura. The trunk channel receives few tributaries in the lower plains, and dwindles in size and becomes increasingly hemmed in by sand dunes. The system also exhibits internal breakdown in that the Bechar no longer floods into the main channel.

Sediment transport in desert streamfloods

The very high turbidities of desert upland floods also occur in the plains. Suspended load concentrations attained between 1000 and 12,000 ppm in the waters of the Todd River at Alice Springs in the central Australian floods of 1967 described earlier (Williams, 1971), and concentrations greater than 5000 ppm are typical of floods throughout the arid western United States (Rainwater, 1962). Much higher values are known, however; for instance a mean daily sediment concentration of 400,000 ppm was maintained during a flood in the Paria River at Lees Ferry, Arizona, approaching the consistency of a mudflow (Beverage and Culbertson, 1964).

Such high load concentrations are largely inherited from the upland flood sources described in Chapter IV or reflect an extension of similar controls, namely the vulnerability of sparsely vegetated desert slopes to erosion by occasional heavy rainfalls and flash runoff. These effects are intensified as flood discharge diminishes downstream due to transmission losses. Turbidity also tends to be maintained by erosion of the channel banks, particularly during the rising flood stage. Desert river channels in alluvium are typically shallow with loose sandy and little-vegetated banks, and flood scour commonly results in excessive widening through bank scour and cavitation, at the expense of adjacent floodplains. Much alluvium is incorporated into the suspended load in this way.

With the flood rise, sands in the floors of large ephemeral channels tend to be dilated and mobilised and the beds are scoured to varying depths. Bed scour has been found to be proportional to the square root of discharge per unit of channel width and can amount to several metres in a large flood, in which it may be a major determinant of increased depth of flow (Emmett and Leopold, 1964). Most of the mobilised coarse sand and gravel is shifted down-channel as bedload, in amounts related to velocity and depth of flow and to particle size. Although the volume of sediment in transport is large, movements of individual particles are intermittent and at velocities much smaller than that of the flow itself; consequently the bedload is shifted only a limited distance downstream, and coarse gravels may move only a few metres (Leopold *et al.*, 1966). Generally, the bed material is re-deposited during the falling stage of the flood, when the bed resumes its former level.

DESERT SAND-BED CHANNELS AND FLOODPLAINS

General characteristics

Over most of their lengths, river channels in desert lowlands tend to be trench-like in section and wide in relation to their depth (Pl. 38), gently winding rather than markedly sinuous, and locally braiding. Deep scour pools at river bends or below rock bars may retain water for periods after flows. Lower and higher sections of the channel cross-profile may be distinguishable, corresponding to the 'minor' and 'major' beds of perennial channels with marked seasonal variations in discharge.

38. Sand-bed channel, central Australia, fairly straight, and shallow in proportion to its width. Photo R. W. Millington, CSIRO, Division of Land Use Research.

Most such channels fall within the range of 'bedload' or 'mixed load' channels as defined by Schumm (1963), with predominantly sandy beds and banks (less than 30 per cent silt-plus-clay) indicating that much of the sediment transported is as bedload, certainly more than 15 per cent and commonly greater than 35 per cent. This reflects the abundance of sand-grade sediment supplied under desert conditions of partial weathering, effective slopewash, and flash-flooding.

Sand-bed channels that carry much bedload adjust to a section in which high shear-stress over the bed is combined with relatively low velocities along the erodible banks. These requirements are met by the wide and relatively shallow desert river channel, because the greater the width-depth ratio the greater the ratio of the velocity acting on the bottom to that acting on the sides; consequently, widening continues until decreasing lateral scour is balanced by the resistance of weak bank materials. These characteristics are shown by the Finke River in its plains sector (Figs. 45, 46a).

Low sinuosities are also required for the efficient transport of bedload, for bedload transport per unit width in a meandering channel may be 20 per cent below that in a straight one (Leopold and Wolman, 1957). This is because of frictional resistance in a winding channel, particularly along the banks. In the loose bank materials typical of a desert channel any tendency to increase in sinuosity is checked by a responsive increase in bank scour. The

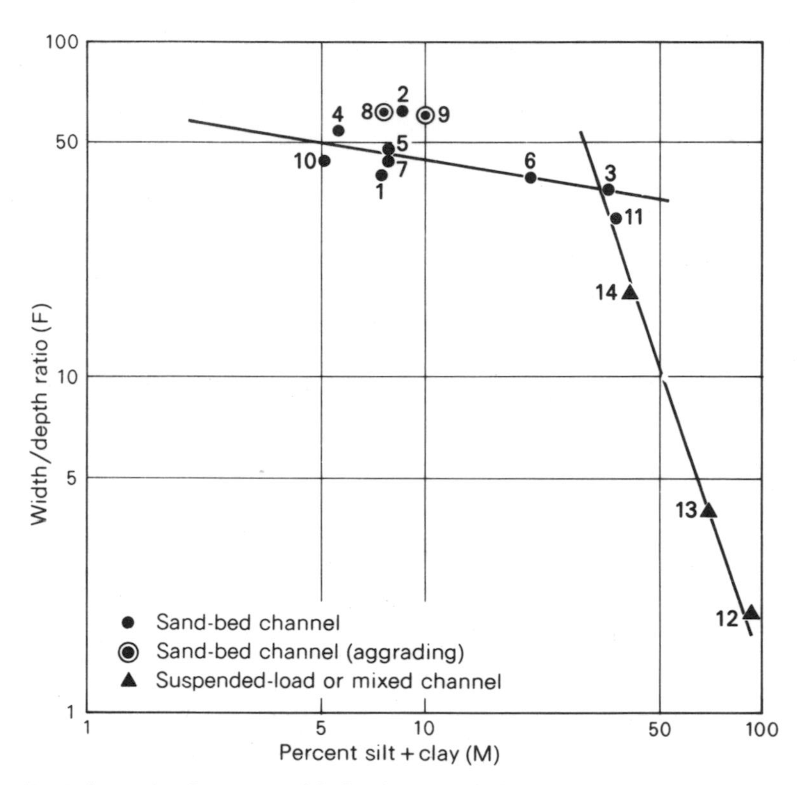

45. Relationship between width-depth ratio of channel and per cent silt-plus-clay in channel perimeter of the Finke River, central Australia. Points are numbered in down-channel sequence.

average sinuosity of 1.35 for the plains course of the Finke River is fairly characteristic of large desert sand-bed channels.

A braiding habit is also common in desert rivers and is expressive of the abundant poorly sorted bedload, the highly variable discharge, and the rapid transit from desert upland to plain. Extreme widening of the channel by erosion of weak banks leads to shoaling through the deposition of coarse bedload particles in a central bar. In the divided sector there is an overall increase in width without corresponding increase in depth, but with increased slope, conveying an ability to transport a larger volume of bedload at the same discharge. Colonisation by vegetation between floods stabilises braid islands, which tend to lengthen downstream. The general distinction established between meandering and braiding channels,

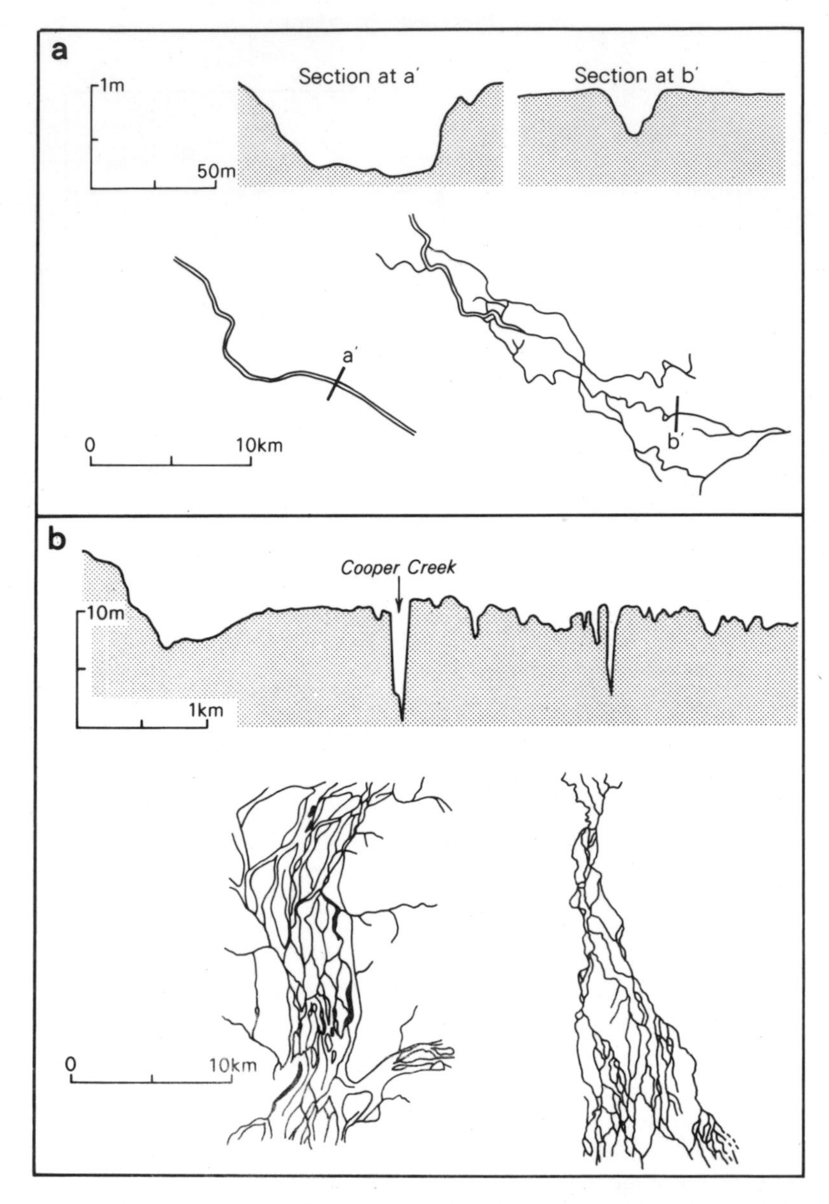

46. a. Cross-profiles and plans of the Finke River as a sand-bed channel in its plains sector and as a suspended-load channel above its terminal floodout; b. (above) cross-profile of the floodplain of Cooper Creek at Windorah, Queensland showing prominent levees and deep channels (below) anastomosing channels of the Thompson River, Queensland, with billabongs (L) and distributary 'deltoid' of the Warrego River, Queensland (R).

that for a given discharge the latter have the steeper slopes, is consistent with the prevalence of braiding in deserts, where coarse sediments, high channel roughness, abundant bedload and high width-depth ratios require steep channel gradients (Leopold and Wolman, 1957).

Relations between flood discharge and channel geometry

Quasi-equilibrium relations between channel geometry and discharge have been demonstrated for ephemeral alluvial channels in New Mexico (Leopold and Miller, 1956). These show important differences from the relations established for perennial channels in humid areas, imposed by the contrasting environment. From Fig. 47 it is seen that the down-channel rate of increase in depth with discharge of constant (bankfull) frequency, given by the exponent f, is smaller for the ephemeral channels than for average perennial channels, and conversely that the rate of increase in velocity given by the exponent n is greater. Since the ratio n/f is related to the rate of increase of suspended-sediment load with discharge (Leopold and Maddock, 1953), it can be anticipated that this will be greater in floods in ephemeral channels ($n/f = .7$) than in perennial streams ($n/f = .25$). The New Mexican studies bear this out. The exponent of increase of suspended load on discharge was found to be 1.3 compared with the average of 0.8 for perennial streams. That the exponent is greater than unity indicates a downstream increase in sediment-concentration with discharge in desert lowland streamfloods due to partial loss of discharge into sandy beds, with negligible settling of sediment because of high velocities of flow.

In consequence of these relationships, and as a result of smaller down-channel decrease in the grade of bed materials — and hence of channel roughness — owing to lack of sorting, the down-channel decrease in gradient with discharge in ephemeral desert channels in New Mexico is less than the average for perennial channels in humid regions. Longitudinal profiles of desert ephemeral channels do generally tend to remain steeper and to be less concave than those of perennial channels of comparable discharge.

The ephemeral channels studied in New Mexico were tributary to the Rio Grande, in a desert region drained to the ocean. Their catchments of a few hundred square kilometres were occupied by integrated channel networks in which the Horton laws relating

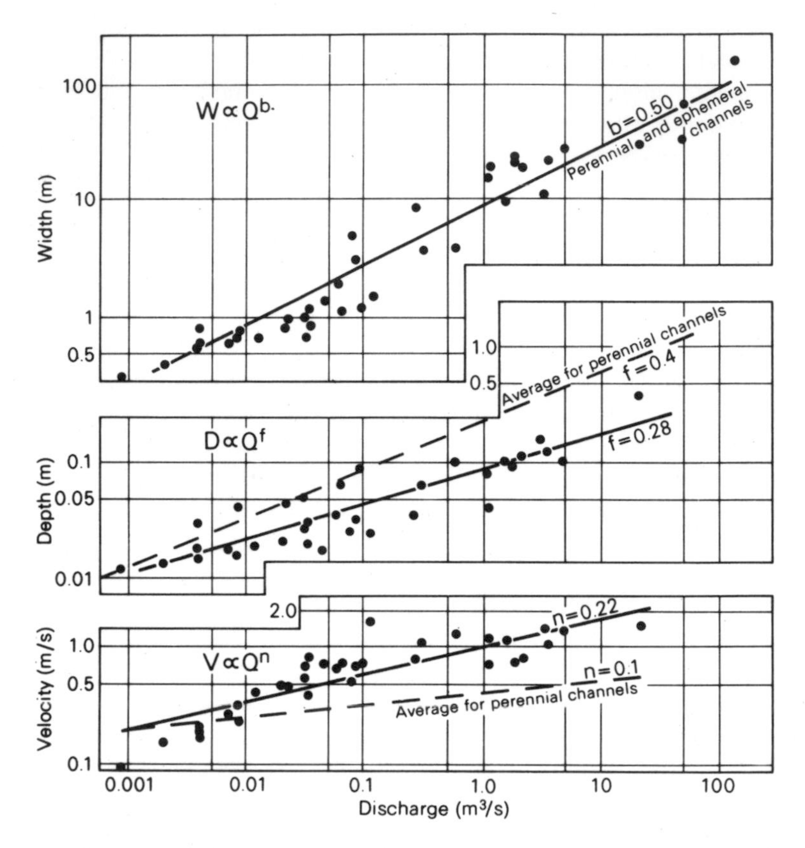

47. Change in width, depth and velocity of ephemeral channels as bankfull discharge increases down-valley with drainage area, based on measurements made in New Mexico by Leopold and Miller, 1956. Regression lines for average perennial channels added for comparison.

stream order to number and lengths of channel segments were found to apply, and in which a down-valley increase in discharge reflected an accompanying increase in effective catchment area. These relationships commonly do not hold for large desert catchments, particularly those in basins of interior drainage and sediment accumulation. In these, flood discharges in trunk channels will diminish beyond a certain distance into the plains, as illustrated by the central Australian and Saharan floods.

One factor contributing to this is the limited extent of desert rainstorms, which rarely cover more than a few hundred square kilometres, whereby part only of a larger catchment may contribute to discharge. Another factor, at least in the lowlands of shield and platform deserts, is that the larger the catchment the greater is likely to be the proportion of included sediment-mantled plains that yield little local runoff. These factors so reinforce the effect of transmission losses of discharge that tributary junctions become increasingly spaced with distance from the uplands and the pro-

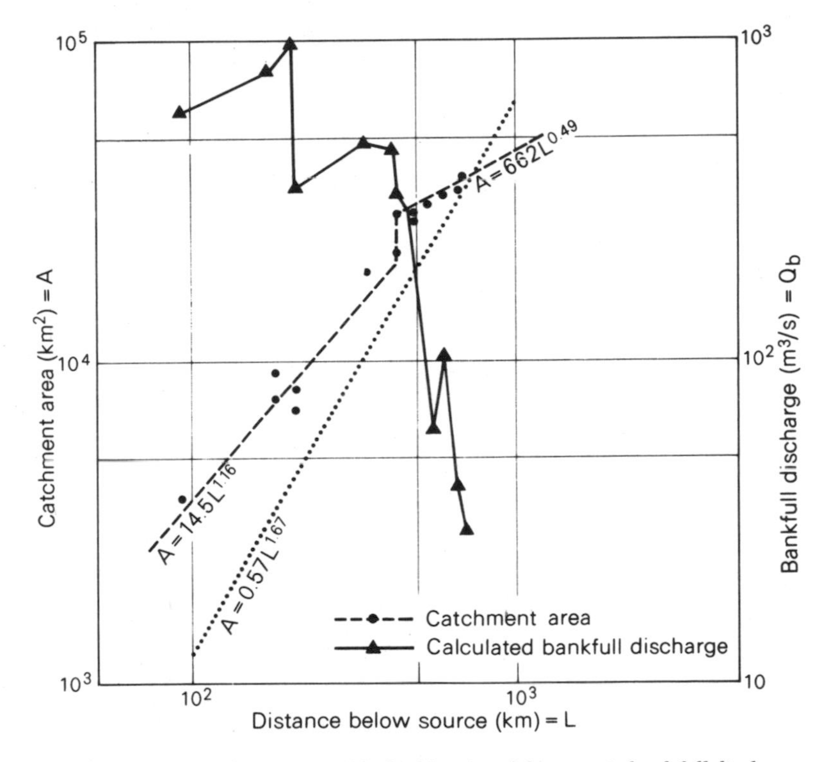

48. Increase in catchment area (dashed line) and decrease in bankfull discharge (solid line), calculated from Manning equation, with increase in distance below source in Finke River, central Australia. Dotted line shows slope of regression of basin area on channel length characteristic of perennial river systems (from Hack, 1957). Discharge calculations include unpublished data from M. E. Sullivan.

portion of the topographic catchment that is hydrologically effective tends to diminish with size.

Under these conditions the rate of increase in effective catchment area with channel length declines from some distance down-channel, as exemplified by the Finke below the confluence of the Hugh River (Fig. 48). There is a concomitant decline in channel dimensions. This is shown by calculated bankfull discharges in the Finke, which begin to diminish some distance upstream but which fall off sharply below the Hugh River confluence. The decline may be checked and possibly reversed at major tributary junctions, as with the Finke at the confluence of the Hugh. This decline in size of desert channels in lower plains sectors tends not to be systematically linked with any change in the shape of its cross-section, for there is little sorting of load and hence little change in proportion of bedload in this tract; for example the width-depth ratios of channel sections of the Finke in Fig. 45 cluster without regard to distance down-channel.

Aggradation in lower sectors of sand-bed channels

Some distance below the fall-off in growth of catchment, aggradation sets in along the sand-bed channel because of increasing sediment concentrations in diminishing flood discharges. This results in shallowing by deposition of bedload, with little change in width. Aggradational instability in such a channel is revealed in an above-average width-depth ratio in relation to the silt-plus-clay content of the channel perimeter (Schumm, 1961), and from Fig. 45 it can be seen that two sections in the lowest sand-bed reaches of the Finke have this form. Aggradation may set in irregularly, reflecting localised hydrologic and sedimentologic controls within the imperfectly adjusted desert channel system. For instance the channel of the Finke River shoals markedly due to 'sand-plugging' both above and for some kilometres below the entry of Goyder Creek, which drains a granitic catchment yielding abundant coarse sand and which in consequence has an extreme width-depth ratio of 63. In major desert sand-bed channels, however, a lower sector of general bedload aggradation is commonly identifiable by progressive narrowing and shoaling of the channel over long reaches and also by steepening. This last feature appears in the lower Guir-Saoura, for which Vanney (1960) cites the following gradients in down-channel sequence:

Guir at Abadla	.0025
Saoura at Beni Abbès	.004
Saoura at Kerzaz	.004

Bedforms of sand-bed channels

Various sandy bed-forms within the channel can yield important evidence of previous floods (Williams, 1971). These remain exposed as minor surface forms in the intervals between flows.

The commonest forms generated by high-stage flow and accordingly formed in the higher parts of the channel are large-scale ripples or dunes between 1 and 5 m wavelength, and longitudinal bars tens of metres wide and 2–3 m high extending up to hundreds of metres down-channel. With higher velocities and intense sand movement, these forms may be erased and replaced by a plane bed. Still further velocity increase into high-energy upper-regime flow, when waves form on the stream surface, results in large anti-dunes facing upstream. As water topping the bank spreads out, it forms flat-crested sand splays of horizontally laminated sands with steep foresetted fronts or rippled tongue-like extensions.

The lower-energy forms typical of the declining stages of the flood and hence found in the lower parts of the channel are large transverse bars with flat, often gravel-veneered crests tens of metres in extent and terminating down-channel in avalanche faces up to 1 m high, and small-scale ripples of 5–15 cm wavelength. In general, small-scale ripples and transverse bars are associated with finer and better-sorted sands, and large-scale ripples with coarser, poorly-sorted sands. The commonest high-stage form left by the central Australian floods of 1967 was large-scale ripples, indicating lower-regime flows with velocities of between 1 and 2.5 m/s, but plane sand beds in the middle reaches of the Finke indicated the attainment of upper regime flows there, with velocities of up to 3.5 m/s.

Channels floored with sand and gravel tend to develop bed patterns due to rhythmic selective scour-and-fill. Commonly there is an alternation of gravel bars and sandy reaches in which the bars show a regular spacing of between 5 and 7 times channel width, similar to the pool-and-riffle sequence in perennial channels with which they appear to be related (Wertz, 1964; Leopold *et al.*, 1966). The gravel does not generally extend in depth and appears to have been brought to the surface through differential dispersive forces produced by grain-to-grain impact during bedload transport. It is the same tendency for larger particles to move to areas of lesser dispersive force that concentrates gravels in mid-channel bars. The gravel moves during floods but the bars tend to remain stationary. Flood scour forms an undulating gravel-floored bed which is then smoothed by selective deposition of sand during the falling flow stage, leaving the gravel surfaces projecting only slightly.

Floodplains of sand-bed channels

Forms typical of these floodplains in central Australia are depicted in Fig. 49a (Perry *et al.*, 1962). The channel is typically flanked by an uneven zone of alternate longitudinal banks and flood furrows rather than by levees. The channel and this innermost floodplain are subject to frequent remoulding, for adjustment to high discharges in sand-bed channels is mainly achieved by bank erosion and widening and by the occupation of lateral arms at higher levels, which may result in shifts of course. Equally, aggradation takes the form of shoaling and braiding, and mid-channel islands stabilised by vegetation can in turn become incorporated into a laterally extending floodplain through the abandonment of braid arms. Schumm and Lichty (1963) have described such a cycle in the Cimarron River of southwest Kansas, in which the channel widened from 15 m in 1870 to 360 m — the whole valley flat — in 1942, in response to a period of variable but below-average rainfall and occasional severe floods, and subsequently narrowed to 165 m by 1954 after a period of above-average rainfall and more regular but less destructive floods.

On the more stable alluvial surfaces beyond the active flood tract the light topsoils are subject to stripping by wind and occasional sheetfloods. The eroded areas commonly form part of regular transverse or longitudinal patterns with wavelengths of tens to hundreds of metres, possibly reflecting original variations in the texture of the alluvium and associated microrelief, reinforced by dependent contrasts in surface salinity which accelerate selective erosion. The stripped areas, known in Australia as 'scalds', may develop into claypans through dispersion of the upper part of an exposed heavier subsoil in standing water, for they are relatively impervious. Limited parts of the inner floodplain may be occupied by small alluvial basins which act as flood sumps, and by their feeder channels.

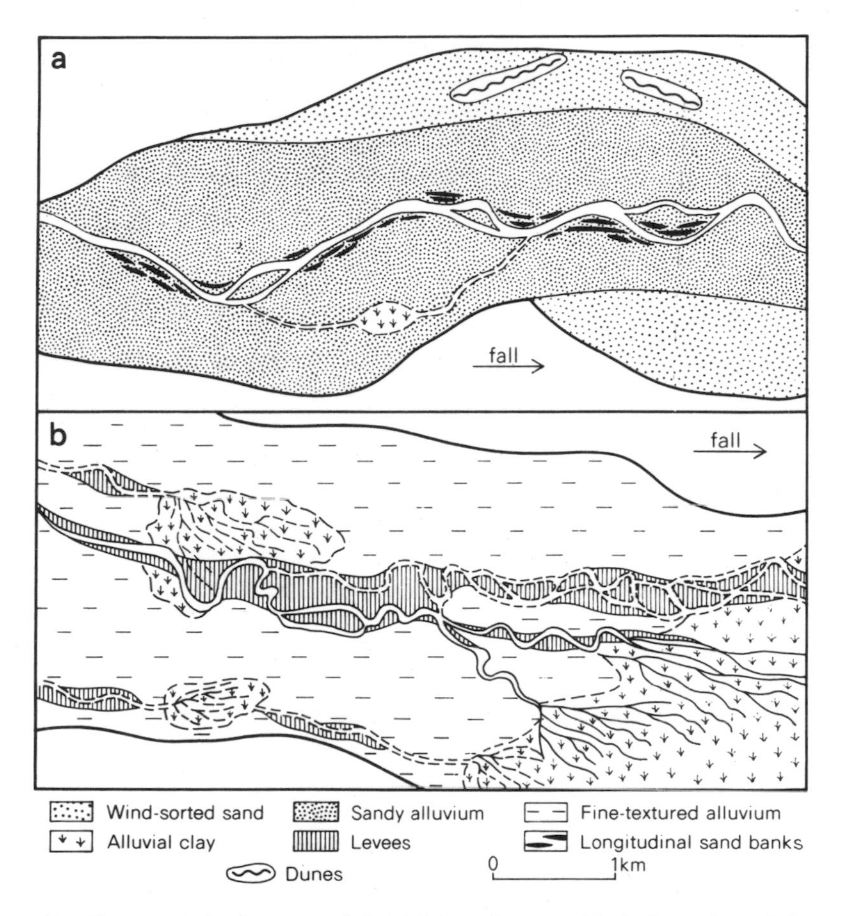

49. Characteristic features of floodplains of a. sand-bed channels and b. suspended-load channels in lowest plains sectors in central Australia. Based on surveys and airphoto interpretation.

The outer floodplain is commonly at a slightly higher level, often bounded by a curved step marking a former course, for unpaired low terraces are the likely result of a shifting channel regime. The fluvial imprint is less strong here and the surface soils are often wind-worked; sand mounds occur about vegetation and low dunes are common.

The floodplains of most desert sand-bed channels are built predominantly of sand. Overbank deposits consist typically of overlapping convex lenses a metre or so thick, with minor silt horizons marking deposition from slack water in swales between sand banks. Channel fills formed within broad shifting channels are similar, with features representing point bars, although the bedding is subject to planation in upper-regime floods. Smaller channel fills are commonly horizontally-layered (Schumm, 1960a, 1961).

SUSPENDED-LOAD CHANNELS AND FLOODOUTS

Channel forms

Continuing aggradation of bedload and relative increase in the percentage of silt and clay in flood sediments eventually lead to changes in the form of the channel. With increase in suspended load and consequently in the proportion of fine cohesive sediments in the banks and bed, the aggrading channel narrows through deposition on the vegetated banks as well as on the bottom, with resultant decrease in width-depth ratio (Schumm, 1961). With these changes comes an increase in sinuosity, commonly to a meandering habit. At the same time, levees form along the now stable banks, above the general level of the floodplain. This can lead to instability, to crevassing of the levee and to the splitting off of a separate channel. The offshoot tends to run parallel with the parent channel and at the foot of the levee backslope, until possibly swinging to re-enter it. The *anastomosing* channel pattern so formed is reticulate and rather rectangular with sharp angles of bifurcation and re-entry. A distributary pattern is common where the channel system enters a flood basin.

In the Finke, these changes as depicted in Figs. 45 and 46b take place over tens of kilometres, but in smaller channels they can be accomplished over shorter distances.

Floodplains in lower channel sectors

Features of floodplains in these lowermost sectors, as shown in Fig. 49b, include anastomosing and distributary systems of narrowing sinuous channels, both active and abandoned, and prominent levees reflecting by their relative development the ages of the channels. Terminal distributaries typically branch out in the form of digitate deltas building out into an alluvial basin or floodout. The channel tract stands higher than the backplain with its lateral flood basins and feeder channels, although abandoned older channel tracts may form low rises within it. Longitudinal gradients are very low (1 to 2 per thousand).

Alluvial deposits in the channel tract consist of layers of fine sand and silt, often in concave lenses because of the tendency to deposition on the banks of suspended-load channels (Schumm, 1960b). These are interbedded with larger bodies of silt and clay deposited from standing water.

Floodouts

A majority of desert streamfloods fail to reach topographic baselevel, whether in the ocean or on the floor of an inland basin. Minor flows may be defeated by infiltration to die out in the sand-bed sectors of the channels, but many floods continue further and empty into lateral or terminal alluvial basins in lower sectors. Such floodouts are termed *sebkha* in the Sahara. Minor floodouts may be limited by alluvial barriers created by the river itself; others are dune-barred, for many major desert streamfloods end at the frontier between the fluvial and aeolian domains (Pl. 37), and others again are located against topographic barriers.

The terminal floodout of the Finke shown in Fig. 41 occurs where the channel is hemmed between the dune ridges of the Simpson Desert and an escarpment, where deposits from local channels flooding from the backing tableland constitute a further impedance. All large floods in the Finke terminate here, from those of 3–4 year intervals to flows with an estimated recurrence interval of decades, as in 1967. An alluvial tract continues southwards through the dunes beyond the floodout and can be traced to the Macumba and so to topographic baselevel in Lake Eyre, but there is no historical record of streamfloods along it.

In the Saoura system the Sebkha el Melah is a lateral basin, trapped between dunes to the west and the strike ridge pierced by the main channel at Foum el Kheneg (Fig. 43). The extent to which the largest floods continue beyond the sebkha down the Messaoud is influenced by dune barriers in the channels.

Various forms of gilgai microrelief occur in floodouts, in swelling clay soils subject to episodic flooding followed by drought and deep cracking. They include round gilgai, often clustered round the ends of distributary channels, reticulate lattice and depression gilgai in the flat floors of basins, and linear wavy gilgai down the marginal slopes (Hallsworth *et al.*, 1955; Harris, 1968).

Clay plains in riverine deserts

Catchments in which source rocks yield mainly fine-textured sediments may contain extensive clay plains drained by suspended-load channels, for example the Channel Country of southwest Queensland in Australia, named from the tracts of anastomosing river channels which cross it (Pl. 39).

The region forms a northeast inner quadrant of the Lake Eyre basin and most of it drains towards Lake Eyre through the Diamantina River and Cooper Creek, but it also includes the similar floodplains of northern tributaries of the Darling River, such as the Warrego. Extensive aggradation by these rivers has resulted from subsidence of the inner basin and has been favoured by rainfall gradients, for the rivers rise on the wetter northeastern upland rim of the arid zone where heavy summer rains cause extensive flooding at intervals of a year or two.

The multiple channel tracts, commonly between 2 and 5 km wide, consist of a regular mesh of sinuous, suspended-load channels which branch at wide angles, the arms pursuing parallel courses separated by the levee rise before linking again downstream in a

39. Anastomosing channels of Cooper Creek, southwest Queensland. Billabongs occupy deep-scoured straight reaches.

similar abrupt angle (Fig. 46b). The term *anabranch* has been applied in Australia to large distributary channels of this type that continue on a separate course. At intervals, several anastomosing arms may link in a single broad straight reach which accords with a slight flattening of the floodplain. This unified sector may be deeply scoured and is then typically occupied by a perennial water hole up to several kilometres long, known as a *billabong*. The channel meshes are crossed by sinuous shallow floodways which feed from and into the deeper channels and which are occupied only at high flood stages.

In the channel tracts, patterns of reddish silt and fine sand outline old meander scrolls as much as 2 m above backplains of dark cracking clay. These featureless surfaces may extend for tens of kilometres, broken only by silty rises marking former levees and by aligned islands of sand dunes also formed in prior channel sediments. Sump basins of circumdeposition in the outer margins of the backplains are fed from the main channel by distributaries; they are regularly flooded and have extensive gilgai.

The floodplains broaden locally in crossing secondary basins of subsidence where gradients fall below 1 in 5000, as at the confluence of Cooper Creek with the Wilson River. The channel system enters such basins in fanshaped *deltoids* (Fig. 46b) (Whitehouse, 1944), and the flat floors of the basins are occupied by regular meshes of interconnecting linear depressions resembling giant reticulate gilgai.

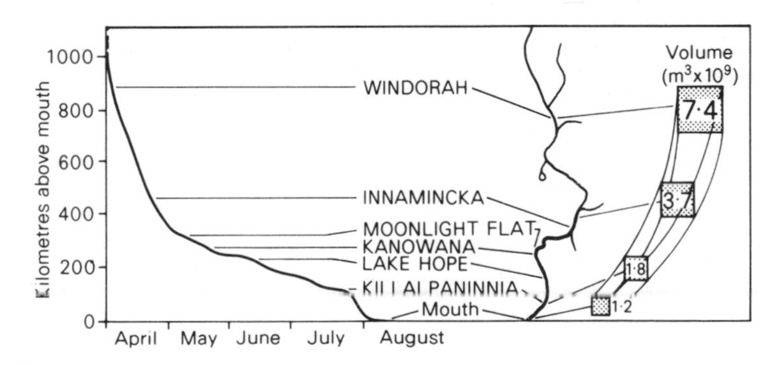

50. Advance of flood down Cooper Creek to Lake Eyre, central Australia, April-August 1963, showing irregularity of advance and decrease in flood discharge due to losses into lateral floodouts in lower sector. After Bonython, 1963.

The various basins along the river courses accept floodwaters on the rising stage and later discharge part of them back into the main channel. These detentions slow the advance of the flood wave in lower reaches, particularly where the channels are also opposed with increasing frequency by dunes, and only the largest floods, at intervals of a decade or so, break through to Lake Eyre (Fig. 50).

DISORGANISATION OF DESERT DRAINAGE

The intermittent functioning of desert river systems has its spatial counterpart in discontinuity or *disorganisation* of channel networks, in forms dictated by structurally-controlled major relief as well as by climate.

In the open shield and platform deserts for example, the limited uplands constitute island sources of connected drainage systems of which only a few trunk channels continue for long distances across the lowlands. Some of these may persist to the ocean or to inland terminal basins, but many more die out at intermediate points on catchment slopes. Maps of existing drainage networks, as in the Sahara (Fig. 64), commonly indicate that these are relics of more extensive and integrated former systems that have retracted towards the upland sources of runoff and have become increasingly disconnected through the isolation of tributary components, and this indication is commonly supported by the distribution of older alluvium or of features such as abandoned water gaps, but over much of the lowermost plains the evidence is now obliterated by dune sands, albeit of fluvial origin! Aridity has had particularly widespread effects on the drainage systems of these deserts, for the vulnerable lowland drainage sectors are here relatively extensive. Examples as evidence of climatic changes will be discussed in Chapter X.

Present-day streamfloods may continue to follow along the lines of former more extensive drainage systems, and their efficacy is enhanced as a result. Elsewhere, some shifts of course have accompanied disorganisation; for example the Hanson and Lander rivers now mainly pursue northward courses from the ranges of central Australia, over young depositional surfaces, but locally intersect and follow northwesterly, structurally-guided Cainozoic valleys revealed by old calcreted valley fills.

In the mountain-and-basin deserts, disorganisation of drainage mainly takes the form of greater compartmentation, each structural and topographic basin having its own centripetal interior system; for example the Great Basin of the western United States has about 150 separate drainage systems. Relief in source areas is generally strong and the distances to be traversed across the piedmont plains are mainly small, and connected systems generally persist to the basin terminals. In past wetter periods a large number of these basins were occupied by permanent lakes, many of which spilled over into adjoining basins, and the drainage became more integrated (Chapter X).

Measures of drainage disorganisation

Martonne (1927) used the relative extent of interior drainage, or *endoreism*, distinguishing this from *exoreism* or external oceanic drainage on the one hand and *areism* or an absence of locally-generated connecting drainage on the other.

Dubief (1953b) employed the *relative extent of connected drainage* expressed as the percentage area of a topographic basin occupied by connected systems of river channels. Since the efficiency of desert catchments has been shown to diminish beyond a certain area, comparisons should be restricted to basins of the same order of extent.

A third measure is the *degree of connectivity* remaining within the channel network of a topographic basin. Dubief (1953b) has distinguished five stages of breakdown:

1. Flows along the trunk channel of the basin system continue to reach baselevel occasionally, but only with the help of tributary floods entering the lowest sector; for example where that sector runs close to an upland source of local runoff.
2. Flows commonly occur along the trunk channel of the basin system as far as it can be traced, but no longer reach baselevel.
3. Only exceptional floods traverse the whole portion of the trunk channel which remains traceable, most ending at intermediate floodouts.
4. The traceable trunk channel is no longer traversed in its entirety by single floods.
5. There is no trace of a trunk channel corresponding to the basin system.

Application of Dubief's criteria of connectivity requires records of flooding which are rarely available in deserts, whilst physiographic or vegetational evidence of flood connections, as from airphotos, may be lacking in advanced stages of drainage breakdown, for example where aeolian sands have covered the lower parts of a basin. Furthermore, the proposed stages are not necessarily successive; breakdown need not proceed regularly up-valley but may be more advanced in the upper part of the basin. For instance, on the debatable assumption that a very large flood in the Finke River might still break through to Lake Eyre, but that local discharge from the Macumba system would be necessary for this, the lower Finke-Macumba system would represent a first stage of breakdown on the above scale; in the western part of the Finke catchment, however, no trunk channel through Lake Amadeus is now traceable and disintegration has reached the fifth stage. This discrepancy reflects the existence within the basin of two contrasted structural and relief provinces with differing vulnerability to drainage breakdown, that of the Western Platform in the west, with island ranges and broad lowlands, and in the southeast the erosional margin of the tectonic basin of Lake Eyre, with favourable centripetal slopes on weak impervious rocks.

The approach is suited to the Sahara for which it was devised, where topographic basins are large and wadi courses remain traceable, although not necessarily functional, over long distances. An appropriate example is the Wadi Igharghar, which rises in the Hoggar massif and runs north towards a topographic baselevel slightly below sealevel in the Chott Melrhir, near the Tunisian border. The system now exists only as far north as the edge of the Erg Oriental, 560 km short of its goal; the lowest sector is occupied by the sand sea and can have operated only in the geologic past. The wadi course now functions as three separate systems (Fig. 51), one of 270 km from the source to the In Tezait floodout, a reach of 80 km below Amguid, and a third of 100 km through the Tinghert Plateau along which extreme floods penetrate to the Erg Oriental. Each active sector depends upon tributaries from an adjacent upland source, and each terminates in a deltaic floodout consisting of a number of ramifying shallow arms, beyond which the former continuity of the channel is indicated by lines of phreatophytes, or must be presumed from topographic evidence. The Igharghar now represents the fourth stage of breakdown on Dubief's scale, but within historical times the first and second sectors were linked and the distance separating the second and third active sectors was much less.

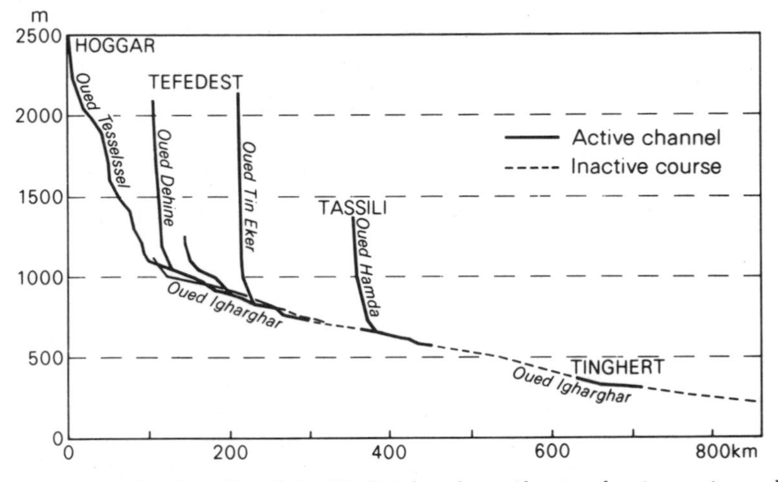

51. Longitudinal profile of the Wadi Igharghar, Algeria, showing active and inactive channel sectors. From Dubief, 1953b.

Status of drainage in arid Australia

Attempts to correlate degree of drainage disorganisation with effective climatic aridity will be discussed in Chapter XI. However, regional differences in drainage disorganisation tend to express tectonic, topographic and surface-geologic influences acting under the general stress of low rainfall, rather than climatic gradients directly. This is illustrated by the differing status of drainage systems within arid Australia as shown in Fig. 52 (Hills, 1953; Mabbutt, 1973).

The close accordance between the limit of the arid zone and the division between endoreic and exoreic drainage is noteworthy. The association is not a simple one, however, for it incorporates the influence of the rainshadow of the Eastern Uplands on the limit of the desert region. Also, the limit of exoreic drainage in the north may post date the aridity of the interior to the extent that it results from an unequal headward contest between an aggressive well-watered coastal drainage and the desert systems to the south.

Within the arid zone the boundaries between drainage classes show little relationship with climatic gradients. The endoreic

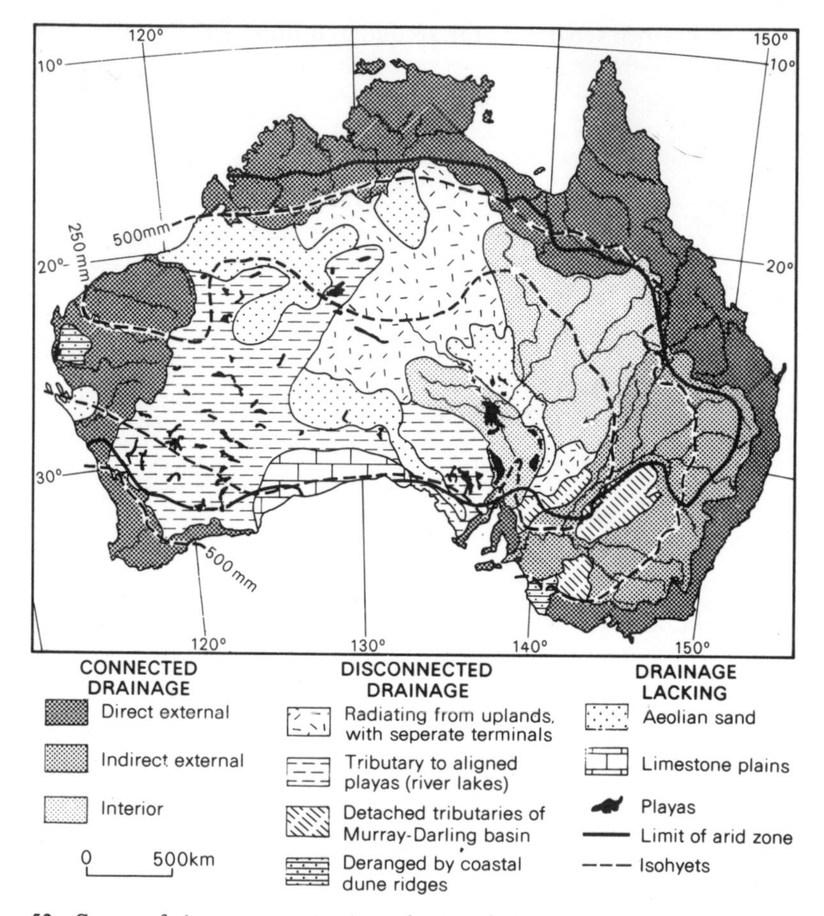

52. Status of river systems in Australia in relation to the extent of the arid zone and to rainfall gradients as shown by annual isohyets. After Mabbutt, 1973.

drainage directed to Lake Eyre is exceptional for its persistence as a connected system despite its inclusion of the driest parts of the continent; this is due to a favourable tectonically-induced combination of better-watered peripheral uplands and centripetal slopes on uniformly weak, relatively impervious rocks. The Murray-Darling system, in a comparable structural setting, owes its exoreic status to the higher rainfall of the Australian Alps and is the only major perennial drainage to enter, traverse and escape seawards from the Australian arid zone. Some of its tributary systems have nevertheless been cut off by barriers of aeolian sand or alluvium.

The two main areas of disconnected endoreic systems occur on the Western Plateau, where restricted upland catchments and extensive plains with pervious sandy mantles have led to the truncation and isolation of drainage elements, many of which are potentially exoreic. Patterns of disorganisation have taken two forms. On shield desert landscapes in the south of Western Australia, where flat lateritic interfluves are mantled with sandplain, systems of broad shallow valleys leading towards the Great Australian Bight have disintegrated into aligned 'river lakes' separated by tracts of aeolian sand but connecting locally at times of heavy flooding; in central Australia, on the other hand, each isolated upland generates radiating drainage systems which die out separately in the adjoining sandy lowlands. The truly areic areas are plains with pervious surfaces, namely the limestone Nullarbor Plain in the south, and the larger areas of sand desert.

VIII

DESERT LAKE BASINS

Maps of deserts show many lakes, but this is not surprising if they are seen as evidence of a disorganised surface drainage. Lakes without outlets, or 'closed lakes', can survive only where net evaporation is at least equal to long-term inflow from surface runoff and discharge from groundwater, and are therefore characteristic of arid regions (Fig. 53A). The association between aridity and closed lake basins has been explained by Langbein (1961) as follows: 'The greater the aridity, the greater is the net evaporation. The greater the net evaporation, the smaller the area of a closed lake. The smaller the lake area, the more numerous are the topographic opportunities for closed lakes'. The majority of the many closed desert lake basins are small; for example a sample of 600 in Western Australia gave a median area of 0.05 km^2 (Killigrew and Gilkes, 1974). Overall, they comprise less than 1 per cent of desert surfaces (Stone, 1956).

Origins of closed desert basins as listed in Table V have been discussed by Stone (1956), Reeves (1968) and Smith (1969, 1972). Primary causes of basin formation, like the subsequent modifying processes, vary in importance between geologic and climatic settings. Block-faulting is characteristic of mountain-and-basin deserts, as in the southwestern United States, whereas the large tectonic lake basins in shield and platform deserts, for example that of Lake Chad, are commonly caused by broad shallow warping. Crater lakes are naturally confined to young volcanic areas, like the Hoggar massif in the central Sahara, and solutional basins to areas of limestone or calcrete, for example the *dayas* of the Algerian Sahara. Deflation basins occur in flat landscapes, commonly of soft alluvium or lacustrine sediments, but in extreme-arid deserts such as the eastern Sahara they are also found to be eroded in subhorizontal rocks. Basins formed by animals, such as the *vloers* of the Kalahari, are naturally more typical of moderate deserts. Basins impounded by barriers of wind-blown sand and alluvium are characteristic of flat plains and valley floors near the terminal sectors of drainage or on the margins of dunefields, where aeolian sand has invaded river systems, and are closely associated with the disorganisation of drainage in riverine deserts, described in Chapter VII.

Desert lake basins vary widely in dimensions according to their origin. They range from animal wallows a few metres across to tectonic basins of many thousand square kilometres. Closed lakes in the larger tectonic depressions may be the complex outcome of more than one process; for example the bed of Lake Eyre in Australia is part-deflational, although the depression was tectonically determined by warping and minor faulting. With the exception of some tectonic lakes in mountain-and-basin deserts, most desert lakes are shallow, rarely exceeding 15 m deep at high-lake stage.

The primary origin of a lake basin is reflected in its *initial shape*. Tectonic basins are commonly elongated along structural axes, those behind depositional barriers will probably share the trend of an impounding form such as a dune or beach ridge, deflation basins may initially be oriented parallel to the dominant wind, whereas those formed by animals, through solution, or as volcanic or meteorite craters tend to be roughly circular. Steeply bounded tectonic basins are naturally more regular in initial outline than those due to shallow warping or other causes.

Where there is little fluvial inflow and input of sediment, as in an isolated basin or in a very arid setting, the outline will remain irregular and the floor uneven. The floors of lake basins largely dependent on groundwater discharge may in fact increase in micro-relief with time. However, where clastic sedimentation dominates, the floor becomes progressively levelled through deposition. More important, to a degree depending on the supply of sediment, the duration of high-lake stages and the size of the basin, the margins are smoothed through erosion and deposition by waves and currents during flooding and by direct wind action in dry-lake stages. The downwind shore is particularly subject to straightening, partly through regression and cliffing by wave attack and more importantly by littoral drifting and the construction of shore terraces, bars and spits. It may also be extended laterally by croding end-currents In a moderate-sized shallow lake the downwind shore will tend to develop a smoothly curved outline with chord normal to the resultant dominant wind during the period of flooding, in equilibrium with patterns of wave-generated currents and associated strengths of littoral drift. Deflation from the dry lake floor and the

TABLE V **Origins of closed lake basins in deserts**

Tectonic basins	
faulted	Lake Natron, northern Rift Valley, Tanzania Death Valley, California
folded	basins of Isfahan Watershed, Iran
down-warped	Lake Eyre, central Australia Chott Djerid, Tunisia
Erosional basins	
aeolian	Qattara Depression, Egypt
fluvial	*charco* and *tinaja* scour-pools, southwest U.S.A.
solution	Sebkha Aarred, C. Bojador, Rio de Oro
animal	*vloers* of Kalahari
Depositional basins	
alluvial	Troy Dry Lake and Lake Tulare, California
dune	Lake Chad Moses Lake, Washington
littoral	coastal lagoons (*aftouts*), Morocco-Mauritania
lava	Alkali Dry Lake, California

Volcanic and meteoric craters	
	Zuni Salt Lake, New Mexico
	Meteor Crater, Arizona (now dry)

building of a transverse lee dune may reinforce this tendency for the development of *oriented lakes* (Price, 1972). Where occasional opposed winds occur the resulting *mature outline* is elliptical, as Stone (1956) found to be the case for two-thirds of desert lake basins in California. The small lakes in Western Australia sampled by Killigrew and Gilkes (1974) are also predominantly elliptical, with a median elongation ratio of 1.5 : 1 and with the major axes oriented normal to the dominant winds of the winter period of high-lake stage.

The evolution towards a mature outline is most rapid in small shallow lakes and is delayed in proportion to the size and depth of the basin and the infrequency of flooding. An upwind shore may maintain or increase its initial irregularity through fluvial progradation where the alluvial input is large — since the intake is commonly sited here — or where winds are markedly unidirectional. It may also be subject to dune transgression in an arid environment. The majority of the oriented lakes of semiarid southeastern Australia exhibit the contrast between a smooth downwind shore and a relatively irregular upwind margin.

Subsequent or *post-mature evolution* involves the filling of the basin with sediment, for all closed basins are in a sense transitory. This stage may include segmentation by growth of beaches under the influence of localised wave fetch, as in coastal lagoons (Bird, 1968). A new cycle of lake formation may be initiated with the erosion of depressions in the surface of the fill.

In many steeply bounded desert lake basins, shoreline changes were accomplished in past wetter periods under a high-lake regime, and many of the forms are now relict and related to water levels above those presently attained. Evidence of climatic change from desert lakes is discussed in Chapter X. Under desert regimes, with basins commonly dry for long periods, the most striking and dynamic features are those of the exposed floors. The three interrelated determinants of the nature of these surfaces are the hydrologic, sediment and salinity regimes, which are in turn dependent on geologic history and past and present climate.

HYDROLOGIC REGIMES OF DESERT LAKES

Flooding regimes

Most desert lakes are markedly ephemeral. Motts (1970) has distinguished between *lakes, playa lakes* and *playas* on the basis of the average proportion of time that they are inundated. The term *playa*, from the Spanish word for 'shore', applies where this *flooding ratio* is one quarter or less, and is an appropriate generic term for the floors of ephemeral desert lakes.

Variations of inflow to a flooding playa in an enclosed basin can be accommodated only by variations in lake volume which bring compensatory changes in lake area and hence in discharge through evaporation. Leakage to groundwater is rarely significant. This balance is expressed in the ratio of basin area to playa area. Fig. 53B shows that this commonly exceeds 5 for closed-basin lakes and that there is an increase in the frequency of such lakes

40. Playas Lake, New Mexico, and flanking bajadas. A playa lake independent of groundwater, here flooded to a maximum depth of 20 cm. Photo J. T. Neal.

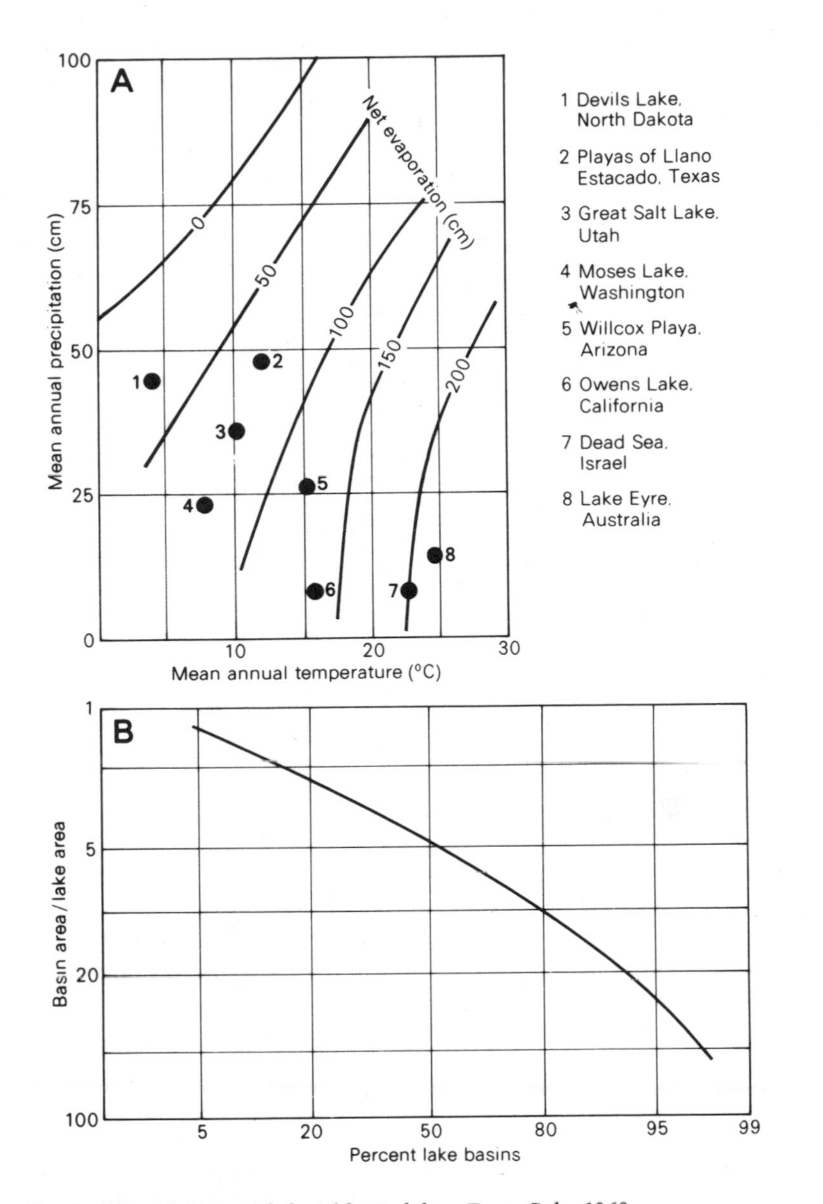

53. A. Climatic range of closed-basin lakes. From Cole, 1968.
B. Percentage distribution of ratios of lake areas at overflow level to catchment areas for closed lakes. From Langbein, 1961.

as the ratio rises. For most playas it exceeds 10; in Iran for example, Krinsley (1970) found a range from 2 to 43, with few values below 10. The ratio also expresses topographic and geologic controls of inflow, and the lowest values for the Iranian playas were in the smaller, steeply-bounded basins close to upland sources of runoff, in which surface geologic conditions also favoured recharge.

Together with excess evaporation, the *basin-playa ratio* largely determines the *response time* of a closed lake, or the drying cycle in a playa. This is because playa-lake area and volume are commonly related in the form

$$A^{0.4} \propto V^{-0.7} \qquad \text{(Langbein, 1961)}$$

For most small playas, the drying cycle is less than one year. Hence there is little carry-over storage; the playa responds to storm episodes rather than to seasonal rainfall and its fluctuations tend to record fully the variability of desert rainfall.

The large playa of Lake Eyre North, for example, has an area of 8000 km^2 and a basin-playa ratio somewhat above 100. The catchment in western Queensland has an annual rainfall of between 100 and 500 mm. The estimated response time is 1.9 y (Langbein, 1961) but the cycle of filling and drying illustrated in Fig. 54a lasted more than three years (Bonython and Mason, 1953). Flooding was initiated by rains of 200–400 mm over most of the catchment in February-March 1949. The major tributary, Cooper Creek, flowed for several weeks, breached the sand dunes in its lower course and entered the lake in June, flooding the southernmost gulfs. Further heavy rain of about 500 mm in January-April 1950 and supplementary falls later in the year caused the lake to fill by September, and the drying phase began in October. The fall in lake level shown in Fig. 54b represents an annual evaporation of 250 cm. The limited records suggest that water can be expected to enter Lake Eyre at intervals of a decade and to fill it completely about twice each century (Bonython, 1955, 1963).

Groundwater discharge

Variability in the hydrologic regime of a playa is diminished where episodic inflow is supplemented by more permanent discharge from groundwater, and this in turn is influenced by the degree of closure of the groundwater system in relation to basin topography.

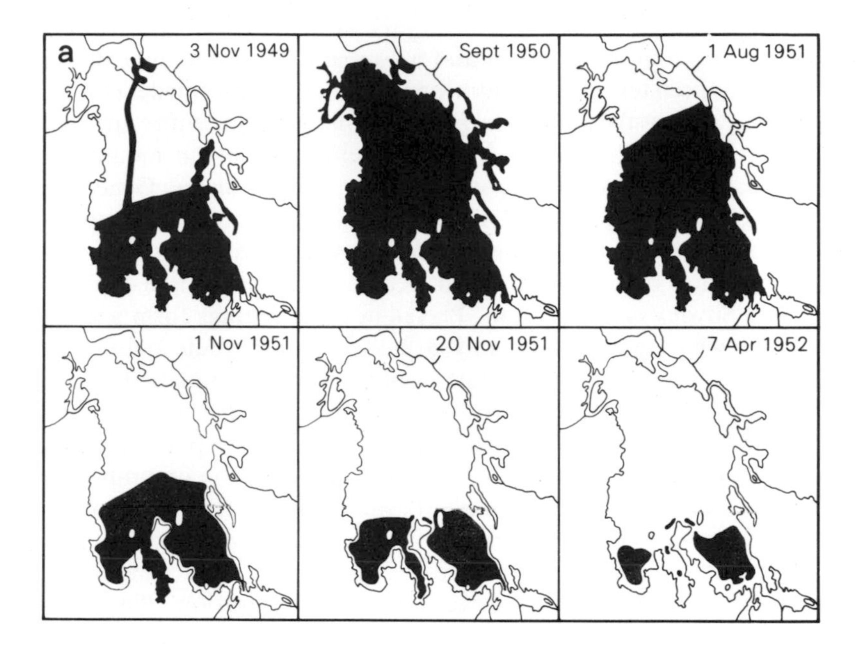

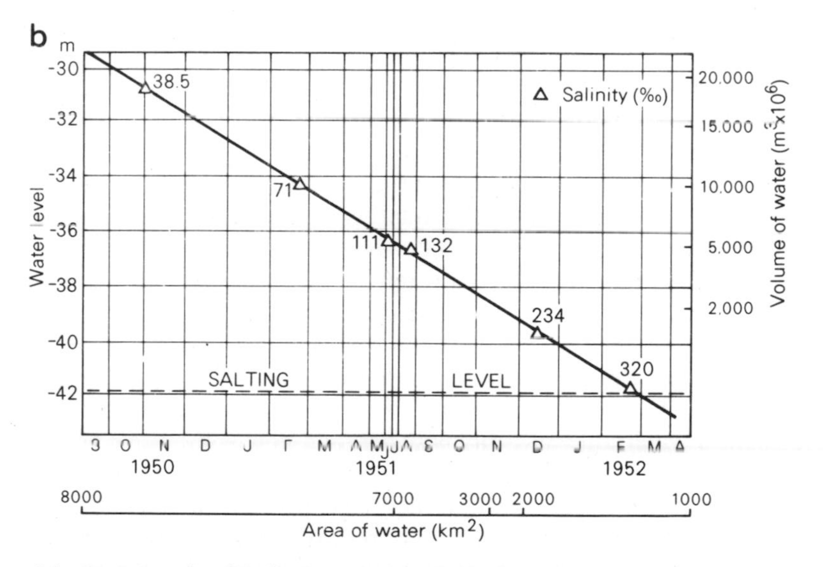

54. *Drying cycle of Lake Eyre North, 1949–52. a. Extent of surface water; b. relationships between level, volume, area and salinity of water. From Bonython and Mason, 1953.*

Where a groundwater basin and a topographic basin coincide, in what has been termed an 'undrained closed valley' by Snyder (1962), the playa forms the point of terminal discharge for the entire groundwater system. This occurs in many tectonic basins in mountain-and-basin deserts, as in the southwestern United States. Generally in shield and platform deserts, and quite commonly also in mountain-and-basin deserts, groundwater provinces are larger than the topographic basins, which then become 'drained closed valleys' under Snyder's classification. Playas in these settings form a large class in which groundwater and surface water contribute in differing amounts depending on geologic, topographic and climatic factors. For example, Krinsley (1970) found in Iran that playas with dominant groundwater discharge occurred in steep catchments with small basin-playa ratios, whereas playas poor in groundwater supplies occurred in larger, open basins with less efficient groundwater systems, in marginal basins from which groundwater drained externally, and in the most arid basins with smallest runoff and highest evaporation and hence with least infiltration. Playas in depressions barred by alluvium or by aeolian sand at intermediate points on the slopes of topographic basins, as is commonly the case in shield and platform deserts, generally lie above the watertable. A playa in a lower-lying setting may be reduced to dependence on groundwater by being cut off from surface drainage, as, for example, Lake Amadeus in the Northern Territory of Australia has become isolated by dunes.

The most vigorous discharge of groundwater occurs in steeply bounded basins with closed groundwater systems, in which the aquifers are coarse basin-fill sediments subject to recharge from upland streamfloods. Groundwater here commonly discharges from marginal springs where bajada gravels are in contact with finer playa sediments. In more open basins, and especially in shield and platform deserts where the important source of groundwater is likely to be bedrock, discharge is through the basin floor from a shallow watertable, possibly with artesian flow.

The distinction between playas fed from groundwater and those dependent on surface inflow largely accords with a fundamental division between 'moist' and 'dry' playas, based on the general state of the playa floor, and as summarised in Table VI has important consequences for the sediment and salinity budgets and hence for surface characteristics in general (Stone, 1956; Motts, 1965; Neal, 1969).

TABLE VI **Classification of playas based on hydrologic regime***

Topographically closed basin		Topographically open basin
No groundwater outlet	External groundwater outlet	
Moist surface	Moist in wet season	Dry surface
Saline groundwater	Saline or non-saline, depending on climate, topography and subsurface drainage	Mainly non-saline, depending on climate and degree of topographic closure
Salt-crust or wet mud flats	Silty-clay surface subject to deflation	Claypan
Mainly large, e.g. Great Salt Lake		Mainly small

→ increasing importance of surface inflow

← increasing importance of groundwater discharge

* After Snyder, 1962; Motts, 1965.

SEDIMENT REGIME AND PLAYA STRATIGRAPHY

The sediment regime as reflected in the youngest playa deposits is dependent in turn on the hydrologic regime. 'Dry' playas subject mainly to deposition from surface floods are floored with silt and clay settled from surface waters which become increasingly brackish during the drying phase. In 'moist' playas the ratio of sediment to precipitated salts diminishes in proportion as discharge from saline groundwater dominates in the hydrologic regime, and the floor deposits range from salt-impregnated sediments to pure evaporites. A single large playa may contain sediment-dominated and solute-dominated parts, the former in a higher outer zone, extending opposite the entrances of river channels, and the latter in the lower parts closer to the watertable.

In a dry playa the rate of net aggradation is determined by the flooding ratio, since deflation loss can occur only when the floor is dry. Most dry playas in full deserts now appear to be undergoing net erosion as indicated by the considerable age of sediments close beneath their floors, for instance 20,000–30,000 y at depths of less than 2 m in Willcox Playa, Arizona (Schreiber *et al.*, 1972).

Former sediment regimes are recorded in the stratigraphy of playa basins, and here it is useful to distinguish between those basins previously occupied by deep lakes and those which have always been playas.

The former are characteristic of tectonic and topographic depressions in mountain-and-basin deserts that contained pluvial lakes during the Pleistocene. As shown in Fig. 55a, they are typically underlain by thick beds of clay deposited in this lacustrine phase, between older bajada sediments and younger saline beds, and the clays then tend to form perched watertables above deeper groundwater. A large number of the playas of the southwestern United States are of this type, represented by Rogers Playa, California, in Fig. 55b.

The more open shallow basins typical of shield and platform deserts normally have a long history as playas and are generally underlain by thin successions of sand, silt and clay with gypsum crystals, and interbedded evaporites, reflecting the ephemeral hydrologic regimes. For example, Miocene dolomite occurs at a depth of only 5 m beneath Lake Eyre North (Fig. 55c). In many playas on the shield desert of Western Australia, thinly veneered or bare rock floors are extensively exposed, particularly along the upwind western margins (Jutson, 1917). In these shallow basins, past wetter episodes have caused extensive shallow flooding of playa margins rather than permanent lacustrine regimes and associated thick deposits. Intercalated aeolian sands may mark drier intervals.

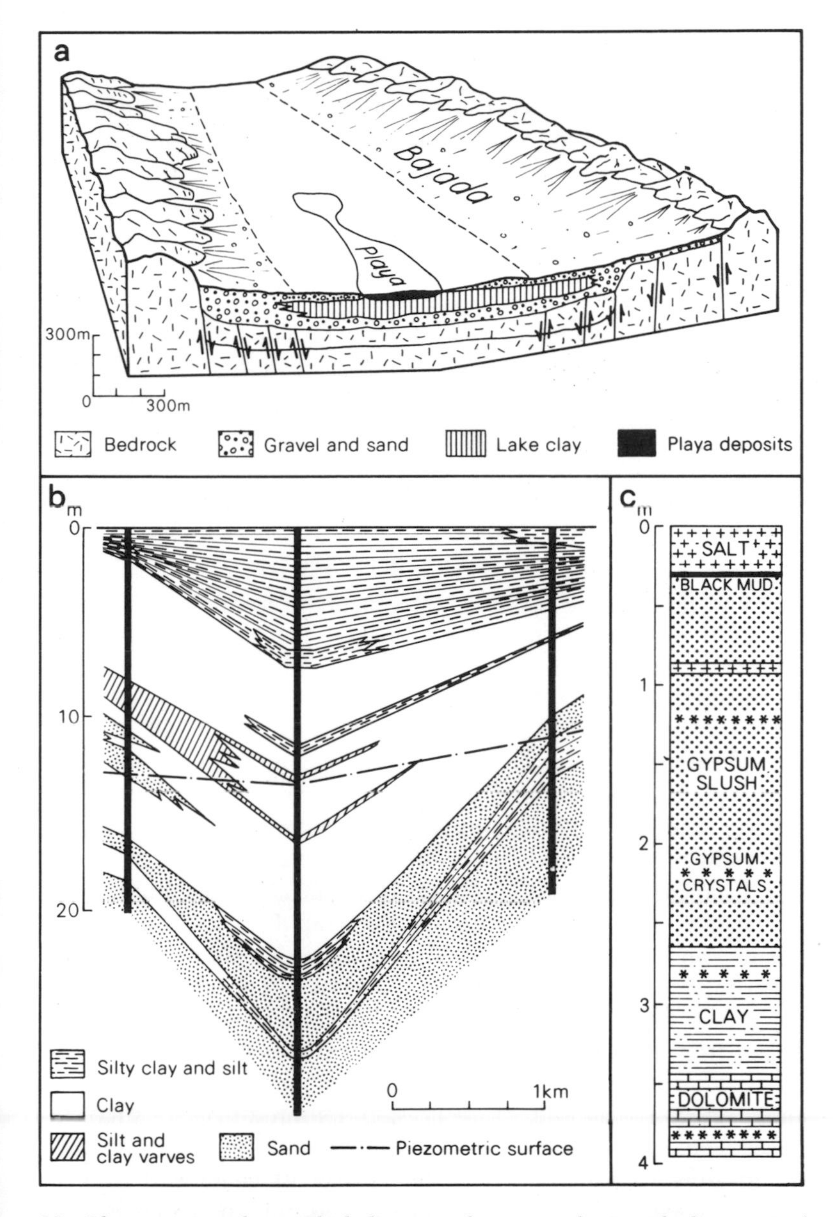

55. Playa stratigraphy. a. Block diagram of a tectonic basin with playa, typical of southwest United States. From Motts, 1970; b. section through Rogers Playa, California. From Motts and Carpenter, 1970; c. section in Madigan Gulf, Lake Eyre North. From Bonython, 1956.

SALINITY REGIMES OF PLAYAS

Excess evaporation in basins of interior surface and groundwater drainage leads to the concentration in playas of soluble salts which in wetter regions are carried to the ocean. These are principally chlorides, sulphates, and carbonates, and to a lesser degree nitrates, of sodium, calcium and magnesium.

Sources of playa salts

Salt relict from a former arm of the sea. This origin has been claimed unsuccessfully for a number of playas in coastal plains, for example those of western Rajasthan including Lake Sambak, but as in this case can often be excluded on the evidence of composition of the salts and known geologic history (cf. Singh *et al.*, 1972). Except in playas of Recent lagunal origin, it has not been shown to be important.

Airborne salts. Salts carried from the oceans through the atmosphere and precipitated in playa catchments in rainfalls — termed *cyclic salts* — are a major source of salts in playas in coastal areas with strong onshore winds. From analyses of rainwater, Jack (1921) demonstrated an annual increment of 169 kg/ha NaCl and 14 kg/ha $CaSO_4$ in the Yorke Peninsula of South Australia, which was alone adequate to account for existing salt and gypsum deposits in local playas; the oceanic source and mode of transport of the salts were indicated by the similarity of composition between the salts in seawater, rainwater, playa brines and in groundwater locally.

At distances greater than 100 km from the coast the amount of oceanic salts in the atmosphere declines rapidly; nevertheless an oceanic source is indicated for salts in large playas considerably further inland. For instance the salts in Lakes Torrens, Gairdner, Frome and Eyre in South Australia, the last-named more than 400 km from the sea, are all similar in composition to seawater, despite differences of local geology (Johns and Ludbrook, 1963; Johns, 1968). The fallout of cyclic salts required may be small if derived from large source areas, and Bonython (1956) has estimated that an annual yield of 1 kg/ha NaCl over its catchment over a period of 3000 y could account for the 400 million tonnes in the existing salt crust of Lake Eyre.

With such low concentrations and large source areas it becomes increasingly likely that salts of oceanic origin have been recycled through soils and saline superficial deposits and taken into atmospheric suspension from bare desert surfaces. That this is generally the case inland is indicated by the relative increase in calcium cations in aerosols (Hutton and Leslie, 1958; Hutton, 1968).

Salt from rocks and sediments in the catchment. This source includes *connate salts* of oceanic or other origin trapped in marine and lacustrine rocks and sediments at the time of deposition. Salts released to a playa in this way are often termed 'second-cycle', and because of their indirect origin they can be expected to have a more localised distribution and a more variable composition. For example Krinsley (1970) found them to dominate in those Iranian playas having Miocene evaporites in their catchments. The waters of the Jordan show a significant increase in chloride content from connate salts where the river traverses the saline marls of the former Lisan Lake north of the Dead Sea. A second-cycle connate origin has been adduced for at least part of the salts of Lake Frome because of unusually low bromide/chloride ratios compared with seawater.

Rock weathering is the main primary source of salts in inland catchments, whence they may reach the playa in streamfloods or through groundwater. As evidence of the adequacy of the source, Bonython (1956) estimated that an annual weathering of 0.25 mm of marine rocks having 1 per cent NaCl over the 250,000 km^2 of Lake Eyre could produce the salts in the lake over a period of 6000 y, or that artesian springs with a daily outflow of 45,000 m^3 and a chloride content of 0.1 per cent would yield the same accumulation in 25,000 y. Ionic ratios in total salts from this source will differ with catchment geology, but in comparison with seawater they will be rich in calcium and deficient in chlorides, as determined by the average concentrations in rocks. However, their journey to the playa floor involves considerable selective concentration.

Salts in playa crusts

In a dry playa where the only source of salts is from the evaporation of floodwaters, the content of soluble sulphates and chlorides in the surface sediments is generally as low as 2–3 per cent, and deflation from the dry floor combined with leaching to groundwater prevents any build-up of these solutes. On the other hand, $CaCO_3$ content commonly exceeds 20 per cent, and even higher concentrations occur in *lime pans* where basin rocks are predominantly limestone, as in parts of the Kalahari. $CaCO_3$, being subject only to partial leaching, tends to concentrate as a subsurface pan beneath dry playas, and so is protected from deflation.

In moist playas salts are supplied from groundwater and may range from 15 to 35 per cent in floor sediments, with higher concentrations in surface crusts, whereas lime content is generally below 5 per cent. The trend of concentration in playa waters with increasing solubility is magnesium → calcium → sodium and carbonate → sulphate → chloride, and accordingly most playa salt crusts consist predominantly of NaCl, although larger amounts of $CaSO_4$ may be contained in the playa sediments. For example, the salt crust of Lake Eyre North consists of between 90 and 99 per cent NaCl, estimated at 400 m tonnes, whereas the gypsum content of the lake sediments may be 4000 m tonnes (Bonython, 1956). Factors such as catchment geology, temperature and inter-

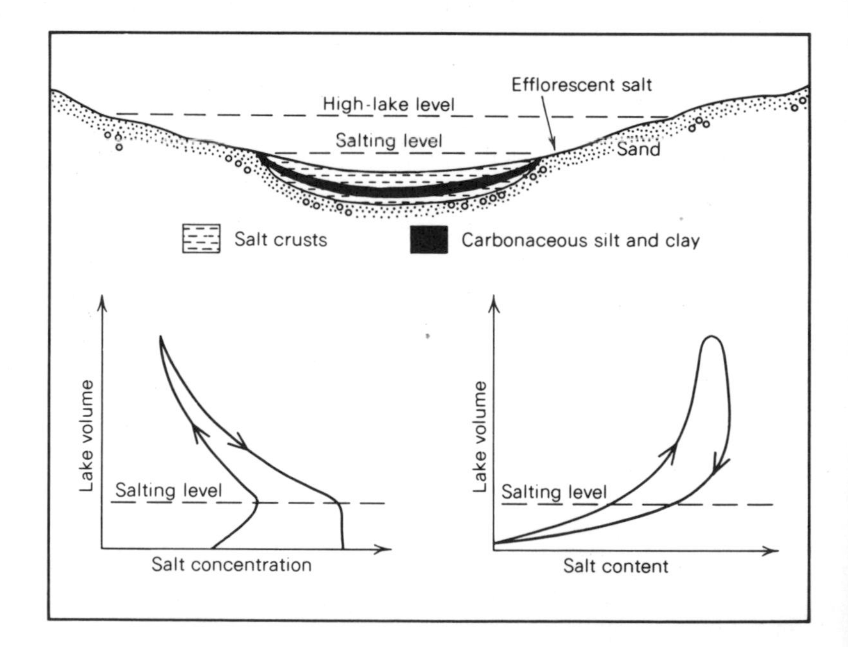

56. Salinity regime of a playa. Relationships between water level, salt concentration, salts in solution, and salting level in a flooding and drying cycle. From Langbein, 1961.

actions of the salts themselves may result in the dominance of other salts, and saline moist playas include *bitter lakes* (Na_2SO_4 dominant) and *alkali lakes* (Na_2CO_3 dominant).

Salinity regimes in filling-and-drying cycles

With evaporation of waters from a flooded playa, NaCl concentration increases until crystals of halite begin to be precipitated at its saturation point (32 %). This is the *salting level* of the playa and it marks the upper limit of an actively forming salt crust as in Fig. 56. With continuing evaporation, precipitation of salt continues and the eventual thickness of the crust should be approximately a quarter of the depth of water in the playa at the salting level. On renewed flooding, the crust begins to dissolve again and salt concentration increases to the salting level, beyond which there is dilution to high-lake stage. However, the mass of salt in solution continues to increase into the drying phase, when concentration again increases to saturation point at the salting level.

Given the estimated inputs of salts from surface inflow or groundwater, the amounts of salts in playa crusts commonly represent remarkably short periods of accumulation as shown in Table VII, and it is clear that losses must occur; furthermore, since the losses have not affected the composition of the salts they must take place at dry stages. The main loss is by deflation from surfaces with efflorescent salts. Another is in the burial of evaporite beds beneath sediments in aggrading playa floors, and in some playas there are losses to groundwater. In view of the small storages, the salts contained in a playa can be represented as expressing a dynamic balance between these losses and gains from inflow (Langbein, 1961).

TABLE VII **Periods of accumulation represented by existing salt crusts**

	Years
Walker Lake, Nevada	90
Devils Lake, N. Dakota	800
Owens Lake, California	1700
Lake Eyre, Australia	5500
Dead Sea, Israel	30,000

Zonation and layering of playa salts

The relative solubilities of salts determine the trends of concentration and precipitation. The least soluble, such as $CaCO_3$ and $MgCO_3$, are precipitated in fringing seepage zones or where river waters meet brackish playa waters, and are not readily redissolved. Their absence from playa waters distinguishes these from the waters of desert streamfloods. Gypsum ($CaSO_4$), being the most readily precipitated of the main constituents of playa waters, precipitates and settles with a fine sediment and occurs mainly as selenite crystals or 'seed gypsum' in saline mud horizons. NaCl, being more mobile, alternates between liquid and solid in the filling and drying phases of playa flooding cycles, whereas $MgCl_2$ and KCl generally remain in solution. Increasing salinity in playa waters is therefore generally associated with enrichment in sodium and with increasing chloride content.

Relative solubilities also determine a zonation of salts on the playa floor as exemplified in Cottonball Basin, Death Valley (Fig. 57). Carbonates predominate in a sandier higher fringe and in delta tracts, and there is transition through a limited sulphate zone to a lowermost inner chloride zone with shallow watertable, which may occupy more than half the playa. Parts of a playa subject to regular wetting by flooding or seepage remain as bare mud flats. Each salt zone also has a vertical layering reflecting the movement of salts in solution. Where deposition is from rising groundwater solutions, as in the lowest part of the playa, chlorides occur at the surface and sulphates below. This layering may be reversed in higher-lying carbonate tracts that are subject to leaching by descending solutions, but the situation is commonly confused, as in Death Valley, by past fluctuations in watertables and flooding levels.

PLAYA SURFACE TYPES

The foregoing account has shown that the hydrologic, sediment, and salinity regimes of playas are interrelated. The key to this lies in the influence of the watertable on the playa surface.

> Virtually every process and feature of playa geology can be directly or indirectly related to the ground-water regimen. The principal surface forms and playa macrofeatures are all controlled to some degree by ground-water depth and composition. Additionally processes of sedimentation and deflation appear to be

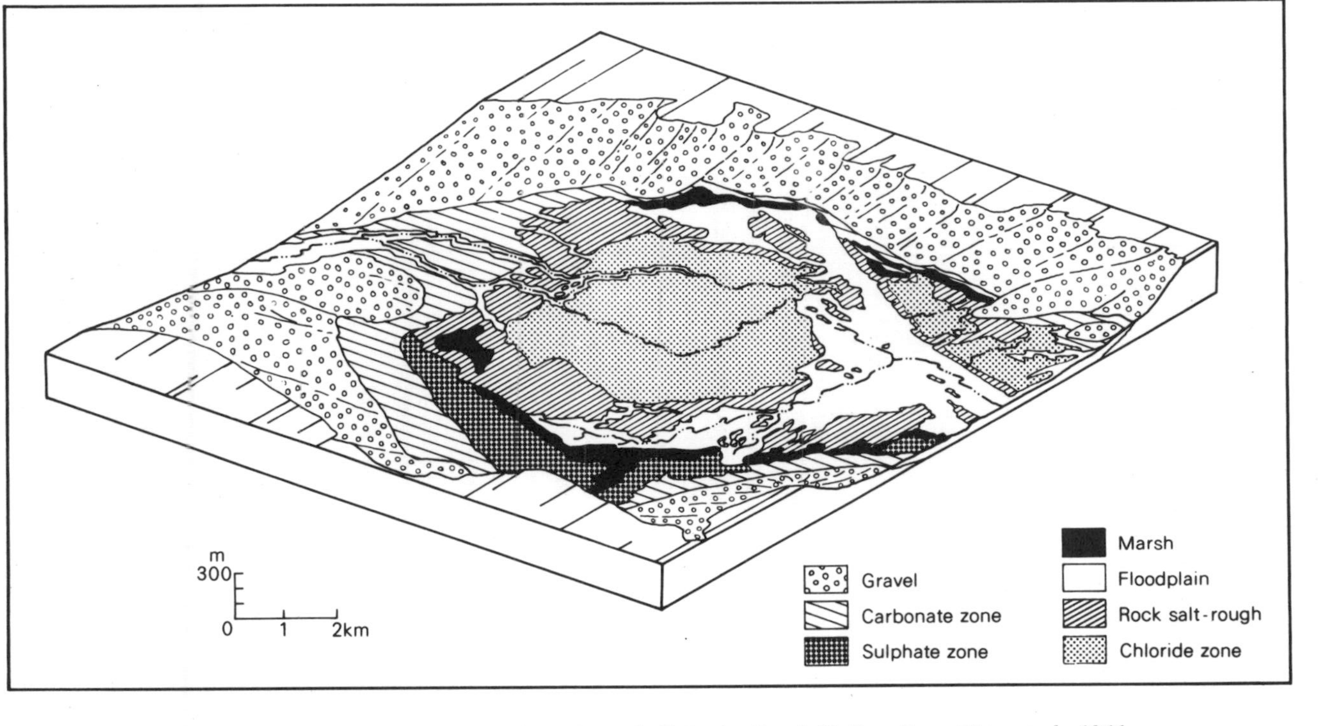

57. *Salinity zones in a playa surface, Cottonball Basin, Death Valley. From Hunt et al., 1966.*

locally controlled by the nature of the watertable. Thus changes in water level will affect these features and processes (Neal, 1965: 28).

This control is expressed in the various types of playa surface. A large playa floor may comprise several of these surface types, in proportions reflecting hydrologic and dependent controls.

Hard dry floors

These are remarkably flat, characteristically of shiny sun-baked light-brown clay with a close polygonal mosaic of fine cracks. They are found on playas not subject to groundwater discharge, either because the floor is above the capillary fringe of the watertable or because the underlying fine-textured sediments are impervious, for clay content of the crust is commonly above 75 per cent. These conditions can exist in topographic basins with an earlier lacustrine history or in lowland playas fed by streamfloods bringing fine alluvium. A hard floor may also occur in a playa receiving groundwater discharge if it is also subject to surface flooding sufficiently frequently. Drift of shallow floodwaters by wind contributes to the extreme smoothness and flatness of the floors, as does creep of saturated mud. Salinity of the floor is generally low (below 5 per cent) but lime may be plentiful at the surface or in a subsurface pan. Vegetation is commonly excluded by the hardness of the surface.

Where the salt content is rather higher, perhaps because the capillary fringe is closer to the surface, the floor may flake after drying and these mud flakes or 'curls' are subject to deflation; otherwise dry floors resist wind erosion. Flaky floors are temporarily smoothed with renewed flooding.

Soft puffy surfaces

This form of 'self-rising ground' occurs where modest groundwater discharge to the playa floor leads to an efflorescence of salts, forming a friable blistery surface of loose granular texture which is vulnerable to deflation. Quite commonly a playa crust of suitable texture lies within 1–3 m of the groundwater table and hence within the capillary fringe, but surfaces of this kind can also result with artesian flow. A soft puffy surface may occupy the lowest part of a playa within a hard dry floor, or form a higher zone in a playa subject to strong groundwater discharge to its lowest part. Again, it may be localised by sediment-type, for instance by the occurrence of silts in a sandy playa that is generally too coarse-textured for capillary movement of groundwater. Colours vary from light brown to white depending on salinity, and the amount of salts and the unevenness of the surface tend to increase with time since flooding. Shallow flooding smoothes puffy surfaces by dissolving the salts, but the puffiness will normally reappear on drying. Part of a groundwater playa subject to very frequent flooding, as near a channel entrance, may never develop a puffy surface.

Salt crusts

These particularly characterise chloride zones of maximum groundwater discharge, where their development is determined by the salinity of the groundwater and by the discharge as controlled by piezometric levels and the transmissivity of underlying sediments. A limiting factor is the flooding ratio, which influences the relative amounts of sediments and solutes in the playa deposits and the liability of a developing salt crust to renewed solution.

Where subject to flooding, salt crusts may remain smooth and silty and relatively thin, and in many playas this is the most extensive type of surface. For example it occupies the lower central part of Cottonball Basin in extension of tributary flats (Fig. 57). The crust here consists of a recently formed veneer of salt with fine polygonal cracks, above a thin crust of salt polygons of about 50 cm diameter. Polygonal cracks form initially because of volume changes in saline muds on alternate wetting and drying. As a salt crust develops it is also subject to cracking owing to thermal expansion and contraction, and in consequence of forces resulting from further crystallisation of halite on the margins of plates.

At higher levels, or wherever flooding is very infrequent, an extensive thick crust may develop and attain an appreciable microrelief through differential upward growth of salt crystals combined with various forms of erosion. A salt crust of this type in the Great Kavir in Iran has an extent of 20,000 km^2 (Krinsley, 1970).

Stages in the development of an irregular salt crust are shown in Fig. 58. An initial form is a network of puffy salt blisters defined by cracks in a drying saline mud beneath. With further crystallisation of salts from the brines below, the intervening crusts thicken into polygonal plates a centimetre or so thick (Pl. 41). The plates grow laterally and thicken and form ramparts at the margins due to enhanced evaporation from brines rising along the mud cracks, particularly on the upwind side where they may overlap the plates to leeward, creating an illusion of overthrusting. Periodically the ramparts are planed by solution following rainfall or flooding, and the plates then thicken further by crystallisation of salts from the brine ponded in the saucers. Smaller plates unite in this way, and the annealed crust may then crack once more into systems of larger polygons. Salt polygons commonly attain 1–2 m diameter, and the giant saucers of Badwater Basin, Death Valley are up to 10 m across, with ramparts up to 1 m wide and 60 cm high.

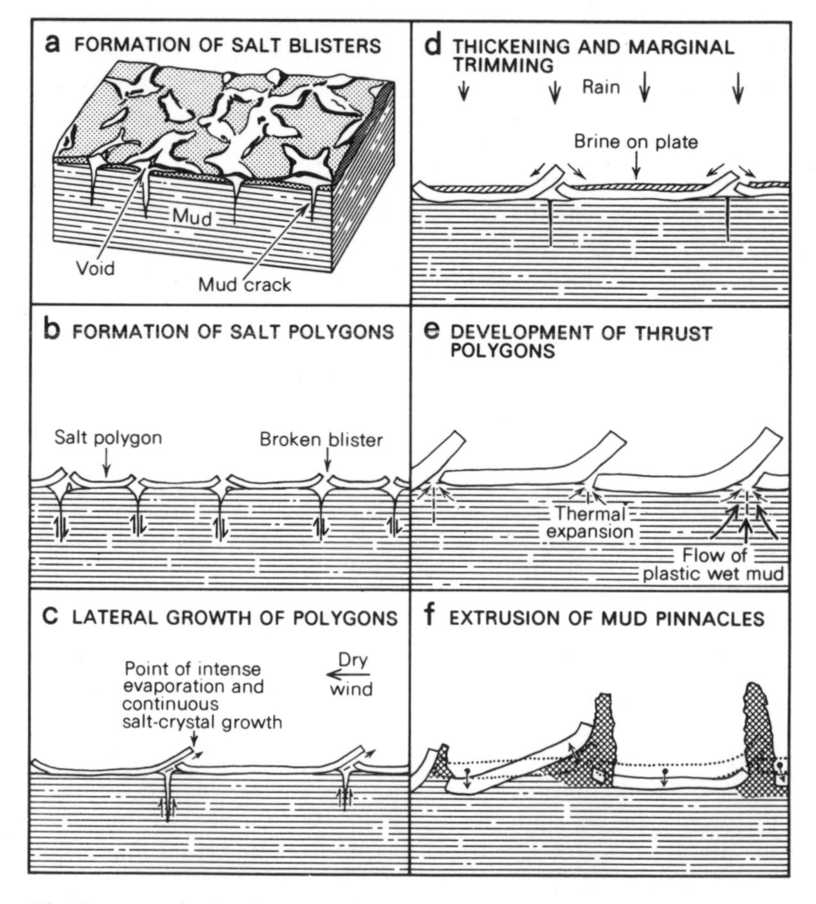

58. Stages in the development of an irregular salt crust. From Krinsley, 1970.

41. Salt crust with polygonal plates up to 150 cm across; Neriz Playa, Iran. Photo D. B. Krinsley.

Salt pinnacles may form along the ramparts of salt polygons by selective crystallisation and solution, and those of the Devils Golf Course in Death Valley are up to 70 cm high. Another possible cause is the extrusion of underlying mud as in Fig. 58. This may result from thermal expansion of dark mud beneath a relatively transparent salt crust, or from the loading effect of salt plates as much as 30 cm thick. On the playas of Iran, the dark pinnacles and honeycomb ridges formed in this way contrast sharply with the white salt surfaces.

Salt pinnacles and ramparts are rendered razor-sharp by solution and recrystallisation of halite. Wind erodes the pillars through abrasion by drifting salt crystals and solution by driven rain, and groovings in the crust are commonly aligned with the dominant wind. Increasing exposure to wind probably sets an upward limit to the microrelief, which rarely exceeds 50 cm. Drain-pits may develop in the floors of the salt saucers by solution of the crust, supplemented by collapse following the leaching of saline mud beneath.

Wet zones

Groundwater playas steeply bounded by fans or otherwise

subject to artesian discharge commonly have a marginal spring or seepage zone which remains moist and soft for much of the time. In their explorations of Lake Eyre for instance, Madigan and Bonython had to devise special methods to cross this zone to the firmer salt crust of the lake floor. The zone tends to be occupied by halophytic phreatophytes, which are excluded from the centre of the playa by excessive salinity.

On the lower part of the playa also, ponding by fine-textured sediment can bring groundwater to the surface, but here salinity is greater and the floors are generally salt-crusted. Playa marshes of this type can occur in the sulphate zone, as in the west of Cottonball Basin (Fig. 57), or in the chloride zone as in other parts of Death Valley. At higher levels the surface may be puffy, and much of Cottonball Marsh consists of cauliflower-like lumps of gypsum crusted with rock salt; in the wetter parts, which may contain shallow pools, the salt crusts are generally smooth, and that of the Badwater Basin marshes further south in Death Valley consists of large salt polygons margined by pressure ridges. The marsh floors tend to be moist and soft at all times.

The wet zones grade into tributary floodplains with surfaces of brine-soaked silt and clay in the most frequently flooded tracts and ephemeral thin salt crusts in the higher parts.

SECONDARY FEATURES OF PLAYAS

Mounds and depressions

A number of secondary features of playas can be grouped in this category. They have been discussed by Stone (1956), Reeves (1968) and Neal (1969, 1972).

Spring mounds. These develop under artesian conditions with the cementation of sediment issuing from the spring outlet by lime and gypsum from groundwater. Tufa in these mounds may be partly of organic origin, as in the 'algal pinnacles' of Mono Lake, California. The mounds along the western fault margin of Lake Eyre are up to 10 m high, and similar forms are reported from many playas in the southwestern United States, including South Panamint and Big Smoky Playas (Neal, 1965). The selenite-capped buttes of Dalby Dry Lake, California, previously held as evidence of deflative erosion of a former higher playa floor (Blackwelder, 1931), are apparently similar, having originated as mushroom-shaped mounds at points of capillary rise of groundwater rich in $CaSO_4$ (Stone, 1956).

Phreatophyte mounds. These form about growing plants by the trapping of wind-borne sediment and salts which can be anchored by a saline crust where moisture discharges around the plant roots. The mound grows to a height limited by the ability of the plant to tap the watertable, and mounds formed by greasewood (*Sarcobatus vermiculatus*) on playas of the southwestern United States have been recorded as high as 3 m. Eventually, growth of the mound carries the plant beyond reach of groundwater, when it dies, and the mound is then eroded away. They are commonest near playa margins where groundwater is fairly shallow but not too saline.

Phreatophytes may play an important part in the water balance of a groundwater playa. Localised lowering of the watertable by a phreatophyte can result in subsidence within encircling fissures in the playa floor.

Sink holes may form where a buried body of rock salt has been subject to solution. Stone (1956) records three natural solution depressions up to 20 m diameter on Dale Dry Lake, California.

Gas rings and gas pits are found mainly on dry playas and are formed by the escape of air trapped during rapid playa flooding, or of methane from organic muds, possibly under the influence of a seasonally rising watertable. Stone (1956) describes pits up to 15 m across on some Californian playas.

Playa grooves form where an inflow channel periodically floods across a playa floor to a low-lying basin. Largest of these is the Warburton Groove, 80 km long, leading southwards across Lake Eyre North from the mouth of the Warburton and possibly reflecting a slight downtilting of the basin to the south.

Patterned ground

Salt-affected playa floors and margins constitute an environment of patterned ground rivalling that of the stony deserts discussed in Chapter VI, and characteristic occurrences in Death Valley have been reviewed by Hunt and Washburn (1966). In addition to the various polygonal patterns on playa floors discussed above, polygonal cracking is characteristic of the saline alluvial margins of playas. It similarly results from volume changes in fine-textured sediments and contained salts on wetting and drying, with associated salt crystallisation, and from thermal expansion and contraction reinforced by changes in the degree of hydration of the salts. Generally, only sorting of salts from clastic sediments is involved in the formation of the polygons, which tend to be fine-grained throughout, but sorted stone polygons can occur on the lowermost margins of gravel fans.

FORMS RESULTING FROM WIND ACTION ON PLAYAS

An absence of vegetation due to salinity exposes the flat floors of desert playas to wind erosion, the course of which tends to be influenced, directly or indirectly, by the hydrologic and salinity regimes. Drying and cracking of a slightly saline playa floor after an episode of flooding render newly-deposited silts susceptible to deflation. Impregnation with salt causes cracking and loosening of surfaces and pelletisation into transportable aggregates of otherwise wind-stable fine-grained materials. The watertable limits wind erosion by acting as a local baselevel. Accordingly, forms expressive of wind action on playas can conveniently be grouped by their respective origins in dry playas, in periodically inundated playas of moderate salinity, and in basins in which a falling watertable has allowed deep excavation.

Wind action on seasonally dry saline playas

Periodic flooding and drying of playa floors of saline silty clay produces aggregates or pellets of clay minerals similar in size to dune sands (0.1–0.3 mm), formed by the efflorescence of salts, particularly of gypsum, or by the mechanical breakdown of saline mud curls and salt blisters (Price, 1963; Bowler, 1973). These pellets are saltated by wind to the downwind margin of the playa, where they are trapped by vegetation or on a pre-existing crust and there accumulate as a non-migratory source-bordering dune. During the next rainy period the layer of clay pellets is stabilised through the coagulatory uptake of moisture by contained hygroscopic salts, and protected by surface crusting. In this way the windward slope of the dune grows upwards through accretion, as reflected in an internal structure of planar sets with upwind dips of about 10°; the pellet layers are too cohesive to drift and avalanche in slip-faces, and cross-bedding is rare. Most of these dunes have moderate windward slopes of up to 15°, rounded hummocky crests, and gentle downwind slopes commonly below 3°. A common height range on moderate-sized playas is 5–15 m. The interaction of waves and currents on the lee shore results in a smooth crescentic dune outline with transverse variation of height controlled by length of wind-fetch across the playa floor (Fig. 59), to which the term *lunette* has been applied (Hills, 1940; Pl. 42).

A combination of conditions is needed for clay dunes to form (Bowler, 1973). Salts are essential for the inhibition of vegetation

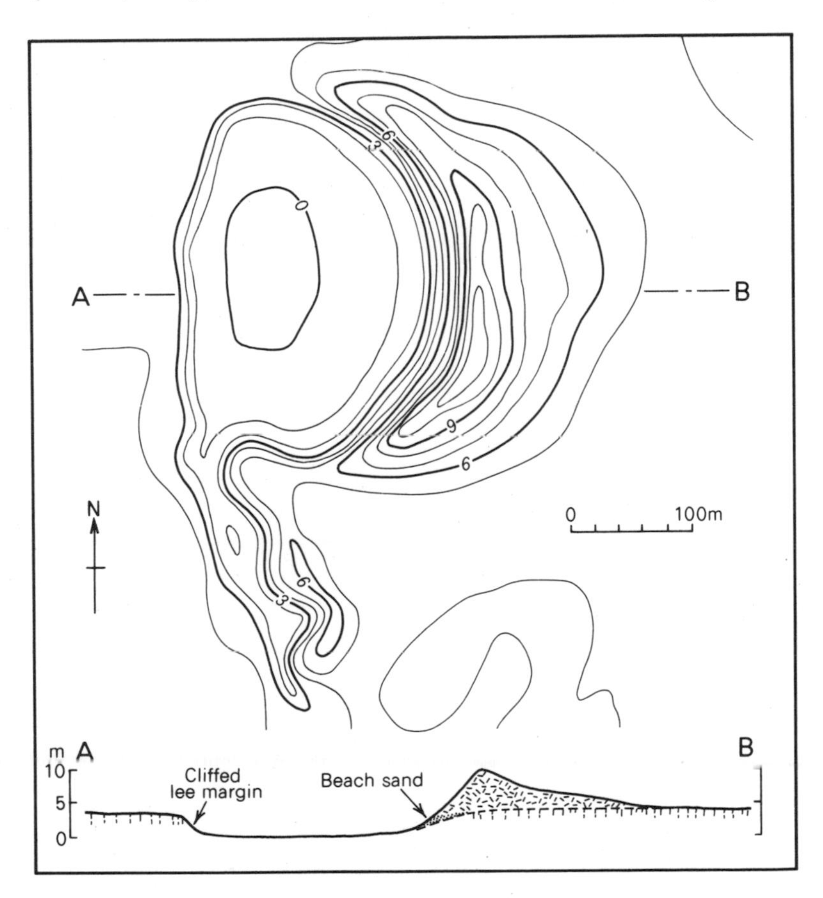

59. Contour plan and section of a playa and clay lunette, Lake Wandella, Victoria, Australia. From Bowler, 1968.

on the playa floor, for the aggregation of fine sediments that might otherwise be exported through deflation, and for dune stabilisation. There must be a continuing supply of silty clay, not too fine-textured for pelletisation; regular flooding is needed for the spreading of sediment and its impregnation by salt; a subsequent thorough drying out of the playa floor must take place, and strong uni-directional winds are needed for dune-building during this dry phase.

These requirements were met in late Pleistocene and Recent times on temperate margins of zonal deserts subject to wet season rains and hot dry summers with strong westerly winds, as shown by the occurrence of clay lunettes of that age in Australia south of 29°S (Fig. 60). The lunettes occur on ENE margins of playas at river terminals and in valley tracts of disorganised drainage,

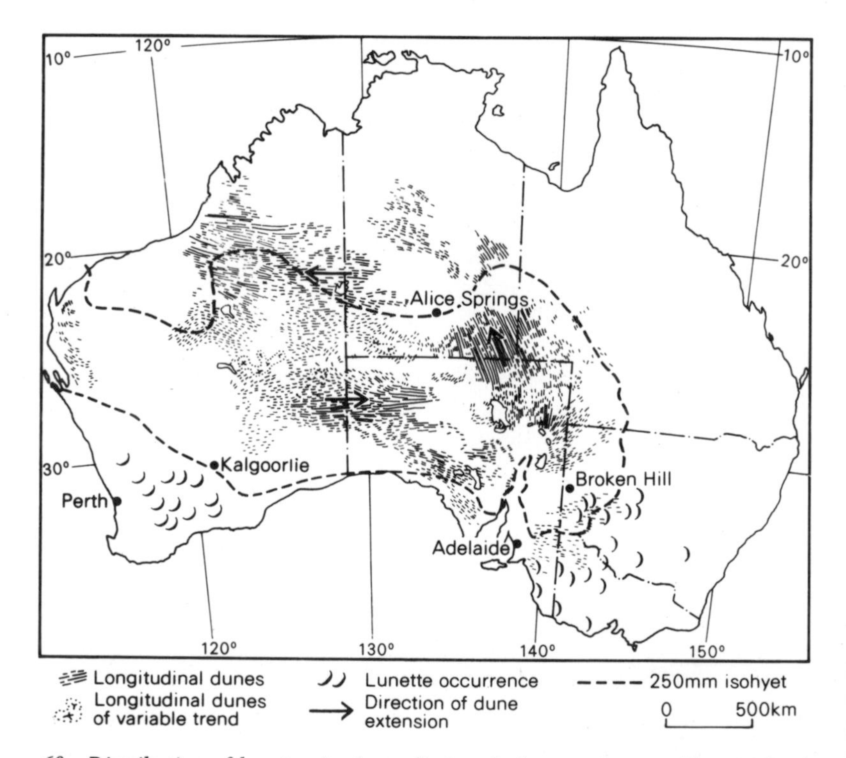

60. Distribution of lunettes in Australia in relation to patterns of longitudinal dunes and present-day rainfall. From Bowler, 1971 and Jennings, 1968.

locally with saline high watertables. There is a corresponding zonal occurrence of lunettes on the northwestern border of the Sahara (Tricart, 1954a).

The lunettes in both these tracts are now almost entirely fossil, as shown by soil profiles and surface crusts. *Active* clay dunes are now largely restricted to tidal lagoons in coastal deserts and semi-deserts of lower latitude, as on the west shores of the Gulf of Mexico (Price and Körnicker, 1961), on the Atlantic coast of Mauritania, and in the delta of the Senegal River (Tricart, 1954b). Boulaine (1954) has also described an active lunette in a basin with artificially enhanced saline groundwater discharge on the Chantrit Plain near Oran.

The sizes of lunettes in relation to their source basins suggest that between 25 and 70 per cent of the wind-excavated material may remain in a lunette. Boulaine (1954) estimated that with an

42. A lunette or crescentic clay dune on the downwind margin of Lake Bulbugaroo, a small playa in western New South Wales. The playa is inactive and the lunette is being eroded. Photo J. M. Bowler.

assumed annual lowering of the playa floor of 1 mm, the lunette of the Sebkha of Ben Ziane near Oran would have taken 15,000 y to form.

For development of these dune forms there should originally have been not less than 20 per cent clay. Chlorides blown with the clays tend to be leached rapidly to the lowest layers and to be recycled to the playa at an early stage, gypsum is concentrated in pipes and as concretions at depth, and lime accumulates more slowly in a near-surface pan. Weathering largely destroys the pellet structure leaving a massive layered buff silty clay. Weathered lunettes are liable to wind stripping and to close gullying, often to the degree of badlands.

Other forms due to the trapping of clay pellets by vegetation or to their immobilisation near playa margins include windrows and drifts in lee of obstacles, mounds around shrubs, such as the smaller *rebdous* and larger *nebkas* of the Sahara, rippled sheets and giant ripples. These tend to occur in arid settings where there is insufficient pelletisation to allow the development of clay dunes (Price, 1963).

Quartz sand lunettes. Many clay lunettes in southern Australia overlie or merge laterally with lunettes of quartz sand, which commonly rest in turn on layers of shell and gravel indicative of high strand-lines. Other sand lunettes occur as older forms outside a clay lunette. They have the form of normal foredunes, with sharp crests, symmetrical cross-sections, and bedding indicative of avalanching on slip faces to leeward. They formed at times of high lake level, for where adjacent bottom sediments consist of clay and silt the floors could not have been exposed at the time of their formation. In South Australia it has been noted that sand lunettes occur on SE playa margins, perhaps in response to wet season winds from north of west, whereas clay lunettes occur on E and NE sides (Campbell, 1968), and there is a comparable overlap in the lunettes of western New South Wales (Bowler, pers. comm.). However, the forms of sand lunettes are less simply related to the extent of playa floor crossed by the formative winds, since beach drift and wave currents influence the relative abundance of beach sand and hence the point of maximum lunette growth. Sand lunettes are less stable than clay lunettes and remain subject to wind erosion, with development of blowouts and parabolic dunes.

Gypsum dunes. Where there is a low input of sediment to the playa, and if the watertable is lowered seasonally sufficient for chlorides to be removed and to allow fractional crystallisation of gypsum, or in by-passed playas that do not form groundwater terminals, crystals of 'seed' gypsum may be blown from the dry crust and form lee dunes. Since seed gypsum drifts more freely than clay pellets, slip faces may form. Hence a gypsum dune resembles a foredune of quartz sand, with a relatively sharp crest and a steep leeward flank. Gypsum dunes readily crust over when stabilised, and weathering destroys the crystalline structure in the upper part, converting it into powdery *kopi.* Gypsum dunes have been reported from some playas in Western Australia, commonly within older lunettes of clay or quartz sand (Bettenay, 1962), but few appear still to be active.

Forms due to wind action on dry playas

Whereas wind erosion on moist playas remains superficial, as dictated by the level of the watertable, in dry playas it can have a greater vertical range and be more selective in its operation. The appreciable age of near-surface sediments on many dry playas points to the efficacy of general deflation.

The most spectacular of the forms resulting from *selective* wind erosion on dry playas are *yardangs* (Blackwelder, 1934), named by Hedin from occurrences in Turkestan. These are parallel ridges of lacustrine sediments between wind-scoured furrows aligned with the dominant wind. They begin to form as broad ridges and narrow wind-scoured troughs, perhaps developed along former gullies, but as erosion progresses these are transformed to wind-moulded narrow ridges on open flats (Pl. 43).

An example of an early stage of yardang formation, albeit on a grand scale, is the landscape of parallel steep-sided ridges and corridors known as *kalut*, cut to a relief of 60 m in soft clayey sandstone in the floor of the Kerman Basin of the Southern Lut, Iran (Dresch, 1968; Fig. 61). The ridges are sheer-sided where composed of sandstone and most exposed to wind erosion; in clays they are steeply convex as in badlands and subject to rilling and mudflows. In the development of the flat-floored corridors and the basal steepening of the ridges, corrasion by wind has combined, now or in the past, with streamfloods, cracking and deflation of clay crusts, and solution in calcareous beds.

43. Yardang cut in old lake sediments, Tibesti, Libya. Photo R. F. Peel.

At such an early stage the ridge crests are often accordant, preserving a former lake floor, but they become narrowed and lower as the corridors open and are then commonly jagged in outline due to gullies on the slopes. At this stage only the basal slopes are smoothed and undercut by wind, but eventually the entire ridge is brought within the vertical range of wind abrasion and takes on an aerodynamic form. The upwind end of such a yardang is prow-shaped, undercut to a height of 2 m at the base and paraboloid above, and the crest is keel-formed, tapering downwind. Such forms do not exceed 20 m high. In the Kerman Basin there is a downwind transition from fields of yardangs to kalut ridges and from these in turn to linear dunes, all of similar trend as shown in Fig. 61.

Wind action in playa basins with a falling watertable

In moist playas the efficacy of wind erosion is enhanced in various ways by saline water at or below the level of the playa floor in combination with episodes of drying. With a falling watertable these effects are given vertical range and deep enclosed basins may result. The combination of processes involved has been termed *hydro-aeolian*, but in large-scale forms developed over considerable

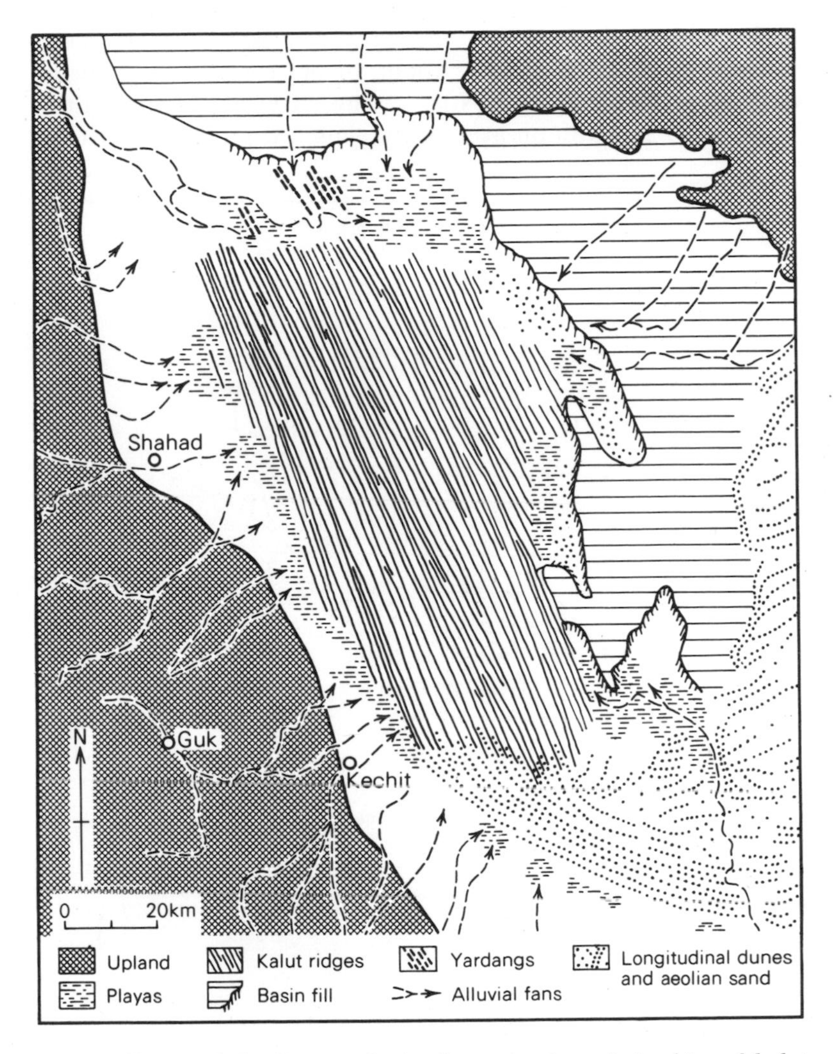

61. Landforms of the Kerman Basin, Iran, showing relationships of kalut ridges and yardangs to the escarpment edge of an undissected basin fill and to a system of longitudinal dunes. From Dresch, 1968 and Krinsley, 1970.

time there generally remains some doubt as to the contribution of other processes, such as karstification, mass movement and fluvial erosion, particularly in early stages of less arid climate.

The oasis depressions of the Libyan Desert provide good examples of this. They cover about 10 per cent of the hamada surfaces west of the lower Nile, having an extent of more than 70,000 km^2, and with an average maximum depth of 250 m they

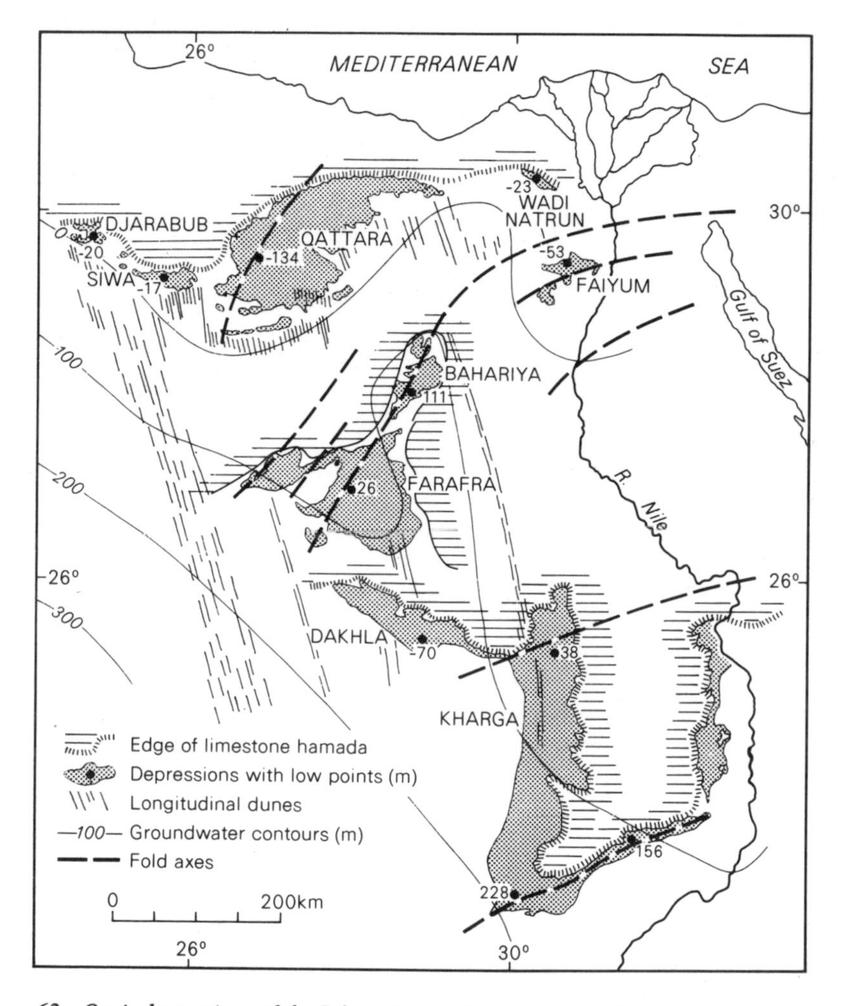

62. *Oasis depressions of the Libyan Desert in relation to fold axes, escarpment margins of limestone hamada, and longitudinal dunes. Floor levels as shown decrease northwards in accordance with the groundwater table. Depressions are defined by the highest closed contour. After Pfannenstiel, 1953 and Ball, 1927, 1933.*

represent a formidable excavation of rock material (Ball, 1927, 1933; Pfannenstiel, 1953). As shown in Fig. 62 they occur in a series of arcuate belts determined by broad anticlines trending WSW to ENE and convex to the north, which have affected the area in more than one phase since the Cretaceous, with minor faulting. The belts of oases occur at or just south of formational boundaries with a marked change of facies; for example the depressions of Dakhla and Kharga occur south of an escarpment of resistant Nubian Sandstone, and the Qattara depression in little-consolidated sands and marls beyond the southern limit of a hard Miocene limestone.

The levels of the depression floors fall from south to north as determined by static groundwater levels, and descend below sea-level in the northernmost belt, where the Qattara depression has been excavated to −134 m.

As exemplified by the Qattara depression (Fig. 63), most have steep escarpments on the north and west, which is the regional direction of gentle dip, and rise more gradually to the general level of the desert tableland to the south. The lowest point of the depression is generally close below the escarpment and typically contains playa deposits at one or more levels determined by past positions of the watertable.

The southern depressions are the oldest and occupy an embayment of an escarpment which may have originated by fluvial erosion as long ago as the Eocene, but although such erosion may have led to the initiation of the depression by the removal of hard caprock and exposure of softer strata in belts of upwarping and tectonic disturbance, any formative river systems must have long since been obliterated under aridity. They can have played no part in the deepening of southern depressions, nor even in the initiation of the northernmost.

Wind excavation to the level of the watertable in soft rocks beneath the hamada caprock has been the dominant process in these later stages. At the same time, the depression floors have been extended at the base of the northern escarpments by spring sapping, solution weathering, hydration and salt weathering, aided by groundwater weathering and flocculation of loosened sediments by contained salts. Mass movement and stream action during pluvial episodes may have aided escarpment retreat. The asymmetry of the depressions and the boldness of the northern escarpments

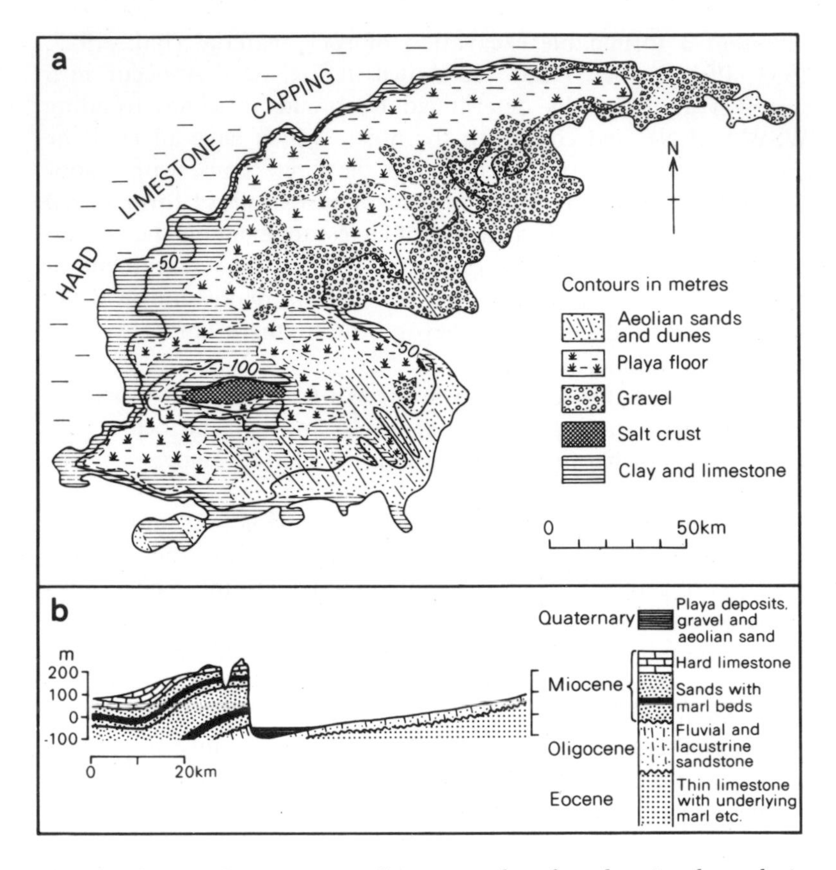

63. The Qattara Depression. a. Contours and surface deposits; b. geologic section. After Ball, 1933.

reflect their northward development. Fluctuations of the water-table in relation to sealevel changes as well as to climatic fluctuations, as recorded in wind-eroded lacustrine deposits above the floors of many of the depressions, could have led to an alternation of water-level weathering and wind erosion. The efficacy of wind erosion is indicated by trains of longitudinal sand dunes which extend from the southern margins of the depressions for hundreds of kilometres SSE across the hamada surfaces.

IX

SAND DESERTS

Sand desert, the popular image of an arid landscape, forms only a minor part of many deserts, for instance less than 2 per cent in North America. Extension of aeolian sands is favoured by sparse vegetation, coarse-textured superficial deposits and infrequent runoff, but area of sand is no measure of aridity; it covers only about 15 per cent of the harsher Sahara but 50 per cent of the moderate Australian deserts. This reflects an importance of factors other than climate, particularly the sand source.

DESERT SANDS

Composition and sources

Desert sands consist largely of quartz grains, with light minerals such as feldspar that are stable under a dry climate. The fine fraction commonly includes a small proportion of heavy minerals including hornblende and ilmenite, iron oxide usually forms a partial coating, and there is a variable admixture of clay minerals.

A main source is *alluvium*, shown by the common occurrence of sand desert in lowlands beyond the river terminals. This is typical of basins of interior drainage in platform deserts, as in the dunefields or *ergs* of the Sahara (Fig. 64), and of the larger tectonic depressions in the mountain-and-basin deserts, for example the Takla Makan of central Asia. An alluvial imprint may persist in the form and composition of the sand, such as links sands in the Simpson Desert with igneous and metamorphic source rocks to the northwest (Carroll, 1944).

Playa sands are typical of basins formerly occupied by lakes, such as that ancestral to Lake Chad. These sands commonly contain evaporites, notably gypsum, and dunefields entirely of gypsum sand occur at White Sands National Monument, New Mexico (McKee, 1966).

Some aeolian sand originates in place from *weathering of bedrock*, as in residual sand sheets on Nubian Sandstone in the eastern

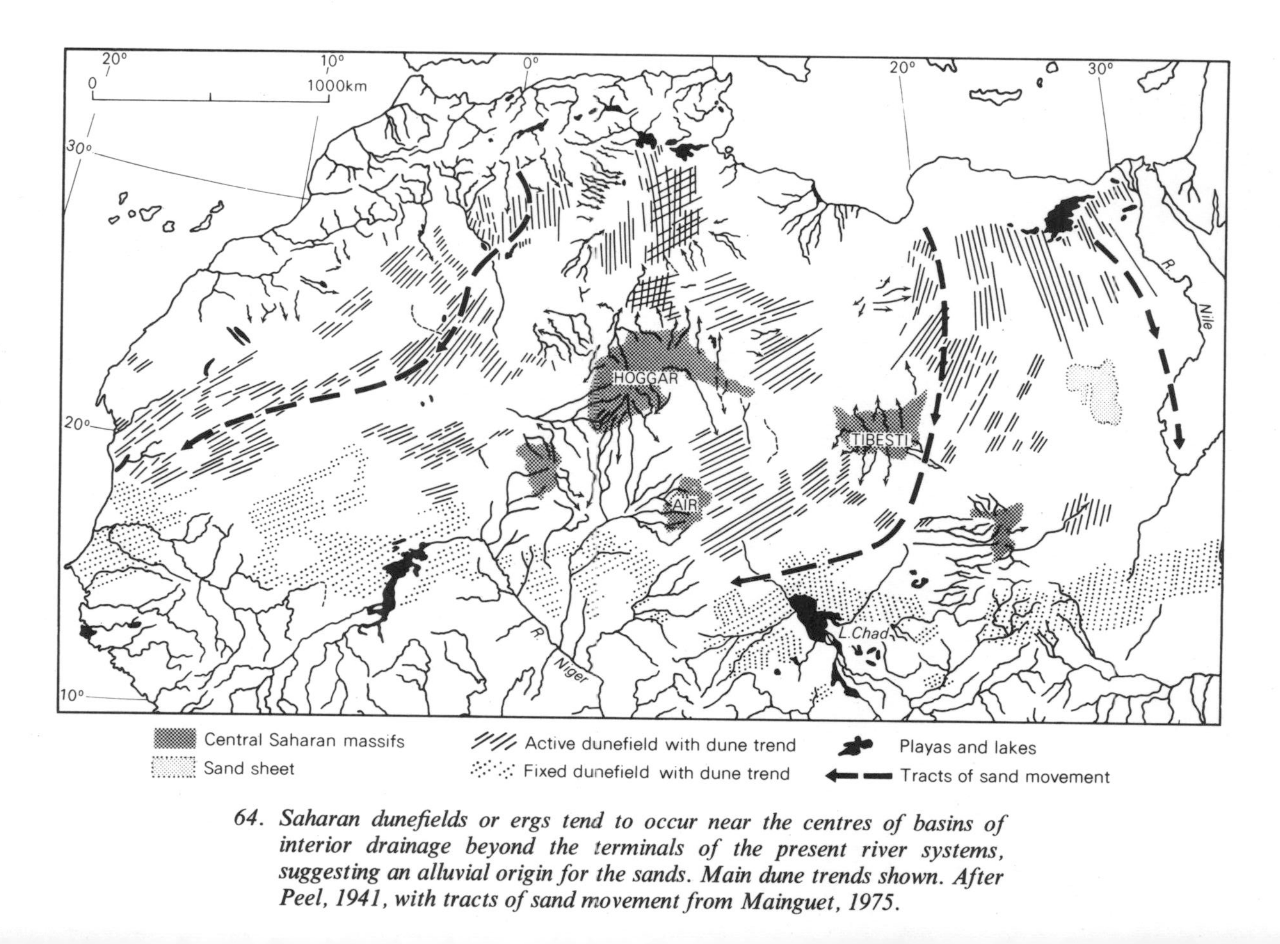

64. Saharan dunefields or ergs tend to occur near the centres of basins of interior drainage beyond the terminals of the present river systems, suggesting an alluvial origin for the sands. Main dune trends shown. After Peel, 1941, with tracts of sand movement from Mainguet, 1975.

Sahara. Dune sands extending south from the Kharga Depression and other oasis depressions in Egypt apparently originated by salt-weathering of sandstones in the depressions (Fig. 62).

Sand beaches and coastal lagoons are an important source of dune sands in coastal deserts with strong onshore winds, for example in the Atlantic Sahara south from Cape Juby. Calcium carbonate grains may predominate near the source but diminish inland relative to the harder quartz.

Dune sands commonly retain characteristics indicating sources nearby, and aeolian transport involved in dune building may be small. Continuity of dune patterns over large areas need not imply corresponding distances of travel of sand grains. Nevertheless, it is possible that sand may be moved over long distances in extreme deserts of low relief. Satellite imagery has confirmed the existence in the Sahara of preferred tracts of sand movement, revealed by reflectance from sand belts or from sand-polished rocks, following the trade winds for thousands of kilometres as shown in Fig. 64 (Mainguet, 1975). Long-distance sand transport may need to be considered in relation to the source, economy and age of dunefields in such areas; for instance Wilson (1971) has postulated an integrated system of sand-flow for the Sahara, with dunefields located in areas of converging sand streams and deflationary regs at points of divergence.

Colour and texture

The yellow to reddish hues of desert sands come from microscopic flakes of iron oxide adhering to the grains. Since these may be removed during saltation, mobile sands are paler than those of less mobile dunes. Regional differences in colour may reflect different sand sources; for example the sands of the Erg Occidental are noticeably darker than those of the Erg Oriental (Capot-Rey and Gremion, 1964). Redness of dune sands tends to increase with age (Norris, 1969), either from continuing accession of dust rich in pigment, or from the weathering of contained ferromagnesian minerals. Sands in shield deserts may inherit their redness from parent laterites.

Desert sands are subject to etching and redeposition of silica by dew or rainwater charged with carbon dioxide in periodically alkaline environments (Kuenen and Perdok, 1962). This results in an overall fine pitting which shows as frosting. The small percentage of lustrous grains includes those with fresh fractures.

TRANSPORT OF SAND BY WIND

Particles classed as *sand* fall within the range of diameters 0.05 to 2.00 mm. The lower limit is close to that at which the velocity of fall becomes less than that of an eddy current in an average wind, when particles can be carried away in suspension as dust. The upper size limit of sand approaches that at which grains are rarely moved directly by wind pressure or by impact of other grains. Sand-sized particles thus have the distinctive property of ready surface transport by wind and with it of accumulation as dunes.

Modes of transport

When sand grains are set in motion by pressure from a turbulent airstream they are initially rolled along the ground, but within a few metres this gives place to a bounding motion termed *saltation*. Grains are taken up a small distance into the airstream and are carried forward by its pressure in flat trajectories as they fall. Saltation is sustained through dislodgment of further particles by the impact of descending grains. The bulk of this transport takes place within a few centimetres of the surface, in hops of a few centimetres. The rate of travel is about half the wind velocity measured at 1 m. Saltation normally accounts for about 95 per cent of the bulk transport of sand in dunefields and involves mainly the fractions between 0.15 and 0.25 mm, with an upper limit of 0.5 mm (Fig. 65).

Larger sand grains mainly between 0.25 and 2 mm are rolled and pushed along the surface in advances of a few millimetres by the impact of saltating grains, a movement termed *surface creep*. The sand in this *traction load* is usually much greater than the saltation load, but since its rate of movement is less than 1 cm/min compared with a few metres per second for saltation, it accounts for only a minor part of the bulk transport.

Particles smaller than 0.15 mm tend to be carried in *suspension*. The very fine sand, above 0.05 mm, is borne above the saltation curtain but within a few metres of the ground and is deposited with the saltation load as soon as the wind slackens. Still smaller particles form a dust cloud which can extend to a great height and persist as a haze for hours after a sand storm.

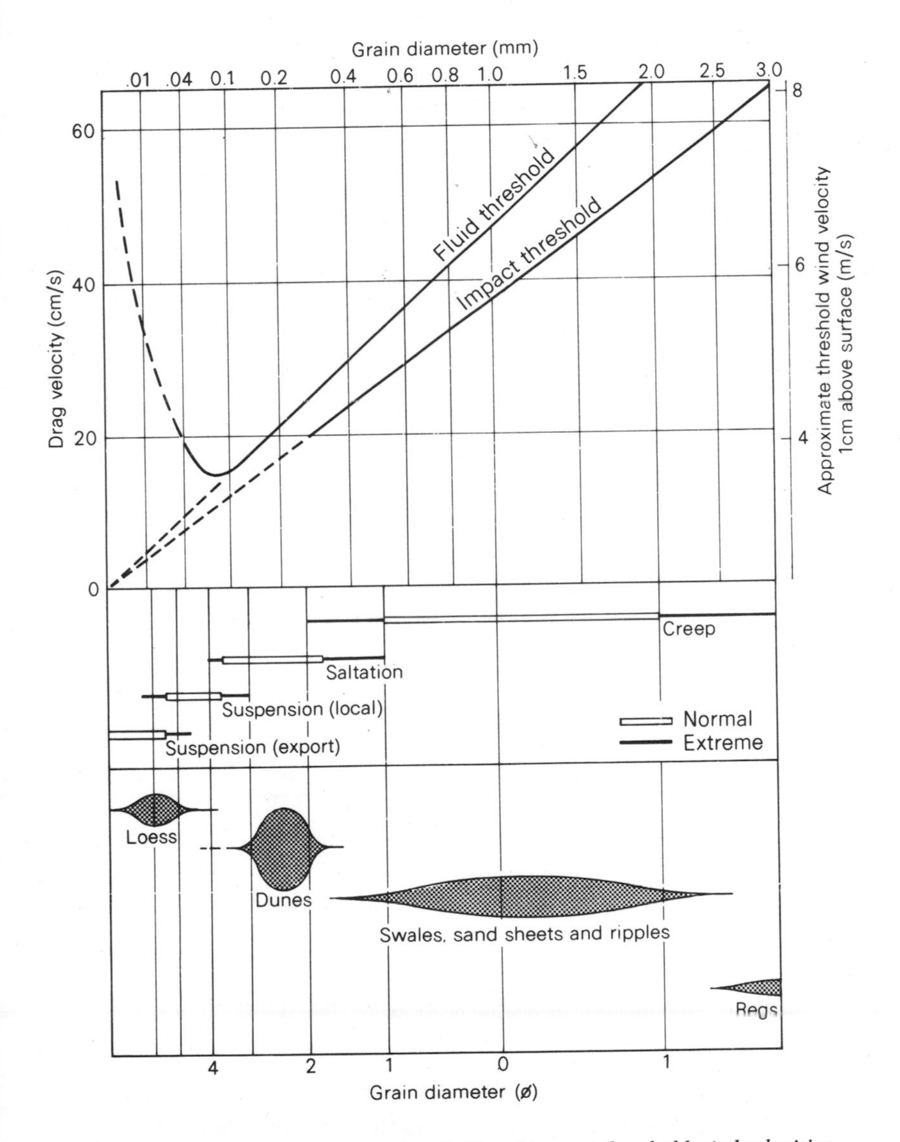

65. Relationships between grain size, fluid and impact threshold wind velocities, characteristic modes of aeolian transport, and resulting size-grading of aeolian sand formations. After Bagnold, 1941, and Folk, 1971a.

Threshold wind velocities

The shear exerted by the wind at the surface is expressed in the *drag velocity* and is proportional to the increase in wind velocity with height. The wind velocity at which sand is set in motion is the *fluid threshold* and the slightly lower velocity required to sustain movement with saltation is the *impact threshold* (Bagnold, 1941). For typical dune sand the impact wind velocity is 4 mps at a height of 1 cm, equivalent to a wind of 16 kph at 1 m (Fig. 65).

Threshold velocity and grain size. With particles above 0.08 mm the threshold velocities increase with the square root of grain diameter, as shown in Fig. 65. The initial fluid threshold is set by the modal diameter of the surface grains, and progressively coarser fractions are set in motion as the drag velocity increases. In dune sands, few grains exceed twice the modal diameter, and the *ultimate threshold* at which all surface grains are set in motion is commonly attained. With grains above 2 mm the impact threshold is reached very infrequently.

TABLE VIII

Threshold wind velocities for sand movement on various types of surface in the Mojave Desert, California*

Type of surface	*Threshold velocity (kph)*
dune	17
coarse sandy wash	21 (creep of granules to 3 mm)
desert pavement with granule veneer	32 (sand only removed) 53 (small granules removed)
alluvial plain	40 (silt chips saltated)
crusted alluvial fan	53
crusted playa surface	stable at 53, threshold not found

* From Clements *et al.*, 1963.

With particles finer than 0.08 mm, threshold velocities increase rapidly with diminishing grain size, with the passage from rough granular surfaces to smooth surfaces sheathed in viscid laminar air, and with increasing cohesion between particles. Surfaces of silt and clay resist wind erosion therefore, but the fine grains are rapidly deflated in suspension if disturbed.

Threshold velocity and surface characteristics. The threshold values stated above are for flat surfaces of dune sand and are altered by dune relief, vegetation, and moisture content. They also take no account of gustiness, known to be important in initiating sand movement (Dubief, 1943). Measurements from the Mojave Desert in Table VIII illustrate the control of threshold velocity exercised by the roughness or cohesion of surfaces.

Wind regime and sand transport

Bagnold (1941, 1951) found the rate of sand movement to vary approximately as the cube of the drag velocity, and to be represented by

$$Q = \frac{10^{-4}}{(\log 100\,z)^3} .t\,(v-16)^3$$

where Q is the mass of sand moved over a 1 m front, in tonnes
t the period in hours
v the wind velocity in kph, and
z the height of the wind recorder in metres (to be halved where the instrument is obstructed by buildings or vegetation).

The effectiveness of occasional strong winds in transporting sand is apparent from this formula; for instance transport by one hour of wind at 50 kph is equivalent to 14.3 hours at 30 kph or 611 hours at 20 kph. Hence the direction of the *dominant* or most effective wind may differ from that of the prevalent wind.

The formula allows the construction of *sand-movement roses* showing vectors of potential sand transport, and the calculation of annual resultants as vector sums. Regional analyses of *effective wind regimes* have been made on this basis by Dubief (1953c) for the Sahara and by Brookfield (1970a) for central Australia. Both

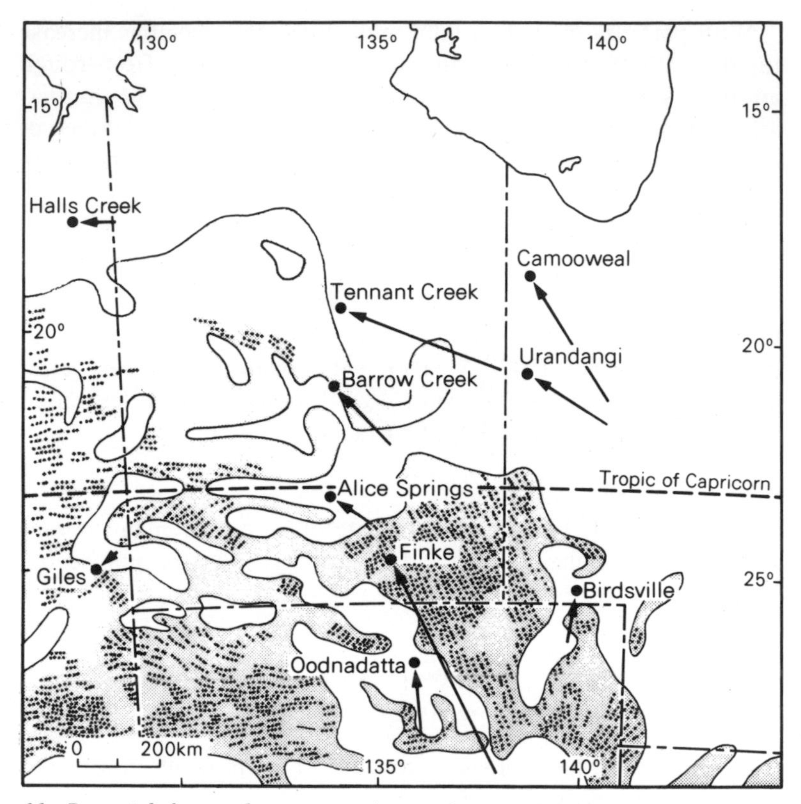

66. Potential for sand movement by wind at stations in central Australia shown relatively by lengths of arrows indicating the directions of weighted resultant winds. Based on data in Brookfield, 1970a.

stress the importance of strong winds as atmospheric depressions converge on the anticyclonic cores. The longest-lasting and most widespread sand storms occur in the warm sectors of depressions, whereas localised but spectacular wall-like clouds of sand and dust mark the passage of fronts (Dubief, 1971).

The zonal deserts are not particularly stormy, although surface winds are stronger across open desert plains. For the Sahara, Dubief (1953c) has shown that sand storms are more frequent along the Atlantic and northwestern margins and less so in the inner dunefields. Persistent northeasterly trade winds give strong weighted resultants in the south, whereas the winds on the stormier midlatitude margin are seasonally variable, with smaller resultants.

Central Australia shows a similar pattern save for the strong southerly wind resultants in the southeast quadrant, where dominant and prevalent winds closely coincide (Fig. 66). The midlatitude deserts are windy in winter and spring, but winds are variable and are also influenced locally by the stronger relief.

Accumulation of sand by wind

The growth of dunes follows from the tendency for saltating sand to accumulate in areas already sand-covered in preference to adjoining sand-free surfaces. This results from the check to a strong wind through intensified sand movement over a sand surface, and from the lowered rate of sand movement where saltating grains splash into loose sand compared with that over firm ground. In the expression for the rate of movement of average dune sand

$q \propto C\, V_*'^3$, where V_*' is the drag velocity and q is the total sand movement (Bagnold, 1941)

the constant C, expressive of surface conditions, is given as

1.8 for dune sand
2.8 for poorly sorted sand, as in a dune corridor
3.5 for a stone pavement.

A strong sand-laden wind will therefore cause accretion on an existing sand patch, whereas a gentle wind may remove sand and extend the patch downwind, because at low intensities of sand movement the greater smoothness of the sand surface may outweigh its disadvantage as a transmitter of saltating sand and induce a slight increase in drag velocity.

Further sand accumulation will be influenced by the nature of the surface upwind and the supply of sand from it. A *stony pavement upwind* will trap sand at times of gentle wind because of its greater surface roughness, and release it in moderate amounts to a strong wind, when accumulation may occur on adjoining sand surfaces. This favours the growth of relatively small but mobile sand forms. A smooth *playa or floodplain upwind* allows rapid migration of sand to its lee margin where it may be trapped in source-bordering dunes due to decreased mobility downwind. A *sand bed in place* combines maximum supply with low mobility, and subsequent dune development may depend upon whether or not a contrasting underlying floor is exposed through sand movement.

SIZE-GRADING AND SHAPE OF AEOLIAN SANDS

Size-grading. This is influenced by the differing modes of transport and deposition. Modal diameters are generally between 0.125 and 0.25 mm and the sands are generally distinguished from beach sands by positive skewness and from river sands by better sorting (Friedman, 1961). Coastal dune sands tend to be better sorted than those of inland dunes because they contain less dust.

Accretion from saltation predominates on dunes and gives well-sorted sands of between 0.125 and 0.25 mm. Admixture of coarser grains with modal diameters of 0.5 mm and above results from *encroachment* by surface creep, mainly in swales and on lower dune flanks. *Sedimentation* of fine particles from suspension introduces a small percentage of grains with modal diameters of less than 0.15 mm in sheltered swales and on lee slopes. Sands on dune crests are generally finer than those of the flanks and swales, but in the Simpson Desert, where the parent alluvium was fine-textured, the crest sands are the coarser (Carroll, 1944).

Sand surfaces subject to deflation normally have a bimodal grading due to the selective removal of grain sizes represented in the sands of nearby dunes, whilst the surface concentration of grains too coarse to be transported by wind shelters finer components. Consequently, they are destined to remain as sand sheets.

Shape. The shapes of aeolian sand grains may reflect a previous history. Except in previously cycled sediments, extremely rounded *millet-seed* grains form only a minority, mainly in the fraction above 1 mm moved in surface creep. Saltation of the mid-size grades produces mainly subrounded to moderately rounded grains through flaking by impact (Kuenen, 1960), but the fine sand grains remain angular because they are carried in suspension and so avoid collisions.

AEOLIAN AND TERRESTRIAL CONTROLS OF SAND FORMS

Desert sand surfaces shaped by wind exhibit regular patterns over a wide range of dimensions, with wavelengths from a few centimetres to several kilometres and with amplitudes from millimetres to hundreds of metres (Wilson, 1972). Linear elements may continue for hundreds of kilometres in systems of continental extent and at the scale of planetary air circulation, for instance the great

TABLE IX **Classification of aeolian sand forms**

- Sand Sheets

- Minor Forms
  - Sand Ripples
  - Granule Ripples

- Free Dunes
  - Simple Dunes
    - Crescentic dune or barkhan
    - Longitudinal dunes
    - Transverse dunes
  - Compound Dunes
    - Linked barkhans
    - Linked longitudinal ridges
    - Reticulate dunes
  - Complex Dunes
    - Barkhanoid forms
    - Longitudinal forms
    - Transverse forms
    - Peaked forms

- Dunes Related to Obstacles
  - Topographic Barriers
    - Leeward accumulations
      - Sand shadow
      - Sand drift
      - Lee dune
    - Windward accumulations
    - Climbing and falling dunes
  - Anchorage by Vegetation
    - Isolated mounds
    - Transverse ridges
    - Parabolic dunes
    - Longitudinal ridges

wheel-round of the Australian sand ridges. These patterns also respond to local obstacles such as topographic barriers. Air photographs and satellite imagery have confirmed the range of these patterns and their similarity in several continents and have invited explanations in terms of fluid mechanics, by analogy with the bedforms of rivers or with cloud patterns (Allen, 1969; Hanna, 1969; Wilson, 1972).

This approach to the study of sand forms must be complemented by an understanding of discontinuities introduced at ground level, as by the localisation of sand supply, differences of sand-grain size, or by contrasts in the roughness of surfaces across which sand moves. The interplay of aeolian and terrestrial controls is shown in any dunefield in contrasts between the exposed crests of looser sand, responsive to regional winds, and depressions in which sand movement is localised by sheltering dunes and possibly hampered by ground moisture.

This interplay is apparent in the following account of aeolian sand forms summarised in Table IX.

SAND SHEETS

Where fine sand occurs with a secondary mode of grains too large to be moved readily in surface creep, wind-moulding of the surface is prevented and a flat sand sheet remains. Coarse grains concentrated at the surface resist arrangement into ripples, prevent sand accumulation by increasing the saltation rate, and protect adjacent finer grains from ejection by saltation impact. The mechanism is that involved in dispersing sand mounds by strewing gravel on them (Lettau and Lettau, 1969). Removal of finer fractions from such surfaces leaves a deflation reg with a grit veneer.

Poorly-sorted alluvia yield sands of this type, as in piedmont deserts. They are also characteristic of primary sands derived from granitic rocks, and in Australia, for example, sandplain is closely linked with shield desert. Sheet-forming sands also originate from weathering of coarse sandstone in place, as in the Selima sand sheet on Nubian Sandstone in southwestern Egypt and the Sudan (Fig. 64).

MINOR SAND FORMS

Sand ripples

These are regular forms, transverse to the wind. Wavelengths are commonly between 10 and 50 cm and amplitudes between 0.5 and 1.0 cm; the ratio between them, the *ripple index*, is generally about 20 (Sharp, 1963; Stone and Summers, 1972). Sand ripples are asymmetrical, with windward slopes of about 10° and leeward slopes near the angle of repose of about 30°. As indicated in Fig. 67a, the crests are built of coarse grains between 0.3 and 0.7 mm, typical of traction load, resting on layered finer sand. As the coarser grains move upslope and eventually over the crest due to the impact of saltating sand, the ripple advances downwind at a rate of a few millimetres per minute; hence it adjusts fairly rapidly to wind changes. Saltating grains trapped in the troughs and overridden by the advancing crest contribute to the finer substrate.

Ripple growth is a function of sand sorting. Stone and Summers (1972) report a relationship between crestal grain diameter and equilibrium ripple wavelength

$$L_a = 63.8\, D^{-0.75}$$

where L_a is the ripple wavelength in cm, and
D the mean diameter of the crestal sand in mm.

An increase in wind strength will normally result in a steady increase in ripple wavelength to a limit set by the diameters of sand grains concentrated in the crests. This supports the view of Sharp (1963) that ripple height is limited by the inertia of the crest grains, rather than the suggestion by Bagnold (1941) that ripple wavelength is determined by the characteristic path of the saltating grains.

Ripples form where removal of sand into saltation is balanced by sand supply. With net accretion, the ripples flatten and may be obliterated by deposition in the troughs: net removal sharpens the ripples, but if the ultimate threshold wind is attained they are swept away leaving a planar surface.

Granule ripples

Ripples may grow further in sands which yield a protective capping of coarse sand and granules (up to 4 mm), and wavelengths of a metre or two and amplitudes of up to 20 cm are reported. Growing over a period of several weeks with varied winds, granule ripples tend to be more symmetrical and sharply peaked than sand ripples and more sinuous and branching (Fig. 67b). They are characteristic of wind-eroded sand sheets, particularly between or adjacent to dune sources of driving sand.

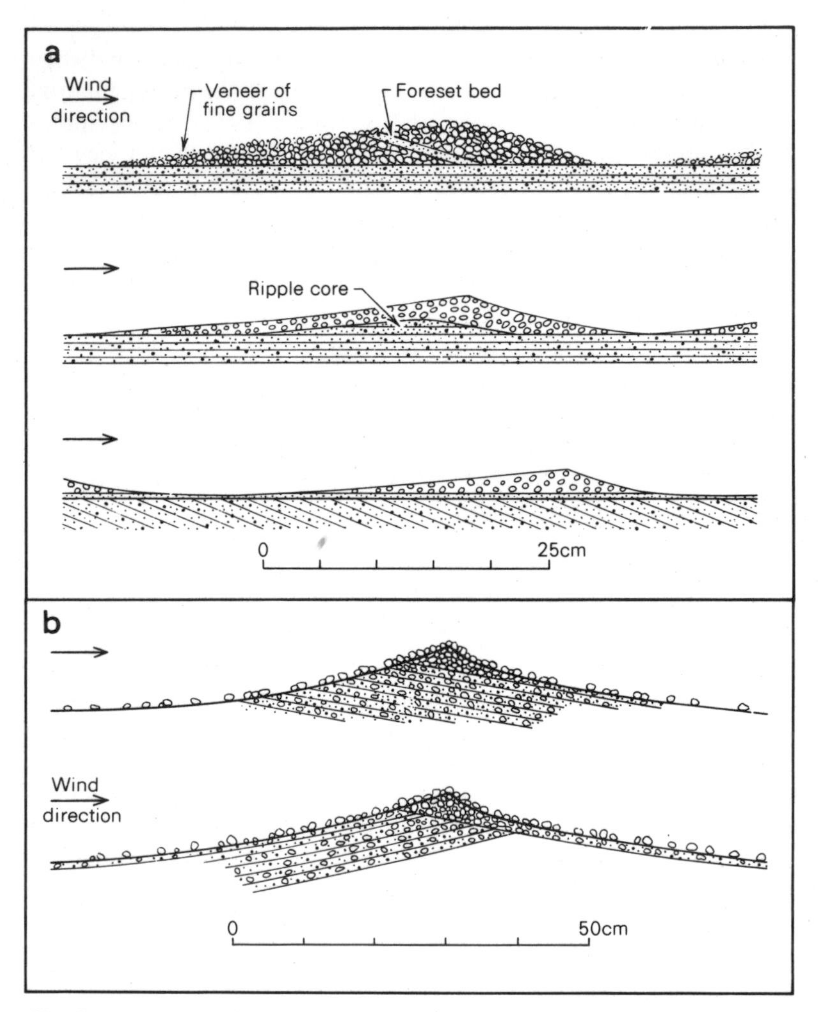

67. Cross-sections of impregnated samples of a. sand ripples, and b. granule ripples from Kelso Dunes, California. The structure of the granule ripple shows an earlier reversed asymmetry due to winds from right to left. From Sharp, 1963.

Differences between minor and major sand forms

Minor sand forms differ from major forms in fundamental ways. They are built by grains in surface creep and are therefore coarse-crested, whereas dunes are formed from the finer saltation load and tend to have finer sands on the crests; further, the transverse form is stable under all conditions, since any gap will be filled by surface creep and any incomplete ridge extended laterally in the same way. Minor and major sand forms are commonly represented as parts of a bedform hierarchy (Allen, 1968; Wilson, 1972; Stone and Summers, 1972), reflecting scales of sand-grain size and airflow, but because of these fundamental differences there can be no direct evolution from one to the other.

SIMPLE AND COMPOUND DUNES

Simple dunes consist of arrangements of form-elements relating directly to one or more wind directions or components within an airstream (Wilson, 1972) and which may be transverse or longitudinal, convex or concave to the effective wind. *Compound dunes* are combinations of simpler forms without change of scale.

Crescentic dune or barkhan

Accretion on a sand patch from strong sand-charged winds may cause it to grow into a sand mound. As the mound increases in height, accretion will be pronounced in the sector of divergent airflow downwind from the crest and the lee slope will steepen to 33°, the angle of repose of dune sand, when an avalanche slope or *slip face* will form. Survival of the dune depends on the maintenance of this sand-trapping slip face in its lee. The face will taper towards the edges of the mound to a minimum height of about 30 cm as set by the intersection of the repose slope and the characteristic saltation path. The wings of the slip face will advance downwind faster than the higher centre until they enter the wind shadow of the dune crest, giving the crescentic form (Fig. 68a). Reverse eddies occur in lee of the slip face, but are not known to shape the dune (Hoyt, 1966).

Barkhans are mainly small; for example the average width in a dunefield in southern Peru was found to be 37 m (Finkel, 1959). The height is consistently about one-tenth of the width. With increasing size they range from elliptical forms with enclosed slip faces as in southern Peru, where downwind length averages three times the width (Fig. 68b), to open forms in which the width may exceed twice the downward length, as in Imperial Valley (Long and Sharp, 1964). The elliptical form has a convex windward profile, with the dune crest separate from the brink of the slip face.

A barkhan advances downwind by the avalanching of sand in

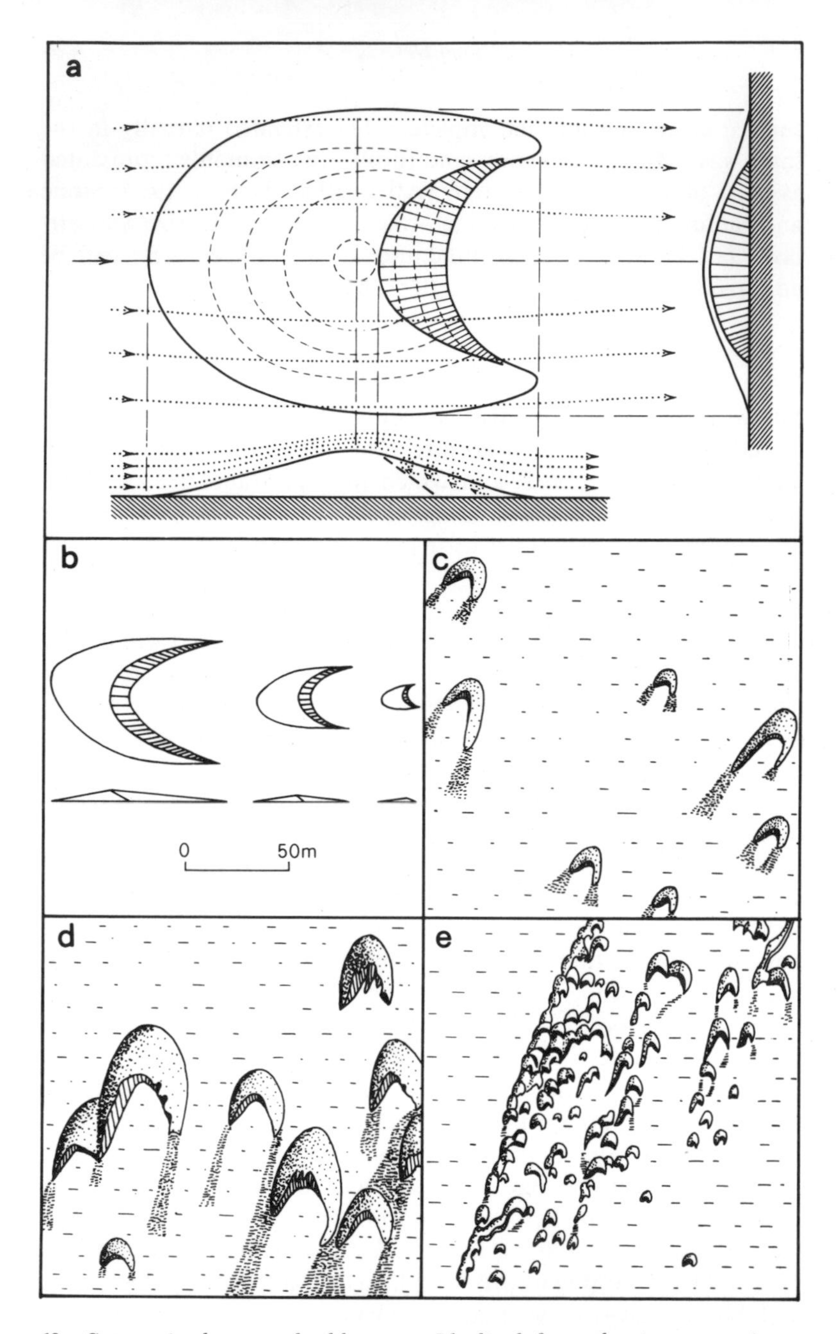

68. Crescentic dunes or barkhans. a. Idealised form showing approximate wind distribution, from Bagnold, 1941; b. relationship between dune size and plan-form exemplified by Peruvian dunes, from Hastenrath, 1967; c. sand streamers, sand shadows and variable symmetry; d. barkhans linked along streamers; e. compound barkhans in a chevron pattern showing greater sand supply from the left. Figs. c–e from Clos-Arceduc, 1969.

tongues down the slip face. The movement is reflected in an internal structure of foreset laminae dipping at 20–40° downwind, overlapped by flat-lying sets with upwind dips of less than 10° representing accretion on the windward face (Fig. 69a). Annual advances commonly range between 5 and 30 m and tend to be inversely

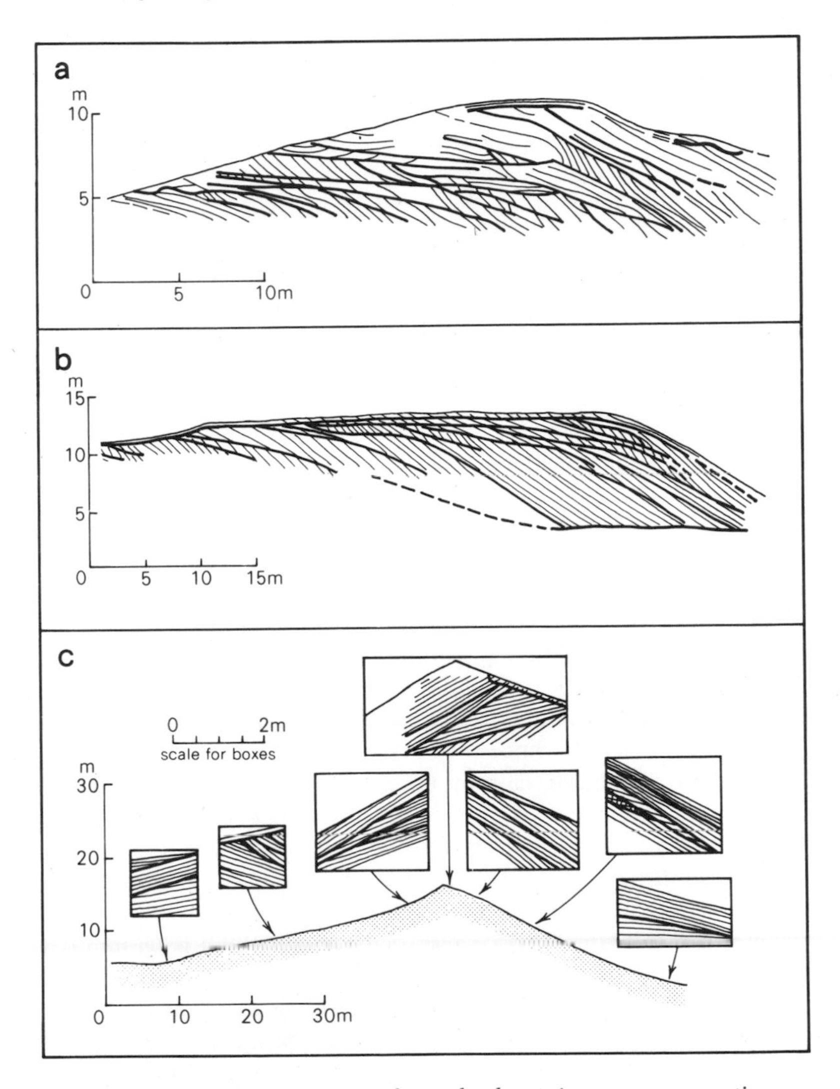

69. Internal bedding structures of simple dunes in transverse section. a. Barkhan, b. transverse dune, White Sands National Monument, New Mexico. From McKee, 1966; c. longitudinal dune, Libya. From McKee and Tibbits, 1964.

related to dune height (Bagnold, 1941). Departures from this rule can often be explained by the stage of development of the dune (McKee and Douglass, 1971), since advance tends to be slowed by growth and increased by shrinkage (Hastenrath, 1967).

Sand transport in a barkhan comprises *bulk movement* in the advance of the dune body and *through-flow* of sand intercepted at the windward toe and discharged as streamer transport from the horns. These sand streamers diverge at 5–20° from the barkhan axis, leaving an intervening sand shadow (Fig. 68c). Asymmetry can result from an obliquely directed supply of sand to the dune; for example the western horns of the south Peruvian barkhans are consistently longer than the eastern (Finkel, 1959).

A smallish barkhan in the Peruvian field with a volume of 700 m^3 represented a bulk transport of 21,000 m^4 in its annual advance of 30 m; in the same period the dune intercepted 18 m^3 of sand along its frontage of 36 m and discharged this from its horns (Lettau and Lettau, 1969). This through-flow is tantamount to a renewal of the dune over 40 years, during an advance of 1.2 km. The Peruvian barkhans had a density of 1 per 7 ha, equivalent to a spread-out sand thickness of 2 cm. Variations in bulk transport within the field required local sand scour of up to 2 mm per annum from the adjacent desert pavement; this could account for the entire field of dunes over about a century, suggesting it as an adequate source.

Barkhans are favoured by a regime of strong sand-supplying and moderate sand-moving winds from the same quarter; for instance 90 per cent of the winds in the barkhan fields of southern Mauritania are from the north, with an average velocity of 25 kph (Clos-Arceduc, 1969). Since they are found with other dune types, however, wind regime cannot be the sole determinant of their occurrence. Barkhans typically occur as isolated bodies on plains with little sand cover, where the floor contrasts with loose sand as a transmitting surface. Hence they commonly occur away from the main dunefields or in firm sand-free corridors within them.

The form tolerates directly opposed winds, which if temporary may smooth the slip face and form a small reverse face on the brink. Seasonally opposed winds may give *reversing dunes* as in Death Valley, where a northeasterly advance in spring and summer is followed by a southeasterly retreat under strong autumn winds (Clements *et al.*, 1963). Where winds are multidirectional in other-

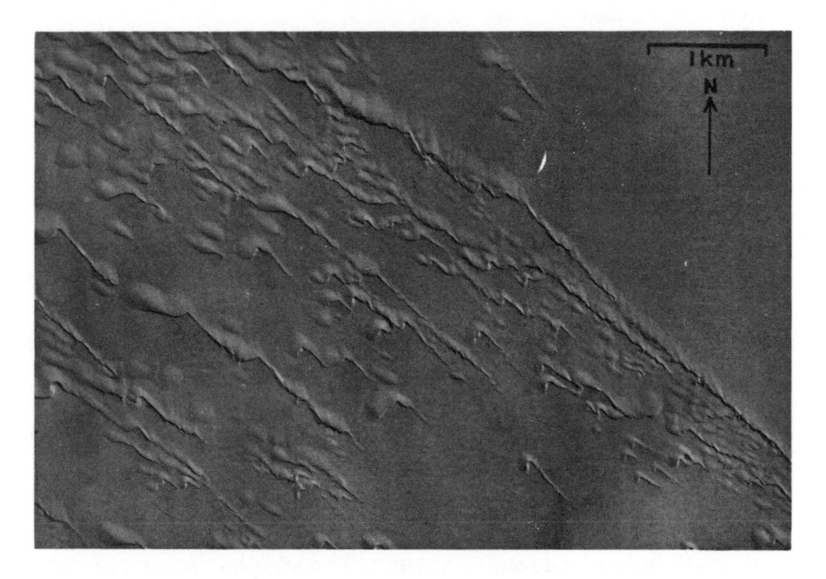

44. Linkage of barkhans to form longitudinal ridges (silk) oblique to the dominant wind, as indicated by the symmetry of the barkhans, north of Tibesti. Photo Institut Geographique National, Paris.

wise favourable conditions, no slip face will form and a sand mound may grow directly into a *domeshaped* dune (McKee, 1966).

Linked barkhans

Where sand supply is abundant, or its transport diminishes downwind, barkhans may link in compound arrangements which are strongly influenced by their own patterns of sand movement. Sand streamers are tracts of barkhan growth and slowing advance and hence of likely capture, whereas accelerating sand-starved dunes are displaced from sand shadows. Clos-Arceduc (1969) has illustrated from southern Mauritania how these differential movements lead to lines and chevrons of linked barkhans trending obliquely at 10–20° from the barkhan axes (Fig. 68 d–e), and relates this to the angle of departure of sand streamers in the originating barkhan; comparable forms are illustrated by Hagedorn (1971) from north of the Tibesti massif (Pl. 44). Wilson (1972) interprets the arrangement as the interaction of longitudinal and transverse components in the airstream, giving progressive lateral shift of the sand form at intervals downwind.

Longitudinal dunes

The ridge elongated in the general direction of the sand-transporting wind is the most widespread simple dune type. Commonly it comprises a broad *plinth* below a narrow *crest*, but these relationships vary. Over most of the Australian and Kalahari dunefields the plinth predominates and is stabilised by vegetation, with mobile sand confined to the crest; however, the type includes the *sloûk* of southern Mauritania (Clos-Arceduc, 1969), which are peaked ridges with limited plinths, as well as rounded forms lacking sharp crests, such as the *alâb* of the Erg Chech (Monod, 1958) and subdued stabilised forms along the margins of the Sahara and the Australian deserts.

A height of between 5 and 30 m is common. In crest-dominated forms the basal width is 5–10 times the height, but in subdued ridges this ratio may exceed 20.

Plinth slopes are concave and generally attain between 10° and 20°, whereas crestal slopes include slip faces at the angle of repose. The crests show considerable variety. The simplest is a razor ridge with a reversing slip face to one side, but there is generally some repetition of form along the crest, for example of the crescentic sharp summits for which the name *sief* (= sword or scimitar) has

45. Longitudinal dune, northern Sinai, showing alternation of lateral slip faces along sinuous crest. Dune height 10 m.

been widely applied (Pl. 45). Multiple and braided crests also occur. In many areas the ridges have a consistent asymmetry; for example the eastern flanks are steeper than the western throughout the western Simpson Desert (Mabbutt, 1968).

Longitudinal dunes generally occur in evenly-spaced systems (Pl. 46). In sandy terrain the spacing ranges between 200 and 500 m, the lower values being characteristic of branching systems and the higher of regular parallel ridges. With spacing above 500 m the catenary swale commonly gives place to a flat corridor, and the widest-spaced dunes generally rest on stony floors. Dune height tends to increase with spacing and with the contrast between dune and swale surface.

Regularity and continuity of pattern go together and the ridges may extend for several hundreds of kilometres. In branching patterns the ridges consist of segments a few kilometres long,

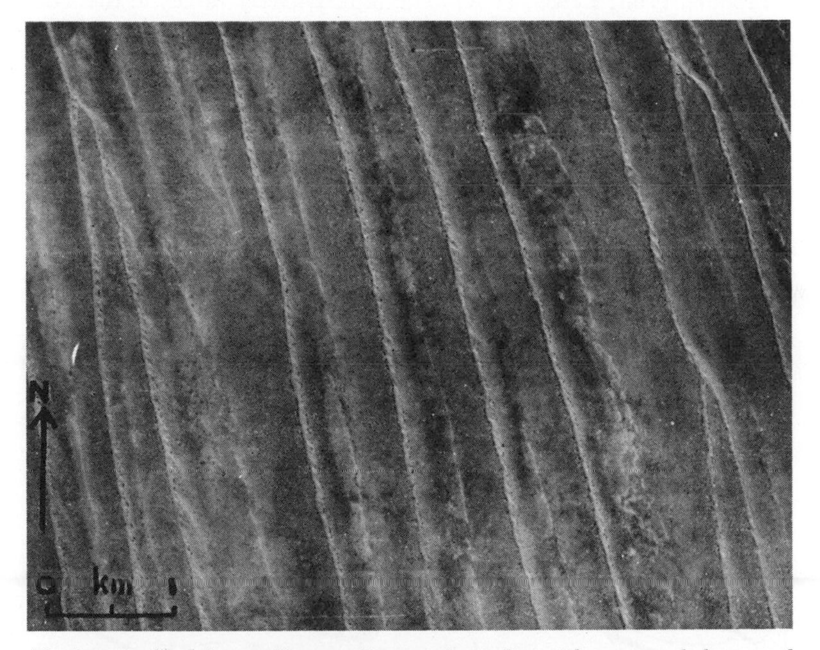

46. Longitudinal dunes, Simpson Desert, Australia, with vegetated slopes and loose crests. Spacing 500–600 m, height 15 m. Junctions point down the formative SSE wind. This photograph has been made available by courtesy of the Director, Division of National Mapping, Department of National Resources.

linking in *junctions* that almost invariably point downwind. Downwind extension of ridges is seen in their abutment against topographic obstacles and their continuation across sand-free terrain.

Over vast areas of the zonal deserts, longitudinal dunes run approximately parallel with the dominant wind, which in the generally persistent regimes is close to the prevalent wind direction. Brookfield (1970b) notes that standard deviations from the mean dune direction in central Australia are seldom above 6.6°. In the trade wind belts, where they have their greatest extent, the ridges trend equatorwards across the tropic and swing westwards in lower latitudes, as in Australia and the southern Sahara. The Australian and Kalahari dune systems also include an easterly midlatitude component in arcuate 'swirls' of anticlockwise growth in response to winds generated by the convergence of low pressure cells on an inner zone of frequent high pressure (Fig. 60) (Jennings, 1968; Mabbutt, 1968; Lewis, 1936). The northeasterly ridges in the Thar Desert are controlled by the southwest monsoon (Verstappen, 1968). Longitudinal dunes are rarer in the midlatitude deserts, but those of the Kara Kum and Kyzyl Kum in Turkestan trend south or southeast, also with the dominant winds. Within these broad patterns, dune trend may respond to local variations of wind direction as induced by topography.

The *exact* relationship between longitudinal dune trend and the formative wind, which is critical to explanations of origin, is obscured in that many systems of longitudinal dunes are stabilised

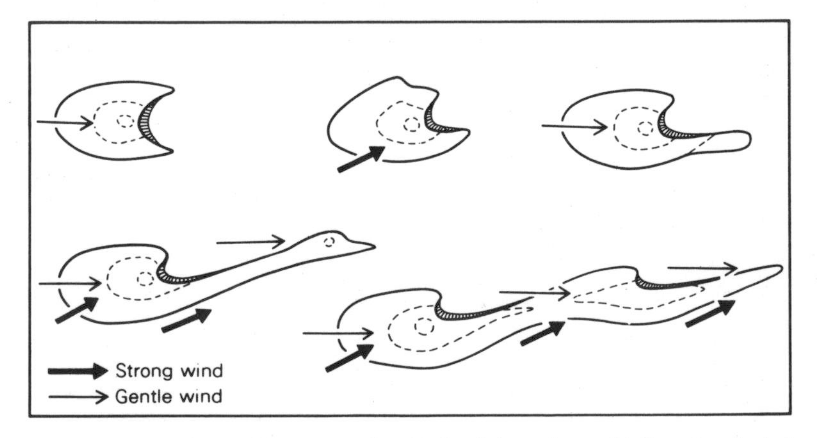

70. Extension of a barkhan into a longitudinal form due to bi-directional wind components. From Bagnold, 1941.

and partly fossil, and may hence express earlier wind regimes, as will be discussed in Chapter X.

Some branching longitudinal dune systems embody more than one directional element, as in parts of the Simpson Desert (Mabbutt, 1968), suggesting that more than one component of an airflow or wind spectrum may be expressed. Contrasting longitudinal forms may intersect discordantly, for example *sloûk* ridges and subdued *alâb* dunes in southern Mauritania, possibly indicating different ages or origins.

Longitudinal dunes have a characteristic internal structure of alternately overlapping avalanche sets indicating the action of cross winds, above a plinth consisting mainly of accretion layers marking the early growth stage of an elongate sand rise (Fig. 69c).

Formation of longitudinal dunes

Extension from barkhans. Bagnold (1941), stressing that barkhans form with relatively unidirectional strong winds, postulated that winds from a second direction might extend one wing to form a ridge with intermediate trend as depicted in Fig. 70.

Clos-Arceduc (1969) accepts a similar explanation for *sloûk* in southern Mauritania which share the trends of chains of linked barkhans, at between 10° and 20° to the dominant winds as expressed in barkhan axes and in the directions of *alâb* ridges in the region. The affinity is strengthened where the *sloûk* have barkhanoid wings. The obliquity of the ridges may express a combination of transverse and longitudinal elements in the airstream as implied by Wilson (1972), rather than the interaction of winds from different quarters as proposed by Bagnold or the displacement of the airflow by the dune itself as suggested by Clos-Arceduc.

This mechanism may apply to the downwind extension of ridges in circumstances generally favouring the formation of isolated dunes.

Wind-rift hypothesis. This postulates that the swales have been fashioned by wind-scour by analogy with parabolic coastal dunes (Medlicott and Blanford, 1879; Aufrère, 1930; Verstappen, 1968, 1972). The parent material is envisaged as alluvial or older aeolian sand and vertical corrasion of the swales is assumed to be accompanied by lateral displacement of sand leading to upward growth of the bounding ridges, and by downwind transport resulting in the extension of ridges and their linkage in continuous systems.

Ridge junctions are interpreted as advancing blowouts (King, 1960; Folk, 1971a), consistent with their general downwind taper. The consistent and inversely related height and spacing characteristic of the dunefields are attributed to an even original 'spread-out thickness' of the parent sand or wind-worked layer, commonly between 5 and 10 m. Folk (1971a) has claimed that the complementary size-grading of dune and swale sands in the western Simpson Desert supports an evolution by differential erosion of a parent sand body.

This explanation is attractive for vast uniform dunefields with no evidence of localised deposition or transport of sand as by river channels. However, the important corollary, that an original surface should occur within the dunes at some level above the swales, appears to be met only exceptionally (cf. King, 1956). The general flatness and common levels of adjoining swales pose problems to this hypothesis, particularly where they coincide with an alluvial horizon which continues beneath the dunes, although it might be postulated that a firm horizon of this type had set a lower limit to wind erosion. Also, an interpretation of the Y-shaped junctions of Australian dunes as blowouts is supported neither by the dimensions nor by the outlines of the arms, and the throats are areas of sand accumulation rather than scour.

Extension downwind from points of sand supply. This suggests that ridges have extended wherever circumstances have localised sand supply on a firm surface. Madigan (1946) postulated that the shifting channels of retracting drainage systems had been loci of sand concentration in the Simpson Desert, and Twidale (1972) has suggested that the ridges may have extended from wind-eroded breaches in source-bordering dunes along sandy playas, scores of which occur in the south of the desert. Under such schemes the dunes are envisaged as growing downwind as the sand supply continues, shifting laterally and growing in height to dimensions and spacing in equilibrium with the airflow. Lengthwise linkages may occur and extension of the ridge downwind assumes greater importance. Junctions have been attributed to deflection of the mobile growing sectors of ridges by cross-winds (Madigan, 1946). This explanation is consistent with the existence in at least part of the Simpson Desert of an alluvial horizon at the level of the swales and continuous beneath the dunes (Mabbutt and Sullivan, 1968), and is more accordant with the characteristic internal structure of longitudinal dunes.

Helicoidal airflow in longitudinal dune systems

Whatever the origin, the forms of longitudinal dunes indicate transverse components within the airstream and hence a vortical circulation. Parallel ridges and swales of comparable, suggestively equilibrium dimensions are consistent with the existence of heli-

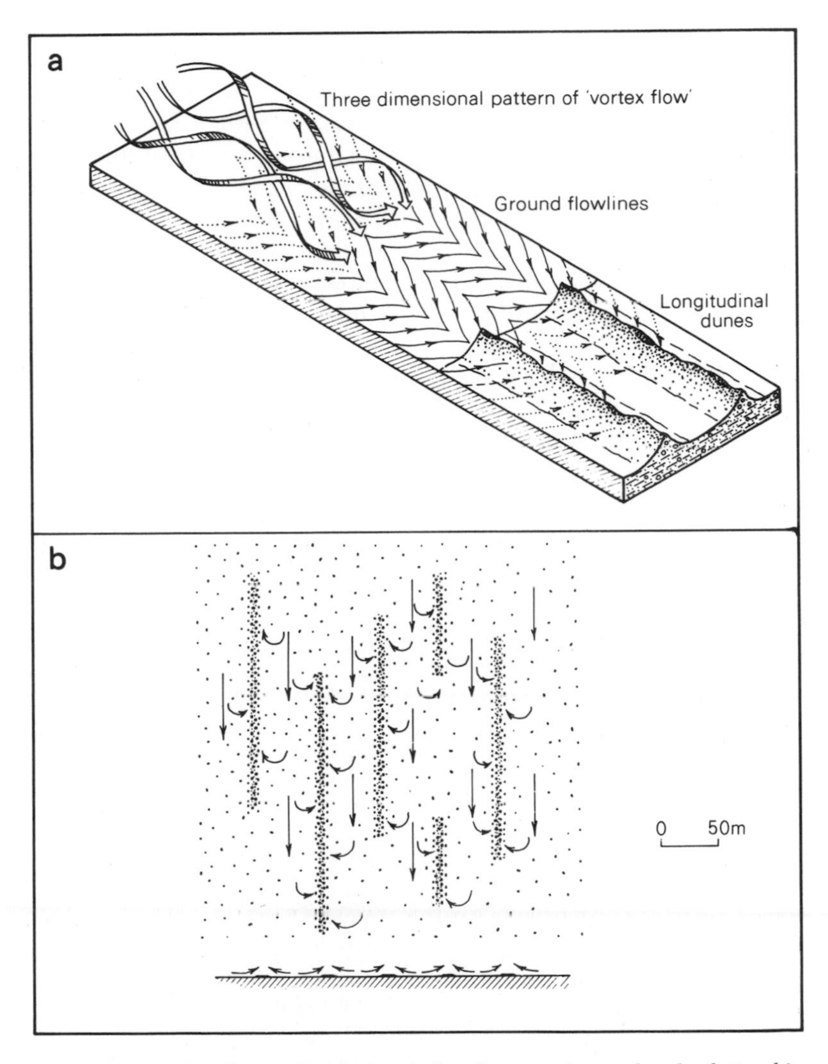

71. a. Vortical airflow, related ground flow lines, and postulated relationship to longitudinal dunes. After Cooke and Warren, 1973; b. longitudinal sand strips with hypothetical wind circulation. From Bagnold, 1941.

coidal airflow in paired opposed vortices with downwind axes, in which the currents rise over the dunes and descend into the swales as in Fig. 71a (Allen, 1969; Folk, 1971a; Wilson, 1972).

Bagnold (1953) and Hanna (1969) have claimed that vortical circulation should develop from turbulence in any heated airstream and would be reinforced by selective solar heating of the dune slopes. However, thermally generated helicoids on the proposed scale have not been shown to exist; no form of thermally generated separation flow or upward eddy has been proved effective in dune building; and the diameters of the vortices should be 2–3 times the thickness of the heated layer, requiring a minimum ridge-spacing of 5 km in the trade-wind belt.

A more plausible cause of vortical air currents is the lateral shear set up within a sand-transporting airstream through the contrast in resistance offered by a sand dune and a firm swale floor, which should induce surface eddies directed inwards towards the ridge. Bagnold (1941) observed in Egypt the concentration of sand into longitudinal strips on a sand sheet, with sufficiently regular spacing to suggest a large-scale rotary circulation as in Fig. 71b. Equilibrium dune height and spacing, as determined by the scale of adjoining vortices, should be set by the contrast in sand transmissivity between dune and swale, explaining why the pattern tends to be more open and the ridges higher where a hard stony floor is exposed. Junctions should occur where a change in surface contrasts leads to a change in the diameters of the vortices and hence in dune spacing, and in Australia they do tend to be aligned along such boundaries, as between stony reg and alluvial plain.

Consistent dune asymmetry might result from a lateral bias in erosion or deposition by vortical airstreams. In the anticlockwise arcuate systems of the Australian and Kalahari deserts the outer dune flank is commonly the steeper, and Coriolis deflection (to the left in the southern hemisphere) has been invoked as a cause of accentuated wind action on the outer flank during dune formation (Mabbutt *et al.*, 1969; Clarke and Priestley, 1970). In the Simpson Desert this outer (eastern) flank is now subject to greater scour by oblique winds, giving more prominent lateral slip faces on the west sides of the crests.

Environmental controls

Longitudinal dune systems require a particular combination

of sand supply and terrain. The sand should not be so plentiful as to mask the downwind-trending surface contrast between ridge and swale, not too scanty to allow the growth of continuous forms. A plains setting offers the uniformity of surface necessary for the evolution of a regular pattern of ridges in response to regional winds. Longitudinal dunes are characteristic of broad lowlands in shield and platform deserts where strong wind resultants are associated with persistent wind regimes, particularly of the trade winds.

Transverse dunes

Linked barkhans or longitudinal dunes may pass into transverse dunes if the sand cover increases sufficiently. These are overlapping wave forms normal to the dominant wind, with wavelengths of 50–200 m and heights commonly up to 10 m (Fig. 72a). They are asymmetrical, with upwind slopes inclined at 5–10° and convex lee slopes which approach the angle of repose but which lack slip faces save where a hard floor is exposed (Pl. 47).

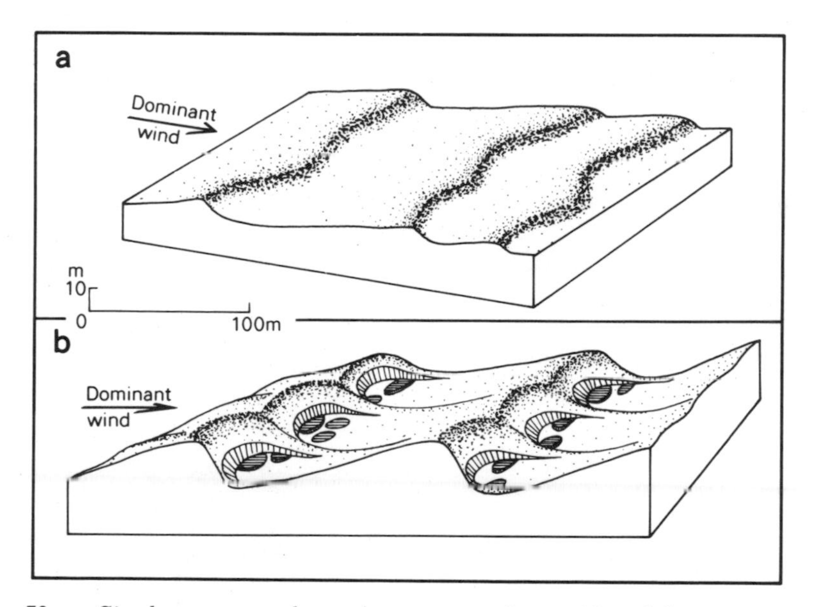

72. *a. Simple transverse dunes; b. transverse dunes with undulating crests, connecting longitudinal tongues, and slip faces with floors of lacustrine silt below. From the Mâjabat al-Koubrâ (Monod, 1958).*

The internal structure in the lower part of the dune in Fig. 6b reflects an encroachment on the dune downwind in the early stages, but the advance of a mature transverse dune is probably very slow. Measurements by McKee and Douglass (1971) at White Sands National Monument indicate rates of advance comparable with those of barkhans in the source areas in the west of the field, but virtual immobility towards the downwind margin. Stability is indicated by the continuity of planar sets over the crest. These accretionary units thin upwards due to periodic scour, suggesting that acceleration of the airstream over the growing dune sets a limit to its height.

Closed systems of transverse dunes are associated with thick sand cover in the lower parts of basins, for example in the Mreyyé of the Majâbat al-Koubrâ in the southwest Sahara (Monod, 1958). They can also result where sand flow is checked by a topographic barrier or by slackening wind velocities (Verstappen, 1972), and vegetation may similarly serve as a constraining factor leading to

47. *Transverse dune, central Australia, with sand-rippled crest, scalloped slip-face and vegetated corridor. Relief 5 m. Photo R. W. Millington, CSIRO, Division of Land Use Research.*

the piling up of transverse dunes, as in the Kara Kum Desert of Turkmenistan. Transverse dunes also occur with coarse sands of limited mobility, as in the gently undulating *zibar* dunes of the Rub-al-Khali (Holm, 1960) and the broad undulations between longitudinal dunes in the central Ténéré desert of Niger (Warren, 1972).

Many transverse dunes have sinuous faces in which salients and re-entrants coincide with crestal saddles and summits respectively (Fig. 72b). By analogy with subaqueous ripples, these have been attributed to transverse components within the air stream (Allen, 1969; Wilson, 1972). They foreshadow the development of longitudinal form elements, and small linking ridges may extend downwind from the salients, up the succeeding crests.

The three main simple dune forms are seen to be differentiated by the amount of available sand and its potential transport, rather than by wind regime, with the progression from barkhan through longitudinal to transverse dunes marking an increase in the thickness of sand and a relative diminution in the extent of surfaces suited for its rapid transmission.

Reticulate dunes

These are networks of ridges with enclosed depressions. A spacing of a few hundred metres and a relief of 10–20 m are characteristic. Commonly the network expresses the intersection of two or more directional components; for example network dunes in the south of central Australia combine the easterly and southwesterly trends of longitudinal dunes on their margins (Mabbutt, 1968). Dune networks have been attributed both to erosional and constructional processes. The former view appeals to standing waves in airstreams from intersecting directions (Clos-Arceduc, 1969), the latter to an interaction of longitudinal and transverse waves in a 'fish-scale pattern' due to the lateral displacement of the longitudinal elements at the points of intersection (Wilson, 1972).

In central Australia, reticulate dunes occupy an area of variable winds at the 'turn-round' of the anticlockwise arc of the longitudinal dune systems. In the Majâbat al-Koubrâ (Monod, 1958), confused networks described as *aklé* dunes occur at the southern rim of the dunefield where sand drifted by northeast winds is also subject to movement by opposed winds, or near escarpments where perturbations of the airstream might be expected.

COMPLEX DUNES

These have dimensions of hundreds of metres and spacings of kilometres, and because of their great bulk appear relatively static. They have been claimed as manifestations of atmospheric turbulence on a distinctly larger scale than in simple dunes, involving interaction between the regional wind and massive, relatively unchanging sand forms, and the name *draa* has been proposed to distinguish them (Wilson, 1972). They are characteristic of thick sand accumulations in extremely arid zonal deserts, notably the Sahara and Arabia.

Barkhanoid forms

Best-known is the Pur-Pur dune in the Peruvian coastal desert (Simons, 1956), with a width of 850 m, axial length of 2 km and height of 55 m. The horns consist of aligned barkhans, which also occur on the windward slope and crest. These move downwind at rates averaging 5 m per year and presumably traverse the larger form, which is relatively stationary with an annual advance of 0.45 m.

Transverse forms

Forms with a wavelength of about 1 km and an amplitude of 50 m have been illustrated from the Erg Oriental (Wilson, 1972, Plate II D). They show a 'fish-scale' arrangement similar to the sinuous transverse dunes of Fig. 72b, with longitudinal rises connecting successive ridges spaced at about 1 km. Their surfaces are covered with simple transverse dunes at one-tenth the scale. Discontinuous flat-floored depressions separate the main transverse ridges.

Longitudinal forms

Linear sand rises about 1 km wide, 100 m high and many tens of kilometres in length are the most widespread complex forms, for example the *draa* of the Saharan dunefields. The flanks are steep but lack slip faces, which are restricted to the combinations of simple longitudinal dunes that run directly or obliquely along the crests. A spacing of 3–5 km is characteristic, and the intervening corridors may be regs, as in the *gassi* of the Erg Oriental, or sandplains like the *feidj* of the Erg Occidental. The *whalebacks* of the Egyptian Sand Sea are similar, up to 3 km wide and 300 km long, and are considered by Bagnold (1941) to have formed over a long period as a composite plinth left by migrating longitudinal dunes on the crest.

Peaked forms

The commonest are pyramidal dunes, known as *ghourd* (pl. *ghroud*, *oghurd* etc) in the Sahara (Pl. 48), with whorled radiating ridges suggesting the name *star-shaped dunes* (McKee, 1966). Up to 150 m high and between 1 and 2 km across, they occur as regular culminations at the junctions of crestal ridges along large-scale longitudinal forms, or at intersections in reticulate systems. With decrease in sand cover they may form regularly spaced isolated mounds on regs, as at the northwest edge of the Erg Occidental (Capot-Rey, 1943).

They have been attributed to the interaction of winds from more than one direction, an interpretation supported by internal structures with alternating avalanche sets (McKee, 1966), or to their situation at nodes in systems of combined longitudinal and transverse wind currents (Wilson, 1972). Once formed, the peaks

48. Pyramidal dune, Grand Erg Oriental, Algeria. Photo R. F. Peel.

doubtless induce their own eddy systems, and Folk (1971b) has suggested that the faces between the arms have been sculpted by convergent vortical currents.

A less common form is the domal summit or *demkha*, which in the Erg Occidental has been attributed to planing by strong winds (Capot-Rey and Capot-Rey, 1948), and in the Murzuk Erg to the inhibition of relief by coarse sands (Capot-Rey, 1947).

DUNES RELATED TO OBSTACLES

Topographic barriers

Leeward accumulations. A *sand shadow* (Fig. 73a) is an elongate mound deposited between airstreams which have diverged to either side of an obstacle. The length and position of the crest downwind are set by the height of the obstacle, and the mound normally tapers downwind.

A *sand drift* (Fig. 73b) is an elliptical mound, less tapered than a sand shadow, deposited in the zone of divergent and decelerating airflow downwind from a gap that has funnelled and accelerated the wind. Drifts can also result from separation flow in a vertical dimension in lee of a steep slope (Fig. 73c).

Should a drift or shadow build up into the main airstream it will extend downwind as a *lee dune* (Fig. 73d), and Melton (1940) has recorded such dunes up to 5 km long below the Moenkopi Plateau in northern Arizona.

Windward accumulations. When a sand-transporting wind meets an obstacle, deposition generally begins some distance upwind and increased turbulence tends to sweep the obstacle clear of sand. The resulting accumulation will have a gentle upwind slope leading to a steep downwind face; if the obstacle is narrow the arcuate face will extend downwind on either side as a *wrap-around dune* (Fig. 73e). The southwest monsoon has built large *windward dunes* or *sand shields* upwind from hills in the Thar Desert (Verstappen, 1970, 1972; Allchin and Goudie, 1971). These are semi-conical and begin as far as 5 km from the hill, rising gently to heights of up to 100 m.

Where sand is less abundant, single or multiple ridges may form parallel with and at distances of up to 3 km upwind from major relief features. Clos-Arceduc (1969) described these as *echo dunes*

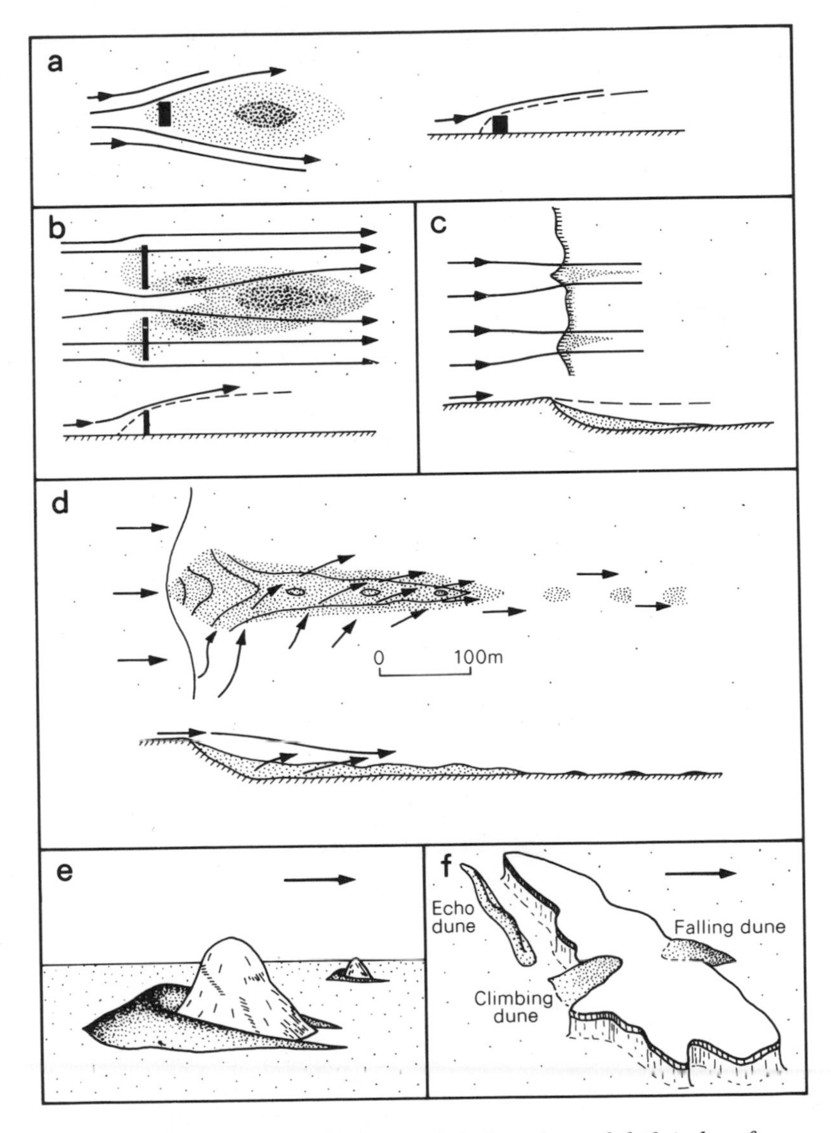

73. *Dunes related to obstacles. a. Sand shadow; b. sand drift in lee of gap; c. sand drift in lee of cliff; d. lee dune; e. wrap-around dune, similar to a sand shield but on a smaller scale; f. falling, climbing and echo dunes. In all cases dominant wind left to right as shown. Figs. a–d after Bagnold, 1941.*

(Fig. 73f) and attributes them to standing oscillations in the climbing airstream. Queney and Dubief (1943) estimate that an escarpment 100 m high should generate a sand-free zone 1 km wide, and such corridors are known in Saharan dunefields bordered by escarpments.

Climbing and falling dunes (Fig. 73f). These form where strong sand-laden winds meet opposing hillslopes. The windward slopes are generally less steep (10–15°) and consist of coarser sands than the leeward slopes (25–34°). The sands are commonly mixed with scree. These forms are widespread in coastal deserts, such as northern Chile and western Sinai, but they also occur inland, for instance in the Mojave Desert (Evans, 1962).

Anchorage by vegetation

Vegetation is unique as a sand-trapper, since it not only acts as a barrier in the path of wind-driven sand and changes surface roughness over large areas, but has the ability to grow up through accumulating sand. It is more important in semiarid than in arid

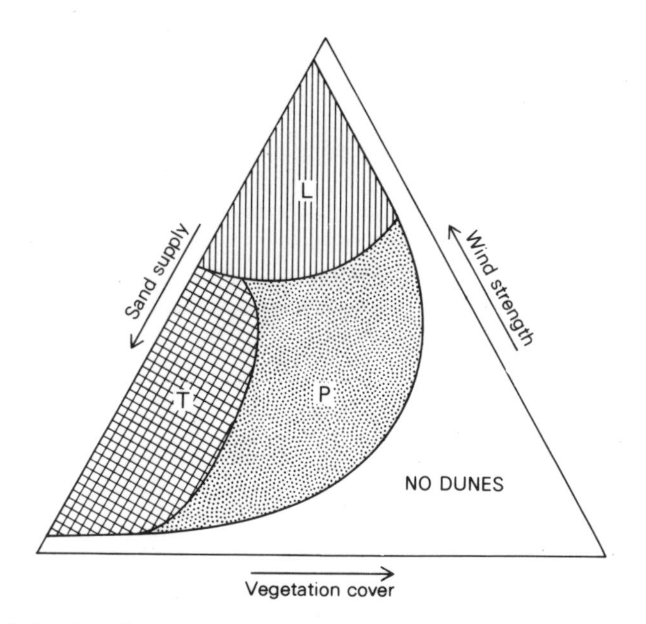

74. Relationships between dune form and the factors of sand supply, wind strength and vegetation in the Great Plains, United States. L = longitudinal, T = transverse, and P = parabolic forms. After Hack, 1941.

regions (Melton, 1940; Hack, 1941), and many of the forms are common to coastal dunes in a range of climates (Bird, 1968). The interaction of sand supply, vegetation cover and wind strength in determining the forms of these dunes is summarised in Fig. 74.

Isolated mounds. These are trapped by shrubs or low trees in open plains with limited sand. They have been reported up to 3 m high and may be prolonged several metres downwind by sand shadows. *Nebka* and *rebdou* are used of smaller and larger forms in the Sahara, equating with *coppice dune* in the United States. The upwind edge of the mound is sandblasted and shifted downwind as the plant dies back, and eventually the plant dies and the mound is blown away.

Transverse ridges. These form with a combination of moderate winds and a renewable sand source, as on playas or river flats, generally as *source-bordering* dunes at the downwind margin.

Parabolic dunes. These arise where anchoring vegetation is locally defeated because of burial through excessive sand supply or corrasion by strong winds, and parabolics of accumulation and of deflation have been distinguished (Hack, 1941). From the initial *blow-out* a convex nose of sand advances downwind, leaving paired wings on which vegetation may resume its hold. The nose is commonly steeper than a normal slip face because it is supported by vegetation.

Depending on wind strength there may be progressive elongation through *hairpin dunes* to *wind-rift dunes* in which the nose is breached, leaving paired longitudinal ridges, but excavation of a parabolic dune may be limited by the watertable. If vegetation is overwhelmed over a broad front, *nested parabolics* may form, with linked wings.

Longitudinal ridges. Hack (1941) associated the development of longitudinal ridges in partly vegetated areas with meagre sand supply — possibly in lee of sand-trapping forms — and strong unidirectional winds. He figures a downwind succession from parabolics of deflation through transverse ridges and parabolics of accumulation to longitudinal dunes.

WIND-BORNE DUST AND ASSOCIATED LANDFORMS

Aeolian dust consists of particles below 0.05 mm. Clay particles are less subject to deflation than silt, because of greater cohesion, whence the predominance of silt in wind-borne dust; however, clay is blown as fine aggregates from salt-affected surfaces as discussed in Chapter VIII. The important sources of dust are floodouts and playas.

The bulk of deflated material is deposited close to the source area. Yaalon and Ginzbourg (1966) describe 'deposits some centimetres thick over an area of 1000 sq miles' in a Negev storm. The finer material remains in suspension and is exported from the desert, to be deposited near the downwind borders through the agency of dewfall and other forms of precipitation, aided by trapping in vegetation. Thickness and particle size diminish with distance from source (Yaalon and Ginzbourg, 1966; Butler and Hutton, 1956).

Loess

This is a yellowish-buff unstratified calcareous silt, well-sorted and with a modal grain size in the range 0.02–0.5 mm, and with a pronounced vertical structure. Sheets of wind-borne clay known as *parna* are widespread in the Riverina of southeastern Australia (Butler, 1956).

The importance of desertic or 'warm' loess relative to 'cold' loess derived from glacial outwash under dry periglacial conditions remains in doubt (Bryan, 1945). Loess is more typical of marginally glaciated mountain-and-basin deserts than of zonal deserts, and it has been claimed that glacial grinding is more effective in producing such fine quartz particles than the saltation-impact of wind-blown grains (Smalley and Vita-Finzi, 1968). But alluvium rather than windblown sand is the main source of desert dust, and comparison of loess with sediment from desert dust storms has confirmed that this source is capable of producing the requisite grades (Swineford and Frye, 1945). In many areas loess lacks glacial sources and *must* be of desertic origin, for instance the loess of the Negev has been mainly traced to the alluvia of the Wadi Arish in northern Sinai on the evidence of mineral content and contained foraminifera (Ginzbourg and Yaalon, 1963).

Loess landscapes. Loess may be deposited as a flat sheet or as a covering on slopes. In the Negev of southern Israel it forms hill-slope mantles a metre or two thick, thinning towards the crest and thicker on leeward slopes. Valley fills commonly exceed 3 m thick and have been washed from the slopes as shown by included stones.

The loess mantles of the Negev and the loess-covered foothills of the Asiatic deserts form smoothly rounded hilly terrain. Shallow valleys are often unchannelled but there can be intense gullying of anthropogenic origin. Such gullies are sheer-walled due to vertical partings in the uniformly massive loess.

X

DESERT LANDFORMS AS EVIDENCE OF CLIMATIC CHANGE

Many desert geomorphic processes, particularly those dependent on rainfall, operate so sporadically that an occasional investigator might suppose them at a standstill and the resultant landforms fossil. Conversely, where landscape evolution is so slow, features produced under an earlier climate can survive little-altered and be attributed erroneously to the existing regime. Accordingly the recognition of palaeoforms in deserts assumes importance, first in elucidating past conditions and second, by elimination, in establishing the morphogenic range of the present.

This chapter deals mainly with landforms expressive of Quaternary fluctuations between heightened and lessened effective aridity, particularly in the warm deserts, and more briefly with survivals from Tertiary climates. The quality of the record varies. The most compelling evidence is that which suggests a fundamental change of process, for instance the widespread invasion of fluvial landscapes by wind-blown sand. Alternations in the mode of operation of a geomorphic agent, between incision and alluviation in a river channel for example, promise a sensitive record of fluctuations although the link with climate may be complex and problematic. Changes in the frequency or range of processes that are in any case episodic, for example mass movements on hillslopes, provide less certain evidence. Least satisfactory are indications from processes that are slow-acting, cumulative in effect or only related with climate in a general way, such as the weathering or varnishing of rock surfaces (Hunt, 1954, 1961).

Landform evidence of climatic change is commonly most marked near the limits of dominance of a particular process. Cut-and-fill sequences are best-developed in piedmont sectors, between uplands of predominant fluvial erosion and lowlands of deposition, and indications of mobilisation or stabilisation of aeolian sands are clearest on the edges of dunefields. Hence desert margins tend to be critical areas, with evidence of desert expansion extending beyond and records of wetter phases within.

Midlatitude mountain-and-basin deserts experienced marked

climatic fluctuations in the Quaternary, involving changes in temperature and precipitation. These factors are difficult to distinguish, and *wetter* and *drier* are here used of contrasts in the effectiveness of rainfall irrespective of cause. Climatic gradients were compressed and their geomorphic consequences enhanced with strong bounding relief, and high-altitude processes such as frost-shattering and solifluction commonly contributed. The most prominent effects of fluctuating aridity were alternating river action in the piedmonts, between erosion and deposition, planation and incision, and the filling and drying of basin lakes.

In the zonal shield and platform deserts comparable effects are limited to upland borders near the midlatitude margins. More widespread evidence is to be found in the lowlands, where extensive dunefields originated near the margins of the expanded arid zones during drier periods and lakes formed in wetter periods, on a large scale where rivers entered from humid regions.

EVIDENCE FROM SAND DUNES

Extension of continental dunes is generally accepted as a sign of an arid climate with dominant wind action, but non-climatic factors such as sand supply are also involved. Dunes can form with only moderate aridity where sand is locally plentiful, as in the lee of a sandy channel, and palaeoclimatic reconstructions should rest on regional evidence. Absence of vegetation is not a prerequisite for dune growth, and parabolic and wind-rift dunes such as predominate in the dunefields of central North America formed with partial anchorage by vegetation. Type of dune is therefore relevant to an assessment of the formative climate.

Inactive dune systems beyond the present climatic limit of wind-moved sands indicate a change to wetter conditions following a drier climate in the past. Stability may be evident from a close cover of vegetation, soil profiles accordant with the present surfaces, modification of dune form or pattern by non-aeolian agencies, particularly by running water, and discordance of trend with the present effective winds. Renewed dune movement, suggesting a further swing towards aridity, may be indicated by wind corrasion of older dune forms and the superposition of younger, possibly discordant dune patterns, or by invasion of non-aeolian surfaces. A study by Smith (1967) in the Mojave Desert of California illustrates the application of many of these criteria in concluding that the dune systems became stabilised and were more recently re-activated on a smaller scale.

Fossil dune systems

Stabilised longitudinal dunes occur along the margins of most of the zonal deserts. The largest tract extends over 5 degrees of latitude along the southern margin of the Sahara as shown in Fig. 64 (Grove and Warren, 1968). Fossil dunes are also found on the equatorward margin of the Kalahari between latitudes 16° and 20°S (Flint and Bond, 1968; Grove, 1969). The Australian arid zone is ringed on three sides by fossil dunes, between 17° and 20°S in the north and between 32° and 36°S in the southeast, but reduced to a width of 150 km by steeper climatic gradients on the eastern margin. In all these areas the ridges are completely vegetated, with some tree cover, and soils are developed in the sands. The fossil status is beyond question although there may be disagreement about the extent of climatic change indicated. The Sahelian dunes have been claimed as indicating a former extension of 'Saharan' conditions southwards by 500 km, into areas now receiving more than 500 mm annual rainfall; but those who support a wind-rift origin for the ridges and hence consider they originated with a partial cover of vegetation would presumably require a smaller change of climate.

Partially stabilised longitudinal dunes occur *within* the moderate zonal deserts of the Kalahari and Australia, where loose sand is confined to the ridge crests and the flanks are mainly vegetated. In Australia there is an increase in the relative amount of moving dune sand towards the drier heart of the southern Simpson Desert, suggesting that the amelioration of climate represented by the fossilisation of dunes on the desert margin also extended, though weakening, into the interior. The factor of sand supply is important locally, however, and completely mobile dunes occur locally along playa margins and river channels, alongside partially stabilised sand ridges.

Eluviated dunes

Modification of a stable dune by non-aeolian processes has been termed the *eluvial* phase of the dune cycle (Smith, 1963). Ridges in the wetter parts of the fossil dune systems mentioned above have been rounded and lowered, their flanks reduced by wash, and the swales filled in (Fig. 75). Slopes on the ancient linear dunes on the southern margin of the Sahara now rarely exceed 5°. The rolling dunes of the northern Kalahari, estimated to have attained 90 m originally, no longer exceed 15 m (Grove, 1969), and low parallel sand rises about 3 m high further east in Rhodesia have also been interpreted as the stumps of linear dunes (Flint and Bond, 1968).

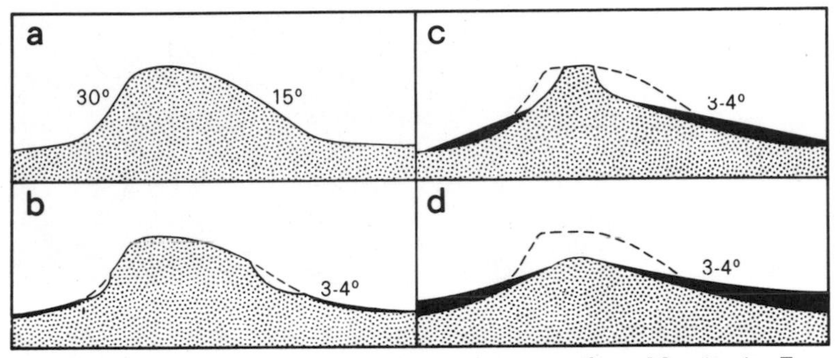

75. Stages in the 'eluviation' of a dune ridge in southern Mauritania. From Daveau, 1965.

Gullying may occur in the initial stages of modification. It is active on the flanks of well-preserved sand ridges in the Mauritanian Sahel south of 17° 30', with 200–250 mm rainfall (Pl. 49), but less important on subdued ridges of the wetter zone to the south (Daveau, 1965). The gullies extend each wet season by piping and the collapse of sand faces through sapping, but their survival depends on the grassing of new sand surfaces before the onset of the dry season, otherwise they become choked by drifting sand. Signs of earlier gullying extend about 100 km further north, to the present limit of mobile dunes, but it seems that this became inactive with a swing back to aridity following wetter conditions in the late Pleistocene and early Holocene.

In southwest Queensland the upwind flanks of formerly stable dunes have been gullied where wash or man-induced sand drift have exposed impervious clay-cemented cores.

Extension of drainage systems into dunefields

Where dune extension has obliterated a river system its re-establishment in a subsequent wetter phase may be prevented by

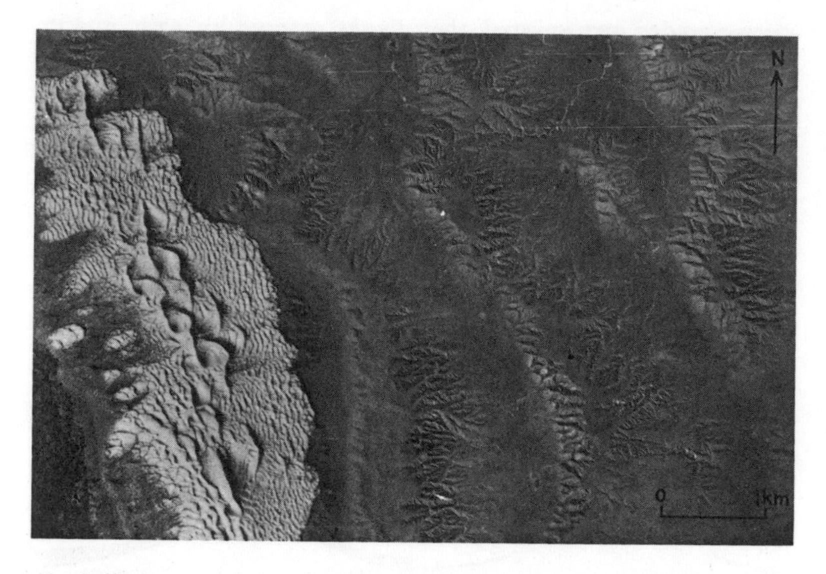

49. Gullied steeper E faces of fossil longitudinal dunes, south Mauritania, an area now receiving 200–250 mm rainfall. An area of active transverse dunes in bottom L. Photo Institut Geographique National, Paris.

the cover of impervious sands. In many cases, however, rivers truncated by dunes have again extended across them, indicating a reversal of the climatic balance. In southern Mauritania the lower Senegal River has breached a dune barrier and one of its upper tributaries, the Karakoro, traverses the large Aouker dunefield (Daveau, 1965). There has been considerable dissection of dunes by rivers south of 16°N in this region, with redeposition of sand in the swales. Connected drainage systems distinguish the older fossil dunes of the Low Qoz of the Sudan from those of the High Qoz to the north (Warren, 1970), and in the northwestern Kalahari older dune ridges have been dissected by tributaries of the Zambesi and Okavango rivers (Flint and Bond, 1968; Grove, 1969).

An example of a changing balance between wind transport of sand and that by running water occurs on granite plains of the Murchison Plateau in the west of the shield desert of Western Australia, where sheetfloods have moulded accumulations of aeolian sands into patterns of sandy banks and alluvial flats known as *wanderrie country* (Mabbutt, 1963b). The name derives from the perennial wanderrie grasses which occur on the banks. The wanderrie banks are up to 1.5 m high, are aligned along the contour for distances of a kilometre or so, and may extend for a few hundred metres down the slope, occupying sectors of steeper fall within slightly stepped profiles. Their outlines are adjusted to the relationship between the direction of the slope, and hence of sheetflooding, and former sand-moving WSW winds. Where these were directly opposed or coincident the banks are arranged in regular series transverse to the slope, whereas they are linked in longitudinal or oblique patterns where water flowed across or oblique to the dominant wind.

The setting is one in which an organised coastward drainage remained undefeated by limited sand drift under aridity, whereas similar physiographic situations further inland developed into unbroken sandplain. The banks are now stable as indicated by a widespread siliceous hardpan, which is deeper beneath the banks and shallower under the flats, reflecting differences of soil texture and infiltration rates. This suggests a trend to wetter conditions since the banks formed.

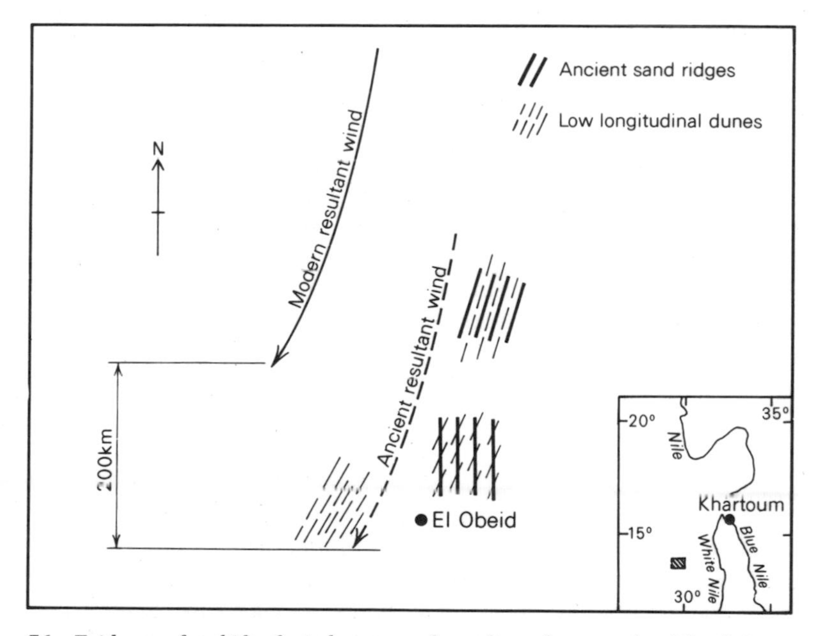

76. Evidence of a shift of wind systems from discordant trends of fossil dunes and modern resultant winds in the Sudan. From Warren, 1970.

Discordance between dune trend and present winds

Where dune trend differs from the effective resultant wind it helps confirm the dunes as fossil and also serves in the reconstruction of a former wind regime, with promise of an understanding of the mechanism of climatic change.

In the Sudan, Warren (1970) has postulated a southward shift of the wind belts by about 200 km to explain the discordance between the trends of fixed sand ridges and the resultant of the present sand-moving winter winds (Fig. 76). Such a shift could explain the occurrence of the stabilised dunes of the High Qoz south of the presently active dunes. An earlier southward displacement of the pressure and wind belts by 450 km may be indicated by the discordance between older longitudinal and lee dunes and the present wind regime, perhaps to be correlated with the formation of the older relict dunes of the Low Qoz.

On the southeast margin of the Australian arid zone the trend of fossil longitudinal dunes and the orientation of lunettes indicate that the dominant summer, or dry season winds of the late Pleistocene were from the west (Fig. 60; Bowler, 1975). The corresponding winds today are southwesterly, and westerlies now dominate in winter. This is held to be consistent with a summer anticyclonic system positioned at least 5° further north than at present. A comparable directional discrepancy of about 60° has been noted between SSE-trending longitudinal and parabolic dunes of late Pleistocene age in the Eyre Peninsula, South Australia and the present southwesterly winds (Bourne *et al.*, 1974). At the centre of the Australian dune swirl, however, there are discrepancies of up to 15° between longitudinal dune trends and weighted resultants of present-day winds which are locally consistent but not regionally in the same sense, as shown in Fig. 77, and are apparently not explicable by mere latitudinal displacement of the wind belts (Brookfield, 1970b).

Discordant dune systems

The evidence is clearer where successive dune systems are discordant in form or trend. This occurs in mid-latitude deserts where arid phases involved only partial destruction of vegetation and incomplete mobilisation of older dunes, leaving superposed dune systems of different ages. Successive systems commonly differ in orientation in these latitudes, where formative wind regimes varied

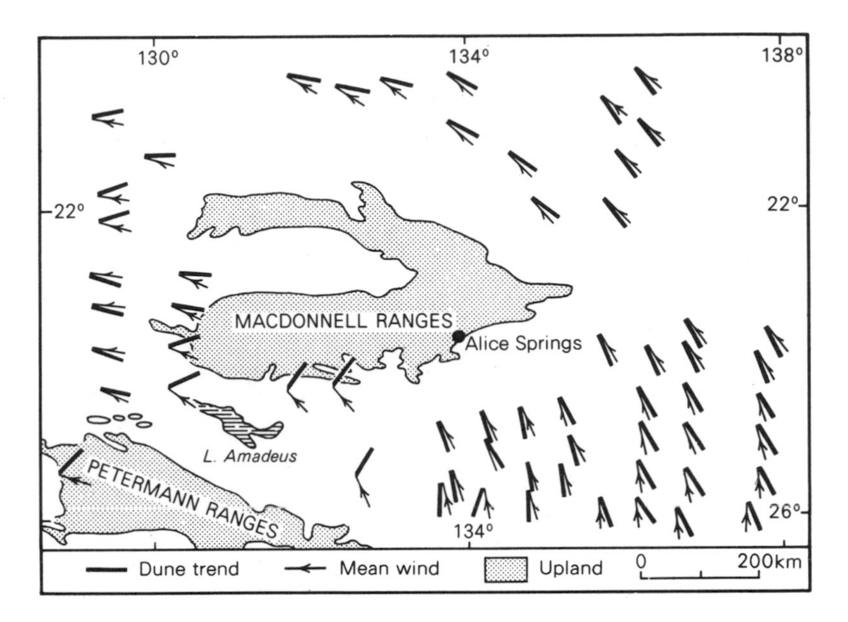

77. Relationship between mean weighted wind direction and dune trend in central Australia. From Brookfield, 1970b.

with the advance and recession of ice sheets. This situation is exemplified in the stable dunefields of the semiarid High Plains of North America, where interaction by vegetation during dune movement has resulted in a predominance of parabolic and related forms.

In Nebraska, Smith (1965) identified three dune series (Fig. 78). The oldest and largest are sinuous compound transverse ridges up to 60 m high and spaced at 1500–2000 m. These were formed by strong and persistent northerly winds with abundant moving sand; associated E-W longitudinal dunes may be due to lighter westerly winds acting on better-sorted sands (Warren, 1967). The second series is mainly a system of parallel low ridges 10–15 m high and with a spacing of 100–200 m; they are thought to have formed as wind-rift dunes with partial vegetation cover under the influence of W to NW winds such as now prevail in the west of the area. They generally occur east of the older ridges but overlap and intersect them further west. The youngest dunes comprise widespread small blowouts and rare U-dunes eroded into older forms by NNW winds, probably under conditions resembling long droughts in the region today. Relations of the dune systems to loess sheets indicate that the oldest and second series of dunes are of early and late Wisconsin age. The strong N winds of the oldest phase may have been associated with a nearby Wisconsin ice lobe, whereas the strong NW winds of the second phase correspond to those of the present winters.

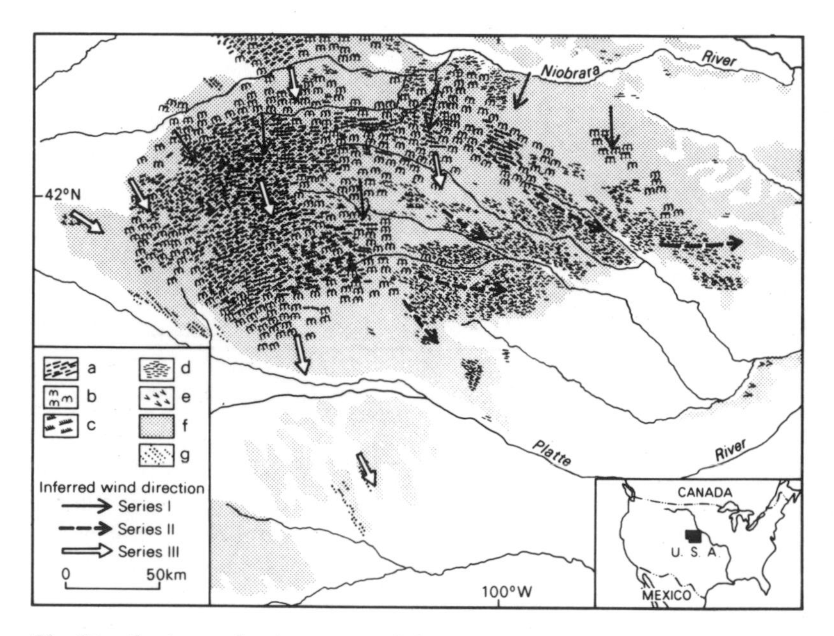

78. Distribution and orientations of dune types in Nebraska, with inferred wind directions: a. large-scale massive dunes, and b. eluviated dunes of Series I; c. dunes of Series II superposed on Series I; d. ridges of Series II with blowouts; e. subdued linear swell-and-swale topography of uncertain age; f. U-dunes of Series III; g. blowouts of Series III and undifferentiated dunes. After Smith, 1965.

Palaeosols in dunes

Alternating stability and instability of dunes may be evidenced by palaeosols, which have proved particularly informative in marginal deserts where sand movements were short-lived and the forms of older dunes survived little-altered. In the Mallee of south-east Australia, a semiarid region of parallel ridges stabilised beneath low woodland, Churchward (1963a, b) has recognised several cycles of sand movement and stabilisation by this means.

Four ground layers are indicated by soil horizons in the calcareous clayey dune sands (Fig. 79). Soil formation, involving obliteration of depositional fabrics and translocation of lime and clay to lower horizons, represents moister episodes of surface stability and leaching. During intervening drier periods the dunes became unstable; in swales and on lower dune flanks an accession of loessial clay or *parna* mixed with dune sand buried the soils, whereas on upwind flanks they were stripped to the lime pan and the winnowed soil and saltated sand formed a new crest and slope to leeward. The amplitudes of the moister and drier phases diminished with time as indicated by depths of leaching and truncation of the layers. The youngest soil, dated at 15,000 BP, followed the final phase of dune-building in this area.

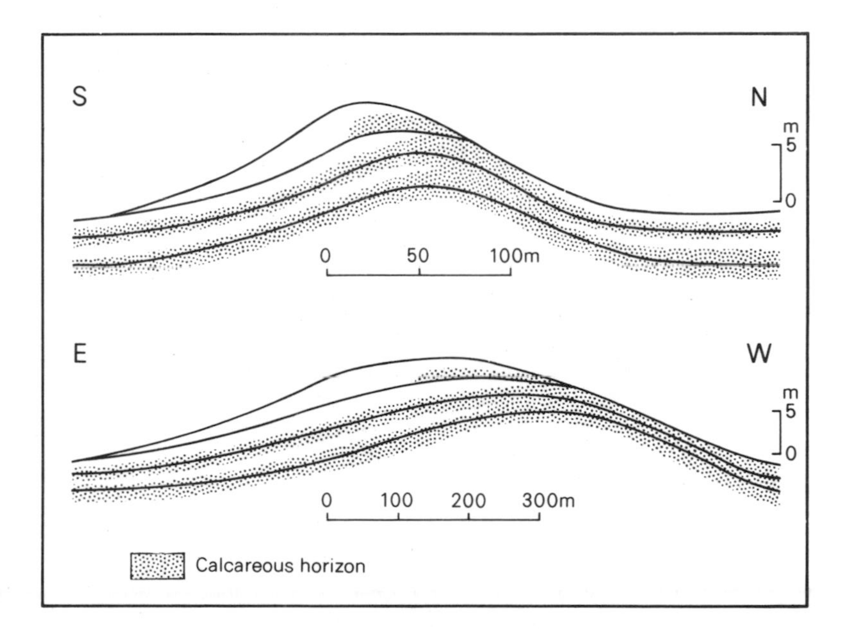

79. Generalised cross-section (above) and longitudinal section (below) of a dune ridge in the Victorian Mallee showing disposition of soil layers. From Churchward, 1963a.

EVIDENCE FROM DESERT LAKES

Growth of lakes is an important consequence of increased flooding in the interior drainage systems of deserts and hence an indication of increased effective humidity. Determination of the dimensions of former lakes also allows reconstruction of water budgets and with this a consideration of related hydrologic regimes as a function of a former climate.

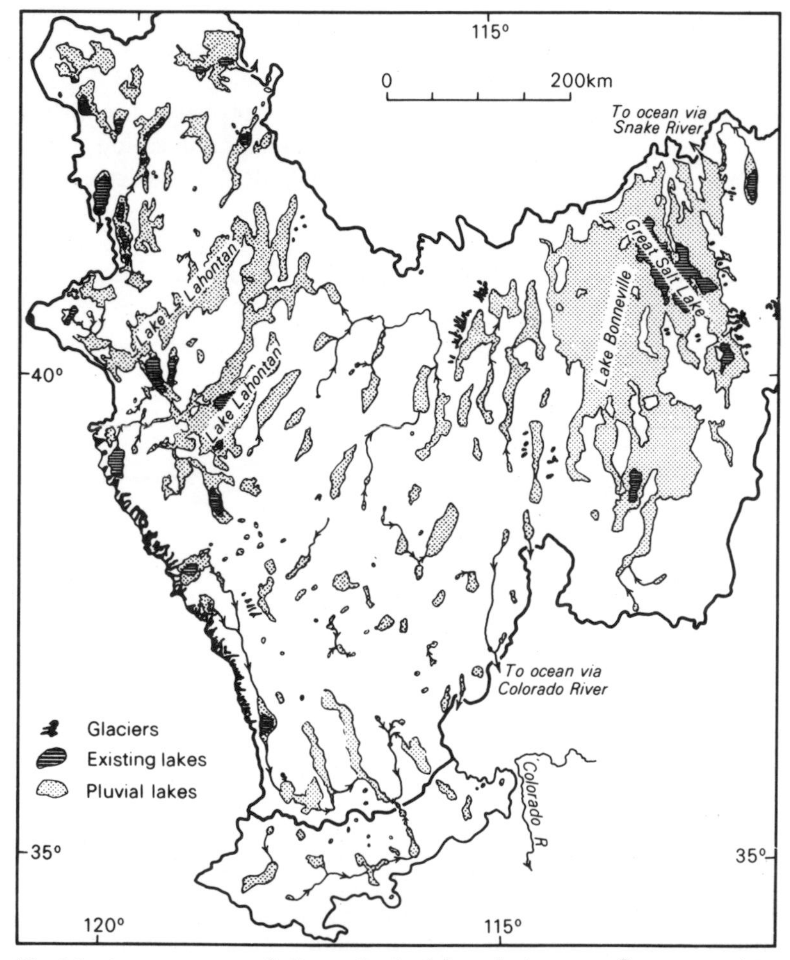

80. Maximum extent of Great Basin lakes during post-Sangamon time. Arrows indicate overflow connections. The heavy line shows the present drainage divide of Great Basin. From Morrison, 1965.

Lakes in intermont basins

In mountain-and-basin deserts many steeply enclosed basins which now contain playas were occupied by permanent deep lakes in the past. The evidence consists of shoreline forms such as wave-cut benches and gravel beach ridges, spits and fan deltas high above present limits of submergence, and thick bodies of non-saline clay and silt beneath the contemporary playa sediments.

These features are widespread in the Basin and Range province of the southwestern United States, as summarised by Morrison (1965). This endoreic region was broken into about 150 separate topographic basins by early Pleistocene block-faulting, most of which were occupied by lakes (Fig. 80). The largest were Lake Bonneville (51,700 km^2) in the basin of the present Great Salt Lake (2000–6500 km^2) in the northeast, and in the northwest the branching Lake Lahontan (22,400 km^2). Many basins were linked by overflow channels and some marginal basins became connected with the ocean.

The shoreline features have generally proved to be composite, with overlapping forms of different ages as shown by degree of weathering and intervening palaeosols. Older and younger lakes have commonly been distinguished on this basis, and the better-preserved suites of younger beaches also indicate major fluctuations of the lakes as well as recessional stages. Many of the lakes were very deep; for example the earlier beaches of the post-Sangamon Lake Bonneville attain 305 m above the Great Salt Lake and the later ones reach 332 m, at which level the lake overflowed northwards. Beach levels may show local warping with isostatic recovery following removal of the water load.

The associated floor deposits provide a supplemental record of lake regime in which clay and silt indicate high-water stages with stream inflow, and intercalated evaporites or palaeosols mark drying phases. At Searles Lake in southeastern California 265 m of deposits have been proved beneath an overburden of playa sediments (Fig. 81), with two lacustrine mud layers poor in salines limited by evaporite-rich horizons (Flint and Gale, 1958; Smith, 1968).

Tectonic desert basins are commonly also closed groundwater basins, and climatically determined fluctuations in recharge may in these settings be accompanied by oscillations of the watertable, although tectonic and other factors may have contributed. Former higher seepage levels may be preserved in tufa terraces and perched

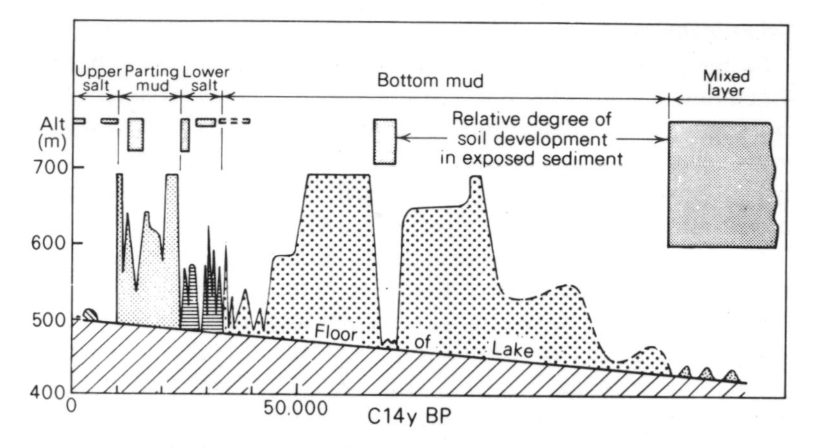

81. Fluctuations of Pleistocene Searles Lake, California, inferred from deep-lake sediments. Time calibration through the latest 45,000 y is based on C14 dates; earlier events are extrapolated. From Flint, 1971, based on Smith, 1968.

spring mounds such as the 'algal pinnacles' of Mono Lake, California. Along the western fault margin of Lake Eyre are flat-topped circular spring mounds up to 30 m in diameter of sand and mud cemented and preserved by travertine and gypsum. They attain heights of 10 m and mark the natural artesian outlets of a watertable some metres above the present. On moist playa floors, former higher phreatic levels may be reflected in evaporite crusts above the present range of flooding; for instance many of the more massive salt crusts in the Death Valley playas are no longer flooded and their formation is correlated with moister conditions in the Holocene (Hunt *et al*., 1966).

Giant desiccation polygons ranging in diameter from 15 to 75 m have been described from several clay playas in the southwestern United States (Neal *et al*., 1968). Near the margins they may grade into parallel stripes several hundreds of metres long, trending normal to the shore. The fractures may be brought into relief where colonising vegetation has trapped sand into ridges. The polygons are associated with underlying thick lacustrine beds rich in expansive clay minerals, which have become desiccated in depth due to change to a drier climate or to artificial lowering of the watertable. Where shrinkage penetrates to the capillary fringe of a deep watertable, rupture occurs and the fracture system then extends to the surface.

Large lakes in tropical sectors of zonal deserts

In the shield and platform deserts, large lakes formed in shallow structural depressions in past wetter phases. Lake Chad is an example and like many others lies towards the equatorward margin of the zonal desert belt, subject to inflow from large exogenic rivers. It occupies a roughly circular basin extending over 15 degrees of latitude from the Congo watershed into the central Sahara (Fig. 82), which subsided and was extensively filled with fluvio-lacustrine

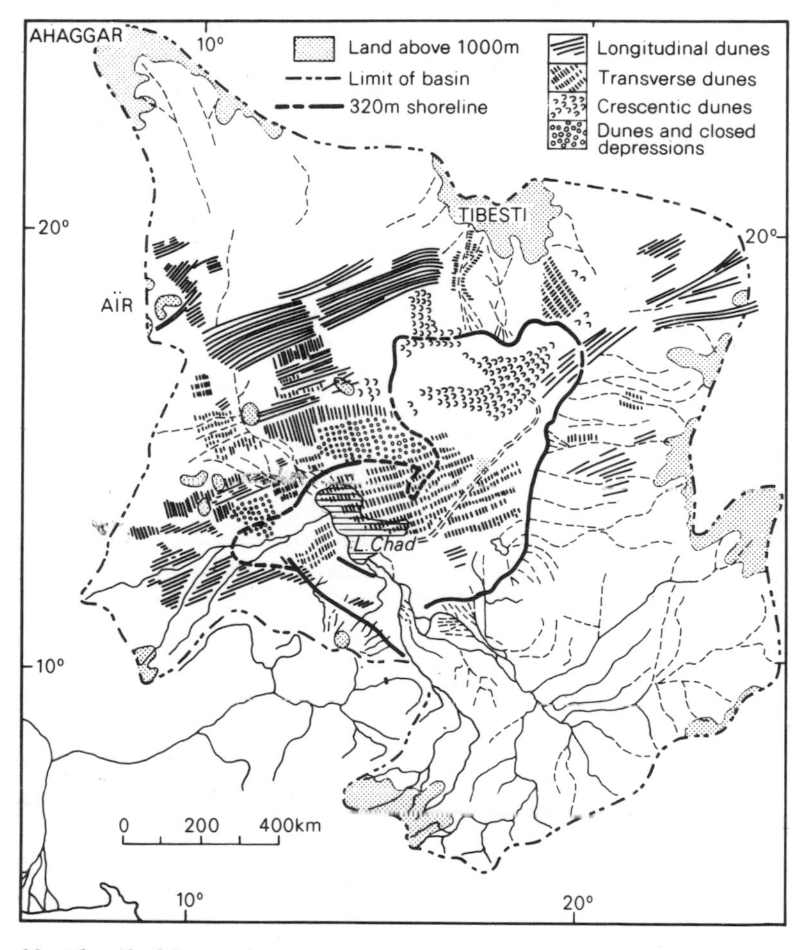

82. *The Chad basin, showing dune systems and the shoreline of 'Megachad' at about 320 m. From Grove and Warren, 1968.*

sediments in the Tertiary and earlier Pleistocene. The flatness of the plains is reflected in the shallowness (3–7 m depth) of the lake over its present area of about 20,000 km^2, and in enormous variations of the lake limits with changes in volume, for example between 8000 and 24,000 km^2 during the last century.

Landforms in the Chad basin indicate an alternation of lake transgressions with arid phases of dune building as set out in Table X (Grove and Warren, 1968; Pias, 1970). With lake recession, dunes extended across the dried-out floor, and lacustrine silts were locally modelled into yardangs, as in the Bourkou lowlands. With lake recovery the dunefields were drowned and silts and clays were deposited between the dunes (Servant, 1970). The most prominent associated shore forms are flat-crested sand beach ridges along the southwestern (lee) margin of the lake, particularly that of the Megachad phase at 320 m, when the lake overflowed southwards via the Benue River to the Atlantic, and the less continuous ridge of a younger shoreline at 287–90 m.

At the 320 m stage, where it stood at 7000 and 5400 BP, the lake had an area of 350,000 km^2 and the basin rainfall may have been between two and three times the present 320 mm, for it may be assumed that temperature change was unimportant in these latitudes. In periods of dune extension the margin of the Sahara was shifted equatorward of its present line by about 5 degrees of latitude and annual precipitation probably fell to below 50 mm.

Desert lakes of this type are characterised by overlapping of lacustrine and dune forms, by ill-defined, generally sandy shore features indicative of enormous fluctuations of lake area, and by modest thicknesses of lake deposits due to shallow depths.

Lunette lakes on midlatitude margins of zonal deserts

The poleward margins of the zonal deserts are stormy with steep rainfall gradients and form zones of potential interaction between fluvial and aeolian processes that have proved sensitive to climatic changes. This is exemplified by the belt of playas and lunettes along the south margin of the Australian arid zone between 30° and 34°S (Fig. 60), an area presently receiving between 250 and 500 mm rainfall (Bowler, 1973). Forms resulting from earlier high-lake stages in effectively wetter periods include smooth arcuate wave-trimmed downwind (E) shores, beaches of sand and gravel with derived foredunes of quartz sand, and cliffed dune ridges on up-

TABLE X **Sequence of lake stages and aeolian episodes in the Lake Chad Basin.***

Stage	Lake level (m)	Lake area ($km^2 \times 10^3$)	Limit of dunes	Dune formations	Age BP
Present	282	24 (24–8)	16°N	Barkhans	
Fourth transgression	287–90	180		Drowned archipelago NE of lake	1800–3200
Arid regression			12°N (3rd erg)	Barkhans and intersecting dunes with enclosed depressions	
Third transgression	320	350 (Megachad)			5400–12,000
Arid regression			12°N (2nd erg)	Transverse dunes, NNW-SSE	
Second transgression	350–400				21,350–30,000
Arid regression			10°N (1st erg)	*Alâb* ridges, NE-SW	
First transgression	400				

* After Pias, 1970.

wind margins. Low salinities associated with frequent flooding and overflow are indicated by freshwater mollusca. The onset of drier conditions and lowering of lake levels to an intermittent moist-playa regime is marked by the entry of brackish-water mollusca and by the burial of lake beaches beneath massive source-bordering clay dunes or lunettes (Chapter VIII) derived from lake-bed deposits exposed and pelletised under increased salinity. With further increase in effective aridity the lakes and their lunette barriers became relict between 17,000 and 15,000 BP. The longitudinal dunes upwind have only locally transgressed on the dry lake floors and are there partially cliffed, confirming that the drying of the lakes and the cessation of dune movement were roughly contemporary, and both indicating the attainment of more arid conditions than have prevailed since. The lunettes and the longitudinal and parabolic dunes of the area, as discussed above, indicate a marked zonal (W-E) air circulation during this late Pleistocene marked aridity (Bowler, 1975).

Bowler (1971) has recorded three such cycles from the Willandra lakes in the northwestern Riverina, on a former distributary of the Lachlan River (Fig. 83). A reduction in the amplitude of successive phases is indicated by the nesting of younger forms within older lake basins, as of the Chibnalwood Lakes within Lake Arumpo, and by a progressive retraction up-river of the extent of younger lake features. The sequence is represented stratigraphically in the lunettes and their palaeosols (Fig. 83).

A comparable record in a sandier setting is provided by numerous isolated pans on flat watersheds in the southern Kalahari, between latitudes 23° and 29°S, an area now receiving between 250 and 400 mm rainfall (Lancaster, 1974). These commonly have two dunes on their downwind SSW margins, an outer parabolic dune of quartz sand and an inner dune of finer calcareous clayey sand. The subcircular outlines of the pans suggest they formed as deflation hollows during a drier period, when the parabolic dune formed at a distance of 1.5 km downwind, which was followed by a wetter phase during which calcareous clays accumulated in the temporary lake from fine materials washed or blown in. On drying once again, the lake sediments provided much of the material for the inner dune which partly resembles the Australian lunettes but which is less rich in clay. Consequently it has a profile resembling a sand foredune, being somewhat steeper away from the pan.

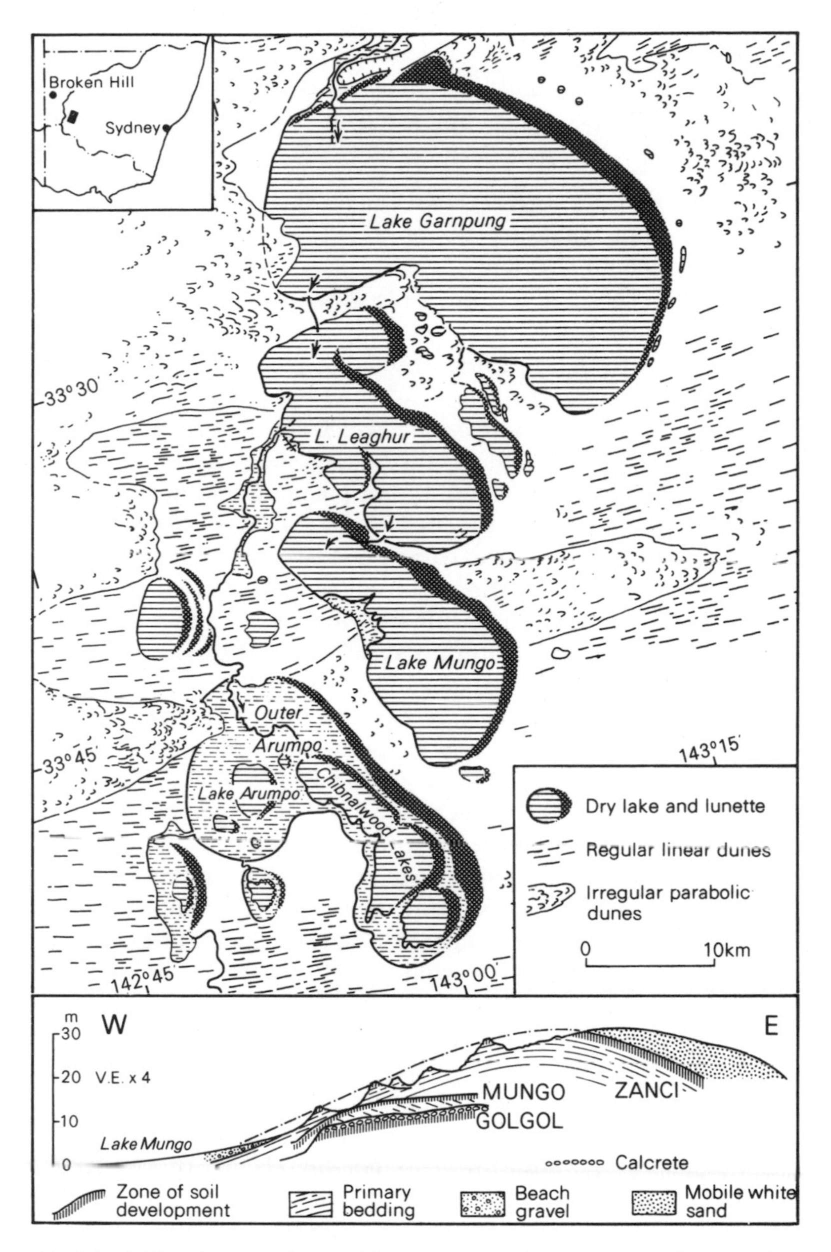

83. The Willandra dry lakes and lunettes in northwestern New South Wales, with adjoining dune systems. Connecting overflow channels shown by arrows. Below, section through the Lake Mungo lunette showing dated ground surfaces. From Bowler, 1971.

Palaeoclimatic reconstructions from closed desert lakes

These involve reconstructions of prior water budgets by estimating former parameters in the equilibrium equation

$R(A_c) + P(A_l) = E(A_l)$ (after Langbein, 1961)

where P = precipitation
R = depth of runoff
E = evaporation from a free water surface
A_l = lake area
A_c = area of lake catchment

Example of an intermont basin lake. Basin floors in the Basin and Range province of the United States now generally receive less than 100 mm annual rainfall, but precipitation increases with altitude on the bordering ranges and may average 250 mm for a catchment, yielding between 10 and 20 mm annual runoff. In the north of the area evaporation and inflow are balanced and small lakes may survive at river terminals, but in the warmer south the excess of potential evaporation is such that only playas now exist. Calculations of the water balance of the former lakes must consider not only that precipitation may have been greater, but also that temperatures were lower and evaporation accordingly reduced and runoff increased. Mollusca in the Searles Lake muds, for example, indicate a lake temperature 5°C below the present mean, in accord with the evidence of lowered Pleistocene snowlines.

Differences of interpretation have arisen largely in the allocation of precipitation and temperature controls. Some authors have concluded that the lakes indicate lowered temperatures but increased precipitation. Snyder and Langbein (1962), for example, in a study of a Pleistocene lake in Spring Valley, Nevada, allocated these factors on a probability basis and considered that the lake could have been maintained with an increase of the present mean basin precipitation of 30 cm to 51 cm in association with a lowering of mean annual temperature by 8°C, which would have reduced mean annual evaporation from its present 112 cm to 80 cm and have increased runoff by one-third. The postulated change is tantamount to an equatorward displacement of the poleward margin of the desert by 8 degrees of latitude. On the other hand, Galloway (1970) has estimated that precipitation at the last full-glacial high stand of closed lakes in the southwestern United States was generally between 80 and 90 per cent of the present.

Shallow lake basins in zonal deserts. Climatic reconstructions are rendered difficult by characteristically large ratios of catchment area to lake area, which introduce great uncertainty in the estimation of former inflows, and by a paucity of lake sediments due to the shallowness of the basins and consequent fluctuations in extent of the lake, which leaves open the question of former lake regimes. These problems are apparent in a palaeoclimatic study of Pleistocene lunette lakes in northwestern New South Wales (Dury, 1973). The former extent of the lakes was determined by mapping lunettes and gypsum-cemented calcareous shore terraces. On the basis that mean temperatures were lowered by 8°–10°C with a virtual halving of present evaporation rates, Dury concluded that a rainfall at least 50 per cent greater than the present 200–250 mm would have been required to sustain the lakes at full stage. These results are in apparent conflict with postulated evidence of drier conditions in the late Pleistocene in southeast Australia as indicated from the Willandra lakes (Bowler, 1971) and from Lake George in the southeastern uplands (Galloway, 1965).

FLUVIAL LANDFORMS AS EVIDENCE OF CLIMATIC CHANGE

Changes in effective precipitation are reflected in the frequency and range of streamfloods in desert rivers and in load-discharge ratios. In the sensitive piedmont zones the response may involve marked alternations between planation and incision. In the lowlands there are corresponding changes in the degree of connectivity of drainage systems, recorded in the alluvial stratigraphy of periodically isolated terminal basins.

Piedmont terraces and staged pediments

Staged piedmont surfaces which converge downslope with distance from the uplands (Pl. 50) have been attributed to climatically-determined fluctuations of river regime, particularly in tectonically stable regions of interior drainage. Among the best-known are the nested mantled pediments (*glacis emboîtés*) in and south of the Saharan Atlas from Morocco to Tunisia, where up to four surfaces of Pleistocene age are distinguished. A suite of three piedmont terraces in strike vales of the western Macdonnell Ranges in central Australia has been attributed to climatic pulsations which have

50. Dissected mantled pediment, Big Horn Mountains, Wyoming. Photo R. P. Sharp.

interrupted general downcutting in the Finke drainage system caused by subsidence to the south (Mabbutt, 1966b). The high terrace mantle is partly iron-cemented, that of the middle terrace has well-organised leached red earth soils, and the low terrace gravels are locally calcreted, suggesting that the climatic fluctuations were imposed on a general desiccation, and in places the dissected terraces have been invaded by dunes.

Such forms indicate episodes of planation and mantling separated by channel incision. All upland channels reaching the piedmont appear to contribute to the former, whereas only the larger channels incise, leaving higher surfaces preserved in inter-stream sectors. Those who stress control over stream action by load-discharge relations regard the planation as indicative of a fully-loaded and incision of an underloaded state. There is no general agreement on the regime changes involved, however, still less on the climatic controls; nor need these be similar under all desert conditions. In extreme desert an increase in the sediment yield of an upland catchment might indicate increased rainfall, whereas in a marginal desert with significant surface control by vegetation a similar result might follow from diminution of rainfall, comparable with the alternation of biostasy and rheostasy in wetter regions (cf. Erhart, 1956).

Where the gravels link upslope with moraines and periglacial mantles, as in Morocco, piedmont planation and mantling have been linked with cooler and wetter episodes and downcutting by the larger channels to sporadic flooding during drier and warmer phases (Raynal, 1961; Mensching and Raynal, 1954; Joly, 1962). In piedmonts unaffected by cold-climate processes the need for preparatory weathering of slope mantles under wetter climates has been stressed, and piedmont mantling has commonly been explained by the onset of hillslope instability with subsequent decrease in rainfall (Coque, 1962).

Dissected alluvial fans

These pose two additional problems of interpretation. First, the effects of earth movements must be separated from those of climatic change, since fans are characteristic of faulted piedmonts. This calls for assessment on a regional scale. Second, *entrenchment* of fan-head washes as part of the normal distribution of alluvium across the fan surface must be distinguished from *incision* of channels leaving the fan, which constitutes a change of regime and shift in the locus of deposition (Chapter V). Those like Denny (1965) who regard erosion as an integral part of the dynamics of a fan in a steady state may see no reason to invoke environmental change to explain it.

Climatic interpretations of fan dynamics vary widely. Lustig (1965) related fan-building in the west of the Great Basin to past wetter conditions, and the current general incision of fan-heads to present aridity and higher sediment-water ratios, leading to scouring by debris flows. Although the fan-head channels are commonly underfit, he considered that the greater tractive force of the flows had shifted the zone of deposition to the middle and lower sectors of the fans. Hunt and Mabey (1966) on the other hand considered that the Death Valley fans were mainly built with the onset of drier conditions and reduction in vegetation cover *after* a wetter phase of preparatory weathering and that dissection was due to earth movements. They claimed that the fans are undergoing little change under the present more arid climate as indicated by the limited extent of active washes and unvarnished gravels.

Williams (1973) attributed aggradation of fans west of the Flinders Ranges, South Australia to cold dry conditions, since the deposits pass headwards into talus breccias interpreted by him as periglacial. Calcification of the fans around 30,000 BP was held to indicate a more temperate climate with rainfall greater than the present 170–300 mm, and subsequent dissection to renewed aridity in the late Pleistocene, with lessened vegetation, reduced

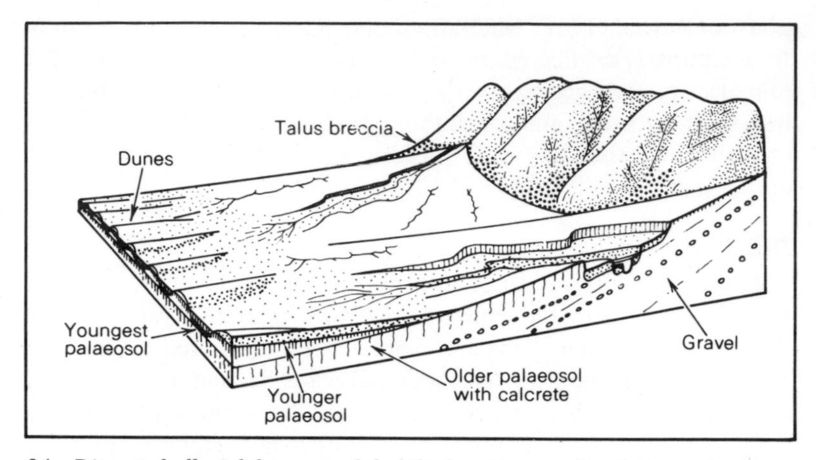

84. Dissected alluvial fan west of the Flinders Ranges, South Australia. From Williams, 1973.

stream discharges and lowered water tables, culminating in dune extension around 16,000 BP (Fig. 84).

Desert river terraces

Alluvial terraces resulting from cycles of cut-and-fill, presumably in response to climatic changes, occur in desert upland valleys and may extend far into the plains along large channels; for example terraces flank the Wadi Saoura in the northwestern Sahara for more than 750 river km from the Atlas Ranges. The Saoura terraces near Beni Abbès reveal two cycles of incision and deposition between the Plio-Pleistocene hamada and the present floodplain (Fig. 85). The sediments of the Upper Pleistocene Saourian cycle are mainly well-sorted sands indicative of a regular fluvial regime (Conrad, 1963). The Saoura appears then to have reached the Touat depression south of the Erg Occidental (Fig. 43). The alluvium in this downstream tract includes much aeolian sand, suggesting that the adjacent sand sea had developed before the Upper Pleistocene. The younger or Guirian terrace, mainly Holocene, has torrent-bedded sediments suggesting sporadic flooding under conditions fluctuating between arid and semiarid.

Arroyo trenching

Many previously unchannelled river flats in the southwest United States are now trenched by discontinuous steep-sided gullies or

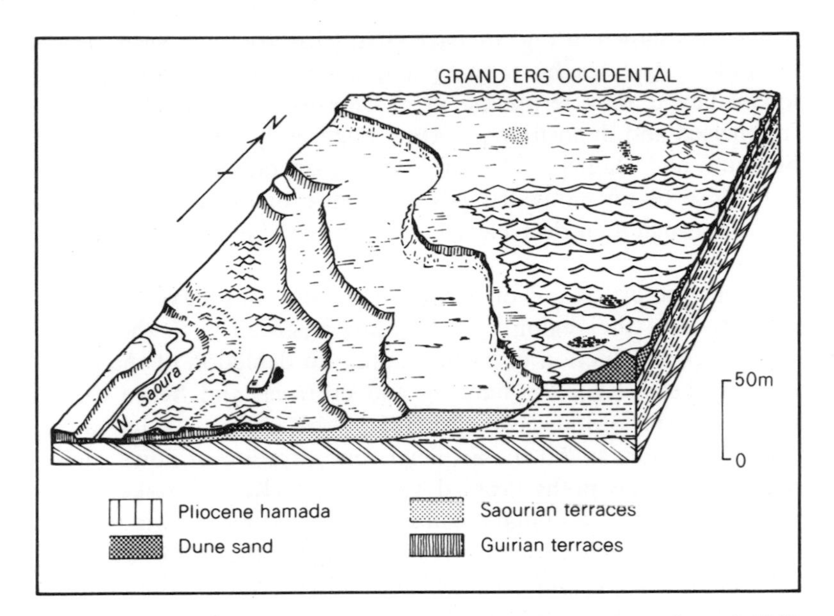

85. Terraces of the Wadi Saoura at Beni Abbès, Algeria. From Conrad, 1963.

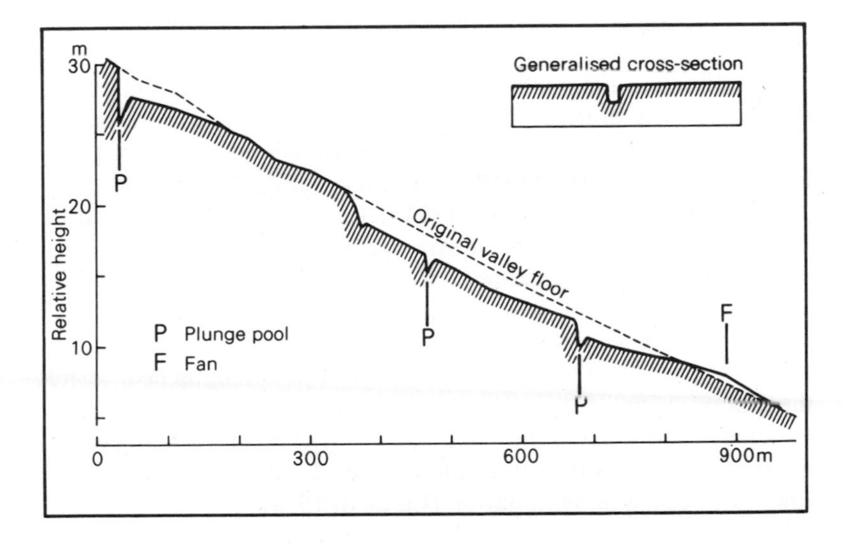

86. Profile along Arroyo Falta, New Mexico. From Leopold and Miller, 1956.

arroyos. As shown in Fig. 86, each channel sector has a steep head-cut which regresses by sapping and plunge-pool erosion, whilst downvalley it may terminate in a steep-fronted alluvial spread. The incised and depositional sectors apparently migrate upstream (Schumm and Hadley, 1957). The gullies operate at gradients less than those of unchannelled floors because of more efficient discharge through their deep clean cross-sections (Leopold and Miller, 1956).

A contributory cause of arroyo trenching may have been the occupation of the region by European pastoralists in the 1870s and the re-introduction of cattle, leading to depletion of the plant cover by heavy grazing, burning and clearing, and hence to increased storm runoff (Duce, 1918). This doubless acted as a localising factor, as did the artificial concentration of flows where tracks and game paths crossed valley flats (R. U. Cooke, pers. comm.). However, a climatic cause is indicated by the widespread occurrence of trenching and by its apparent periodicity. For example the trenching that began between 1860 and 1890 was predominantly into fine-textured alluvial valley-fills or *cienegas* considered to have been shed under wetter conditions from well-vegetated catchments with regular stream regimes, in several episodes more than 200 years ago (Melton, 1965b). Two earlier cycles of cut and fill by streams in the region are indicated, at approximately 2000 and 700 BP (Emmett, 1974).

The climatic causes are not readily detectable, however. Annual rainfall records indicate no significant change to a drier climate in the southwest during the period of trenching, as claimed by Bryan (1925b) and Antevs (1952), nor to generally wetter conditions as postulated by Huntington (1914). Both in New Mexico and in Arizona, however, arroyo cutting appears to have begun during periods of summers with more frequent heavy storms and fewer light rains than at present, which would probably have reduced the cover of perennial grasses and so increased runoff (Leopold, 1951; Cooke, 1974). This tendency would be strengthened where the heavy storms followed a drought such as preceded arroyo trenching in Arizona in the 1880s and 1890s. The subsequent period of more equable summer rainfall seems to have ushered in an epicycle of renewed aggradation (Emmett, 1974).

EVIDENCE OF CLIMATIC CHANGE ON HILLSLOPES

There is a widely held view that the present period, regarded as relatively arid, is one of diminished geomorphic activity on desert hillslopes, that rapid mass movements have temporarily ceased, that erosion is more localised and linear than in the past, and that the hillsides now yield less and finer detritus than previously. This has been stated of the lower slopes of the Tibesti Massif for instance (Pachur, 1970). Ahnert (1960) attributes sheer joint-controlled cliffs and amphiteatral alcoves in escarpments of the Colorado plateaux to sapping and mass movements in past wetter episodes, and regards present activity as dominated by granular disintegration, with regrading and rounding of bluff angles. On the other hand, Schumm and Chorley (1966) have assembled reports of recent rockfalls and slides in the region and would explain the few records by the sparsity of observers (see Chapter III).

Evidence from landslide blocks

Along escarpments of the Colorado plateaux where massive sandstone caps thick shale there is evidence of large-scale slumping and sliding which are held to be no longer active to the same degree. Under this view the mass movements have been attributed to cooler and wetter Pleistocene episodes when the shale became liable to saturation and failure due to percolating groundwater. Enormous block glides, some measuring several kilometres, occur in the Chuska Mountains of northwest New Mexico (Watson and Wright, 1963) and form a belt of ridges and troughs parallel with and up to 13 km from the escarpments. Three generations of ridges are recognised, each more weathered than the next inner one. The Toreva blocks of the Black Mesa region of northeast Arizona are masses of Mesaverde Sandstone up to 600 m long that have rotated backwards in moving above the Mancos Shale and now occur up to 300 m from the escarpments (Reiche, 1937).

Talus flatirons

Where a hard capping above soft rocks breaks down into boulders and finer grades, hillslopes may reflect an alternation of planar fashioning and gully erosion. On duricrusted mesas in central Australia the fossil planar elements are detrital slopes between

20° and 35°, with protective mantles up to 2 m thick of boulders cemented in a red clay-sand matrix. Under gully attack the slope mantles are breached and eventually detached from the hill as erosion extends rapidly in the soft pallid rock beneath, and the former slope element is left as a triangular abutment or *talus flatiron* (Koons, 1955). In Cyprus, flatirons have been preserved by calcrete (Everard, 1964), and in the Negev by dark flints which protect soft white chalk.

In the interpretation of flatirons it is commonly assumed that linear dissection, as at present, coincides with drier conditions and a diminished supply of colluvium, which in turn allows the concentration of runoff, perhaps accentuated by sporadic heavy rainfalls on slopes less protected by vegetation. The processes are analogous to those postulated for the dissection of piedmont profiles, and talus flatirons may continue downslope as mantled pediments or terraces.

WEATHERING CRUSTS AS EVIDENCE OF CLIMATIC CHANGE

A weathering crust may indicate a change of climate if it exists outside its formative climatic range, if there is evidence of climate-determined periodicity of crusting, or where breakdown suggests that it is out of equilibrium with the present climate. However, factors such as lithology and relief may also influence the nature and movement of weathering solutions, and different processes with different climatic significance may produce similar crusts — for example the source of a gypsum crust may be sheetflow, ground-water or wind-borne dust.

Calcrete

Calcrete crusts typify semiarid to moderately arid climates with between 100 and 400 mm annual rainfall, particularly those with cool-season rainfall as on the midlatitude margins of zonal deserts (Chapter VI). Within this range they grade from massive crusts in the wetter parts to rubbly or thin lime-cemented horizons near the dry limits, as in the northwestern Sahara. Where they occur in extreme deserts, as in the Tibesti, they have been interpreted as evidence of past lessened aridity (Meckelein, 1959). Calcretes form at rates which make them particularly relevant to Pleistocene climatic changes, and the possibility of radiocarbon dating of younger Pleistocene crusts adds to their potential significance.

Periodicity of calcrete formation. This is suggested by superposition of calcretes in thick deposits or by the occurrence of distinctive crusts on successive land surfaces, as on staged pediments in and south of the Moroccan Atlas (Beaudet *et al.*, 1967; Ruellan, 1967). The crusts are more evolved on the higher surfaces and are better developed on the flatter parts (Fig. 87A). Calcrete is considered to have formed early in the onset of drier conditions following a wetter phase of planation and mantling, with climates still wetter than the present, under which crusting is very slow.

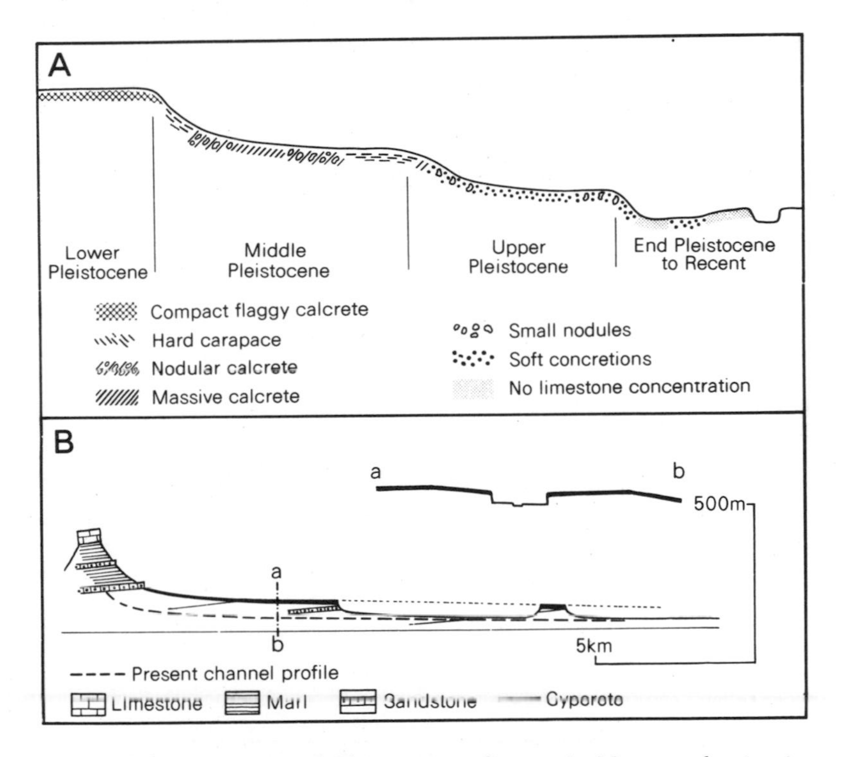

87. A. Calcretes on staged Pleistocene pediments in Morocco showing increasing development with age and with position on the terrace. After Ruellan, 1967.
B. Gypcrete capping a dissected pediment flanking the east end of the Atlas ranges in Tunisia. From Coque, 1955.

However, all the changes are considered to have been within the range of a dry Mediterranean regime. From the relative development of the crusts it is claimed that the earlier climatic cycles were longer and of greater amplitude, and that older crusts have continued to evolve in subsequent cycles. The crusts are considered to provide independent pedogenic evidence which supports the

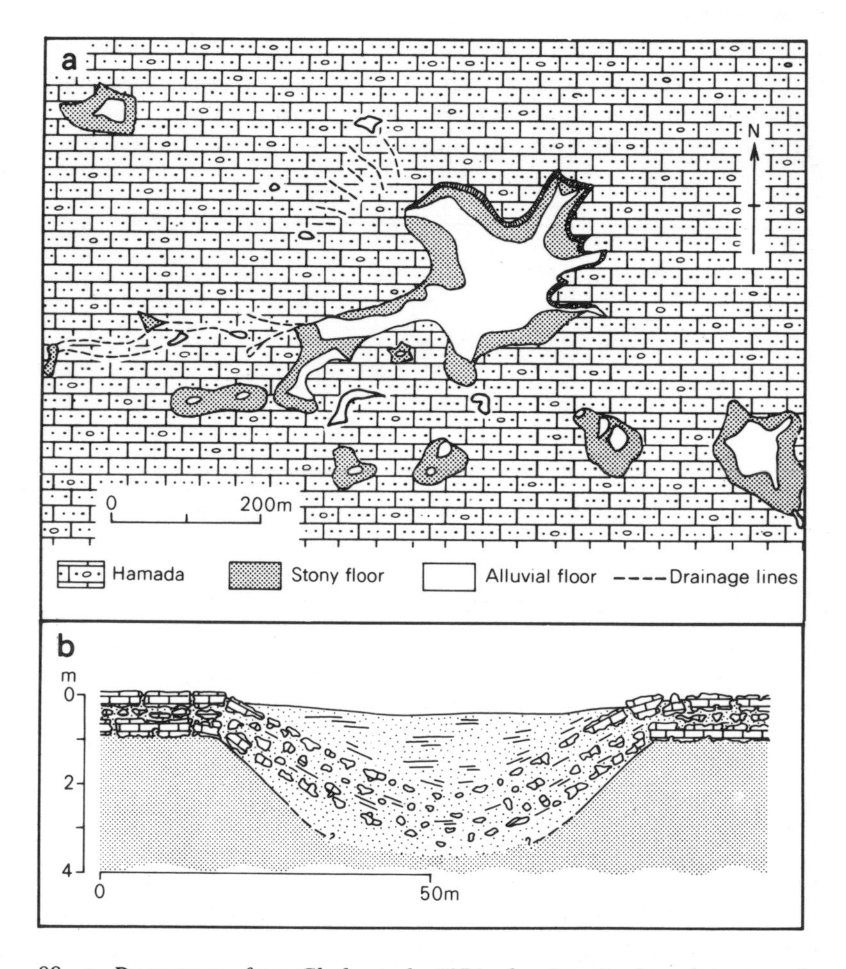

88. a. Daya types, from Clark et al., 1974, showing circular, elongate and star-shaped forms; b. section through a small daya on the Hamada du Guir, formed as a collapse doline following solution of the calcrete crust and subsequently filled. From Conrad, 1969.

palaeoclimatic interpretations of the associated land surfaces made on geomorphic grounds.

Subsequent modification of calcretes. Hard flaggy upper surfaces of calcretes, particularly those with rhythmic layering, have been attributed to solution and redeposition of crusts exposed at the surface and subject to sheetflow. Reduction in rainfall has been claimed as a cause, but similar effects might arise where a thickening calcrete becomes impervious to descending solutions, independent of climate (Flach *et al.*, 1969). Rubbly calcretes in dry settings have been explained as crusts fragmented by desiccation.

More convincing evidence of a change to wetter conditions is afforded by solution pipes in calcrete, commonly associated with penetration by plant roots. Karstification on a larger scale can result in closed depressions such as the *dayas** (Fig. 88a) on hamadas in the northwestern Sahara and on the high intermont plains of the Atlas (Estorges, 1959–61; Conrad, 1969; Clark *et al.*, 1974; Mitchell and Willimott, 1974). The smallest are saucer-shaped, tens of metres across and less than a metre deep, are floored with clay and sand and may hold water after rains; their origin as dolines is indicated by remnants of calcrete beneath their floors (Fig. 88b). Elongate dayas are found aligned along drainage tracts or joint lines, and circular forms occur isolated on the plateaux. The latter include large star-shaped forms, hundreds of metres across and several metres deep, bounded by scarplets of calcrete and with stony margins leading to an inner alluvial depression. Dayas are commonly attributed to solution of calcrete in wet phases following a drier period of calcification. Subsequent deflation of floor sediments, perhaps in later drier periods, may have contributed to deepening, leading to channel incision in turn. Deflation is claimed to have played a major part in forming the *tinajitas* or solution pans in the *caliche* of the southwestern High Plains of the United States (Judson, 1950; Reeves, 1966).

Gypcrete

Coque (1955) has linked with climatic change the formation of gypcrete which cements thin gravel on a pediment at the east end of the Atlas Mountains in Tunisia, adjacent to the saline Chott Fedjedj. The pediment is now dissected but the gypcrete does not extend on the younger surface (Fig. 87B), suggesting it relates to

* Daya is also used of larger karst basins in the subhumid Middle Atlas of Morocco (cf. Martin, 1965).

a past episode. The gypsum is considered to have been blown from the nearby Chott with the drying of the floor, and the encrustation is placed at the onset of a drier phase of incipient dissection of the pediment; its fracture by arid weathering and subsequent recementation are attributed to an ensuing swing to a wetter climate.

Silcrete

Massive fossil silcrete crusts occur in several deserts (Goudie, 1972) but their palaeoclimatic significance remains uncertain because no exact equivalents are known to be forming today. In Australia they commonly occur towards the continental interior from fossil laterites, and on the assumption that they are of the same age it has been concluded that the silcretes reflect drier contemporary climates with climatic gradients similar to the present (Langford-Smith and Dury, 1965; Stephens, 1971).

It now appears that, in the Lake Eyre basin at least, silcrete has formed *either* on a surface planed in an older weathered profile *or* within overlying sands, and is accordingly younger than the laterite of arid Australia (Senior and Senior, 1972). This relationship may explain why the silcrete is restricted to sedimentary basins (Fig. 36): in these less stable settings the older profiles tended to be eroded, there were arenaceous rocks suited for silicification, and the lowland topography favoured the concentration of silica into surface crusts and their subsequent survival. In contrast to the older laterites, which appear discordant with present climatic patterns, silcrete indicates an early stage in the establishment of a dry interior Australia in the early-middle Cainozoic as evidenced by dating of silcrete in north-central New South Wales as older than 22 million y (Exon *et al.*, 1970). This interpretation also reconciles the association of silcrete with a profile of deep weathering, explains the occurrence of silicified laterite on the shield of Western Australia (Stephens, 1971), and accounts for the silicification of terrestrial and freshwater limestones younger than laterite in much of dry Australia.

FEATURES INHERITED FROM HUMID CLIMATES

Forms older than the desert itself extend the climatic record and may yield evidence of the ingress of aridity. They are naturally more common in stable shield and platform desert settings, more particularly in the zonal deserts of the southern hemisphere, where drift of the Gondwanaland plates combined with increasing differentiation of the zonal climatic belts from the Miocene onwards resulted in a sequence from equable wetter climates in the Cretaceous and earlier Tertiary, via hotter semiarid regimes of the Miocene, to desert by Pleistocene times. In contrast, the heart of the Sahara has been arid since the early Mesozoic.

Given a slow rate of desert landscape evolution, it must be presumed that many major landforms of the tectonically stable deserts have persisted from non-desert climates, for example that the extensive plains and inselbergs of the shield and platform deserts may date from former savana regimes. The present state of climatic geomorphology, however, does not permit the interpretation of entire landscapes in palaeoclimatic terms, and the examples cited below, taken mainly from Australia, are specific features out of keeping with their present desert settings.

Weathered land surfaces and related duricrusts

Lateritic surfaces. Ferruginous crusts and underlying pallid zones indicate strong leaching under warm conditions, probably with a seasonally distributed annual rainfall of at least 750 mm. Where these occur in deserts they must be regarded as relict from wetter regimes, possibly of savana type. Lateritic crusts of Cretaceous to early Cainozoic age define and preserve the Old Plateau level in the shield desert of Western Australia (Jutson, 1934) and occur sporadically across the granitic plains of central Australia (Mabbutt, 1963a, 1965a). Lateritic crusts within the Australian arid zone are mainly confined to ancient surfaces on granitic shield blocks, for these offered the most suitable lithologies for the development of laterite and the most stable settings for its survival. There is little accordance between the distribution of the oldest laterites and present-day climates, but the more limited distribution of younger lateritic crusts suggests that the contrast between a wetter north and a drier interior was established by the Miocene. A major inheritance is the low relief of these ancient lateritic surfaces, which has profoundly influenced the geomorphic expression of aridity in Australia.

In West Africa, dissected laterites extend north into the Sahelian zone almost to the middle Niger, where rainfall is now only 200 mm. They are younger than the Australian laterites, are distributed in closer accordance with present zonal climates, and indicate that the southern boundary of the Sahara may have run about 5° further north at the beginning of the Pleistocene (Tricart, 1959).

Etchplains and related features. Where these result from the stripping of an older weathering profile under a desert regime they can be regarded as an indirect inheritance from the past. For example the granitic plains and tors of the New Plateau of Western Australia have been exposed by the stripping of adjacent lateritic profiles (Mabbutt, 1961). In central Australia, remnants of laterite of pre-Miocene age form low platforms on pediments and slope up gently towards nearby granitic hills as though part of former concave profiles. The present angular hill-foot junctions cut into the laterite have been influenced strongly by compartmentation in the former weathering front (Mabbutt, 1965).

Old drainage features

With the onset of aridity the plains sectors of drainage systems became progressively disorganised as described in Chapter VII. In the Sahara, they were eventually obliterated by windblown sand, giving the pattern of lowland dunefields and peripheral wadi systems depicted in Fig. 64. In moderate shield and platform deserts, however, pre-desertic drainage features have survived extensively, albeit now fossilised in place.

In Western Australia, for example, shallow valleys cut below the laterite of the Old Plateau are traceable as former connected systems, now choked with calcrete, alluvium and windblown sand and dismembered into lines of playas or 'river lakes'. Reconstructions of the original drainage by Gregory (1914) and later workers (cf. Mulcahy and Bettenay, 1972) show that some of the valleys led southeastwards to a former gulf marked by marine limestones of the Nullarbor Plain, whilst others linked westwards to the Indian Ocean (Fig. 89). Eocene marine sediments have encroached on the southeastern outlets, suggesting that the old drainage may be at least Palaeocene. The closeness of the southern divide to the coast indicates that disruption of this drainage began with the marginal rifting of the Australian plate in the Eocene and was overtaken by aridity with drift into lower latitudes during the early Cainozoic and with increasing differentiation of the zonal dry belt by the Miocene. The drainage lines have remained unchanged since the Tertiary save for minor westward displacements of some playa floors due to accumulation of windblown sand on

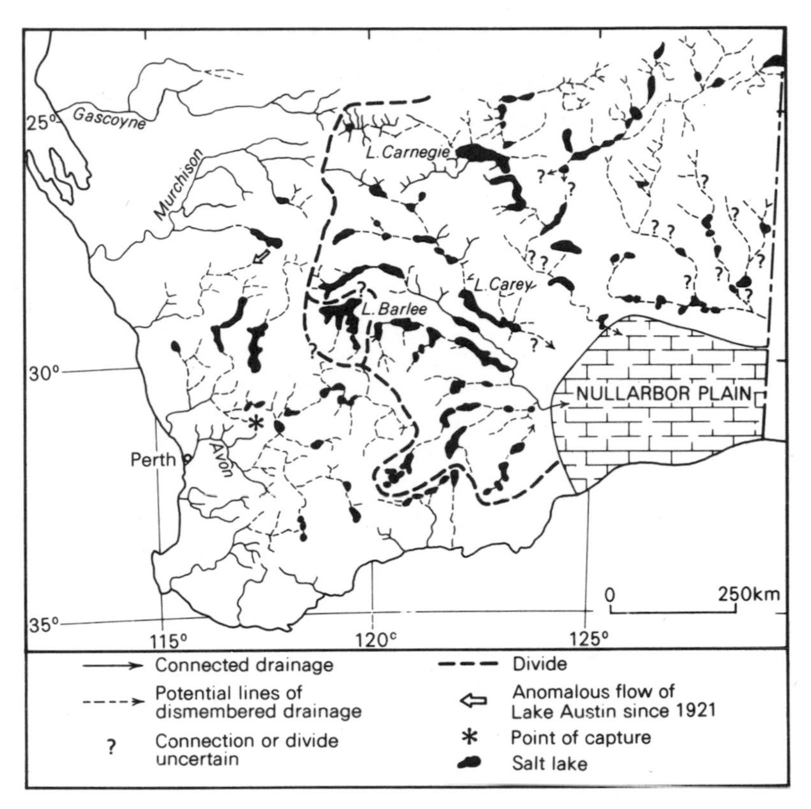

89. Ancient valleys and 'river lakes' of Western Australia. After Beard, 1973.

the east and extension by saline water-level weathering on the west (Jutson, 1934).

A record of out-going drainage in the early Cainozoic, breaking down by the Oligocene into separate interior drainage systems and lake basins, is common to much of arid Australia, for instance the lakes ancestral to Lake Eyre (cf. Wopfner and Twidale, 1967). Although these lakes may have been initiated by tectonic movements, their survival as closed basins and their characteristic deposits of fine-textured and calcareous lacustrine and fluviatile sediments suggest a combination of dry climates and low catchment relief.

XI

THE SYSTEMATIC GEOMORPHOLOGY OF DESERTS

EARLY VIEWS ON DESERT LANDFORMS

Impressions of the Old World deserts

Faced with the unfamiliar landforms of the Old World deserts and the apparent lack of any active agent capable of shaping them, a reaction of the earliest observers was to deny their subaerial origin and to regard them as relict marine forms. Many features combined to encourage this view: vast plains with sharply up-standing island mountains, spreads of sand and gravel in areas devoid of rivers, salt-crusted surfaces in the lower parts, and the chaotic relief of the dunefields. The desert appeared to be a dead landscape no longer worked upon, and abandoned by the agents which had formed it.

In the later nineteenth century the concept of a 'Saharan Ocean' gave ground as the desert became better-known, and homologues of humid landforms were recognised, such as the systems of wadis. But whilst the forms were now acknowledged to be subaerial, they were still judged to be largely relict and were attributed to a past 'pluvial' period linked with the lately recognised Ice Age. They were held to be fossil or to be undergoing only slight modification under the present aridity.

When in due course many desert landforms *were* recognised as products of their present climatic setting, the dominant processes were held to be those peculiar to the desert (cf. Walther, 1900). Wind action was particularly stressed. To the action of wind armed with corrading sand was attributed not only the isolation and undercutting of 'mushroom' or 'monument' rocks, but on a larger scale the basal steepening of hillslopes. *Deflation*, a term introduced by Walther to describe the export of fines from the desert by wind, was held to be of major importance in the excavation of enclosed depressions and the levelling and lowering of desert plains. So prevalent was this latter view that the inselberg savana landscapes in East Africa and around the Kalahari were interpreted as ancient desert plains (Passarge, 1904).

Similarly, the breakdown of rocks and the comminution of rock debris were attributed to thermal expansion and contraction under strong solar heating, as being consistent with the clear skies, high temperatures and large diurnal temperature ranges of the hot deserts, and the stone-strewn surfaces of the desert uplands and hamadas were held as evidence of its efficacy.

Impact of the New World deserts

In North America, alternative views acknowledged a greater similarity between the landforming processes of deserts and of humid regions and placed greater stress on the work of water and less on wind and the effects of insolation. They reflected the impact of the mountain deserts of the southwest United States, with more accidented relief and less extreme aridity, on geologists like Powell and Gilbert who were less bound by traditional views. They brought a first appreciation of the balance and tempo of land-forming processes in deserts of this type. The recognition that moisture was important in desert weathering implied such slow progress that running water, however sporadic its action, could move the limited supply of detritus across little-vegetated slopes. This is evident from Gilbert's explanation of the structural control of desert relief:

> where vegetation is scant or absent, transportation and corrasion are favored, while weathering is retarded. There is no accumulation of disintegrated material. The rate of erosion is limited by the rate of weathering and that varies with the diversity of rock texture. The soft are eaten away faster than the hard; and the structure is embodied in the topographic forms (Gilbert, 1877: 119).

An 'arid cycle of erosion'

When Davis (1905) attempted to synthesise knowledge of desert landforms in an evolutionary scheme he combined views from the New World and from the Old about the relative importance of desert land-forming processes, postulated an initial relief characteristic of North America but deduced an end-form more typical of the desert plains of Africa and Australia.

The postulated controls in the Davisian cycle under the 'climatic accident' of aridity are scant rainfall and sparse plant cover, superficial weathering of exposed rock surfaces by physical rather than by chemical processes, and an interior or dismembered drainage. The original forms are assumed to be tectonic uplands and basins; to a degree dependent on the strength of this initial relief, desert floods are held to dominate in erosion and deposition in the youthful and mature stages. The scheme envisages a progressive reduction in relief owing to the accumulation of the products of upland erosion in the basins, and an eventual integration of drainage with the capture of higher-lying basins by drainage in basins at lower levels. Aeolian corrasion and deflation, although always active, come increasingly to the fore in the slower stages of late maturity and old age, as the upland sources of drainage diminish and the open landscape drowns in its own detritus (Udden, 1894). Wind is held to be the only agent capable of exporting material from areas of interior drainage and hence effective in the final lowering and smoothing of desert plains.

The Davisian arid cycle demonstrates the fallibility of the comparative approach in that it mistakenly places landforms of the mountain-and-basin deserts alongside those of shield and platform settings as evidence of different stages in a common development; it was by no means the last failure to appreciate the differentiating effect of geologic structure and consequent major relief on the physiographic response to aridity.

RESPONSE TO GEOMORPHIC DIFFERENCES BETWEEN DESERTS

More specific studies of landforms and landforming processes were carried out in a number of deserts during the first forty years of this century, and the climatic data necessary for a first appreciation of the magnitude and frequency of desert geomorphic 'events' became increasingly available. The various regional groups of publications from this period continue to show differences of emphasis and interpretation reflecting the variety of desert geomorphological settings.

Most numerous are the writings on the *North American deserts*, which are dominated by accounts of the fashioning of pediments by running water, mainly on the granite terrains of the Mojave and Sonoran deserts. These involve detailed consideration of the relative effectiveness of rills, sheetfloods, and stream channels, as in the later works of W. M. Davis (1938). Processes on the fans and bajadas of the Basin and Range province are neglected by comparison. Pedimentation is linked with hillslope retreat, and there are outstanding analyses of the backwearing of slopes under the control of structure and debris cover (Lawson, 1915; Bryan, 1922). These North American studies are remarkable for their time in their close attention to evidence of processes, presumably reflecting advantages conveyed by the simplifying conditions of aridity, and the greater accessibility of the American deserts.

There is a natural tendency to stress analogues between the landforms of the moderate North American deserts and those of humid areas, for example between pediments and valley-side floors (Davis, 1930), maintaining the Davisian concept of desert morphogenesis as a deviant of the 'normal' sequence of wetter climates. This attitude led to useful re-interpretations of features previously attributed to processes peculiar to the desert environment. For example, pedestal rocks are now explained by differential weathering rather than by undercutting through sandblast, and the effectiveness of insolation weathering is questioned on the field evidence of desert varnishing and from laboratory tests. The view of rare cloudbursts as the essential feature of desert rainfall and morphogenesis is also queried from the evidence of the semiarid Great Plains (Russell, 1936).

There is a natural neglect of features characteristic of more arid shield and platform deserts, such as sand dunes, and wind sculpture is seen to be important only in soft playa sediments.

Landform studies in the *Egyptian and Libyan deserts* at this time, as represented particularly in the works of Hume (1925) and Ball (1933), reflect the greater aridity compared with the North American deserts and the contrasted platform setting and generally flat relief. There is a much greater preoccupation with wind action, as shown in the deflative excavation of the newly-discovered oasis depressions and in the extension of sief dune chains downwind from them. The Great Sand Sea was explored late in the period, initiating fundamental studies by Bagnold (1941) of sand transport by wind and of the resulting aeolian sand forms. The extensive reg and hamada pavements flanking the Nile Valley were also held as evidence of widespread and effective deflation.

Fragmented boulders on stone pavements were accepted as incontrovertible evidence of shattering by solar heating, although studies of weathering on monuments (Barton, 1916, 1938) had established the importance of chemical weathering and its control by available moisture.

The geomorphic effectiveness of occasional streamfloods was accepted for the uplands of Sinai and the Eastern Desert, but not in the tablelands of the Western Desert, where the wadis were held to be relict and subject only to modification by sapping at the base of their cliffs (Peel, 1941).

Shield and platform deserts, as in the *southwestern Sahara, southern Africa and Australia*, yielded accounts of the great systems of linear dunes which were now recognised to run parallel with, not across the wind (Aufrère, 1930; Madigan, 1936; Lewis, 1936). Suggested origins included erosion by dominant trade winds as well as extension downwind. The dunes were seen to be partly fossil and a sign of the former extension of aridity, in contrast to the eastern Sahara with its evidence of landforms relict from a more humid past.

There is a natural emphasis on the extensive lowland forms of these deserts. The work of Jutson (1934) on 'levelling without base-levelling' in the shield desert of Western Australia typifies this; a younger plateau here extends at the expense of an older laterite-capped surface by escarpment retreat through a combination of salt-weathering, waste removal by sheetfloods, and playa migration at groundwater level due to salt-weathering deflation and the growth of lee dunes.

Accessible literature on the deserts of *central Asia* for this period stems mainly from reconnaissance expeditions by investigators from Europe and the United States (cf. Berkey and Morris, 1927). Earlier explorations, for instance by Hedin, had established the unique complex of landforms in these cold mountain-and-basin deserts, locally of extreme aridity: gravel plains or *gobis*, partly vegetated dune tracts, often of confused relief, and saline alluvial plains or *kavirs* at the terminals of upland drainage. In the later systematic geomorphological accounts the emphasis is on wind action on fine basin sediments, in the corrasion of yardangs, the deflation of depressions, and the mantling of foothills with loess.

In studies of coastal deserts at this time, publications by German geomorphologists are preeminent and are of particular interest in their treatment of extreme aridity in these special environments. In his account of the *Namib Desert*, Kaiser (1926) linked his studies of weathering and deposition with the Cainozoic stratigraphy to establish a record of climatic change. In the *Chilean Desert* Mortensen (1927) used the gradations of desert climates, processes and landforms with distance inland and with altitude as the basis of a climate-based morphogenic classification of deserts which was to influence much later German work on climatic geomorphology, and which is discussed below. His account stressed the importance of salt-weathering in areas of negligible rainfall, and the suppression of wind action where lack of streamfloods results in an absence of corrading sand.

THE GEOMORPHIC EXPRESSION OF ARIDITY

Although the regional studies summarised briefly above clearly demonstrated the significance of differences of structure and major relief in the distribution of desert landforms, attempts to systematise this experience tended to rely exclusively on the climatic factor.

Desert geomorphic types as a measure of aridity

German geomorphological writings stand out for their attempt to interpret combinations of geomorphic processes in terms of degree of aridity, as part of a climatic geomorphology of deserts. From his observations in the Namib Desert Kaiser (1926) had claimed that *extreme-arid* areas with minimal and random rainfall were characterised by predominant wind action and particularly by the deflation of fine sediment, that the rare streamfloods were geomorphologically unimportant, and that limited weathering included surface silicification. Under a *normal-arid* regime with some seasonal expectation of rainfall, streamfloods occurred in most years and were the predominant geomorphic agents. They deposited unsorted alluvium which was subject to calcification (fanglomerate). Under *semiaridity* the rivers flowed seasonally, wind action was relatively minor, and weathering involved leaching.

These views on the relative efficacy of wind and water action in deserts were later challenged by Mortensen (1927, 1930a) on evidence from northern Chile. He argued that in the extreme-arid desert cores (*Kernwüste*), as exemplified by the inland valley of northern Chile, corrasion by wind was inhibited by the thin protective crusts formed by moistening of fine-textured saline soils comminuted by hydration weathering, particularly with the lack of sand due to restricted weathering and fluvial deposition. Under such conditions channel incision prevails on the hillslopes and the imprint of depositing water rills on the plains persists unmodified between the rare episodes of rainfall. Both wind and water action are considered to increase in effectiveness with the transition to the inner and marginal deserts (*Mittel-* and *Randwüste*), categories which comprise most full desert regions, with increase in the frequency of runoff and the availability of sand. However, the increase in wind action is commonly the more marked, although both wind-dominated and water-dominated landscapes are found. On the semiarid desert margins (*Halbwüste*) the roles are reversed, for wind action is here hindered by a closer vegetation whereas rainfall and water erosion continue to increase. These changing relationships are shown graphically in Fig. 90.

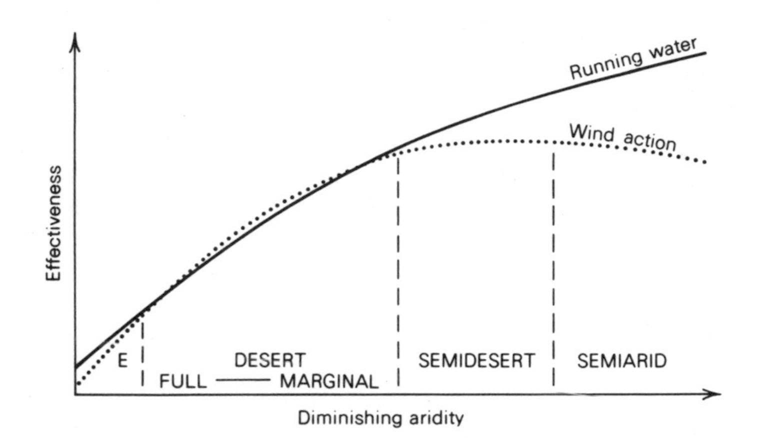

90. Relative importance of wind and water action in relation to degree of aridity in deserts. From Mortensen, 1950.

Mortensen's views were subsequently supported by Meckelein (1959) on the basis of field studies in the east-central Sahara. From the extreme-arid reg plains of Tibesti he reported powdery soils like those of northern Chile, including soft gypseous *fech fech* dust, attributed to salt weathering. These may overlie subsoils with polygonal patterns, regarded as fossil solonetzes presently being broken down by the prevalent salt-weathering. The soils are protected from wind erosion by crusts a millimetre or so thick above thin vesicular topsoils, and by reg pavements. Meckelein notes the prevalent marks of water rills on these surfaces despite the aridity, and cites the persistence of wheel tracks over several decades as corroboration of the inadequacy of wind erosion. These reg plains lack the fossil wadis of the adjacent Sahara and the area has been interpreted as an ancient and persistent desert core which did not experience wetter periods during the Pleistocene.

In contrast, the full desert to the north, equivalent to Mortensen's *Mittelwüste*, shows the complete range of desert landscapes including regs, dunefields, and wadis with floodouts. Many of these landscapes are held to be fossil from past wetter conditions, however. The wadis are mainly inactive and are now being widened by basal sapping of their walls and their terminal alluvial plains are being eroded by wind and rills; even the extensive dunefields are relatively immobile and are characterised by mature or old-age forms. Deflation, although more active than in the extreme-desert, is nevertheless limited by the light winds of the Saharan anticyclonic belt, and rill patterns suggest a prevalence of water action on the stony plains. The widespread desert varnish and saline coatings on stone pavements indicate active but slow chemical weathering under the present regime.

It is in the marginal desert (*Randwüste*) still further north that Meckelein finds the main areas of dune growth, with more frequent sand storms and a more abundant supply of alluvial sand; youthful mobile dune forms such as the barkhan are more typical here. The gravel plains are also clean-swept, pointing to effective removal of fine weathering products by wind and sheetwash. Here the wadis flow in most years and reg plains are actively extending at their terminals.

Status of surface drainage as a measure of aridity

As discussed in Chapter VII, Martonne (1927, 1942) used *endoreism*, or interior drainage, and *areism*, or an absence of locally generated connected channels, as expressions of the disorganisation of surface drainage. He claimed a correlation between the relative extent of these conditions and degree of aridity as measured by his own index (see Chapter I). Endoreism was held to be characteristic of semiarid climates with indices between 5 and 10, and areism with truly arid climates with indices of aridity below 5. The correlation was based on measurements within latitudinal zones, as shown in Fig. 91.

However, the associations are complex and the correlation breaks down at a regional level, as was demonstrated by the account of Australian desert drainage in Chapter VII. In some deserts

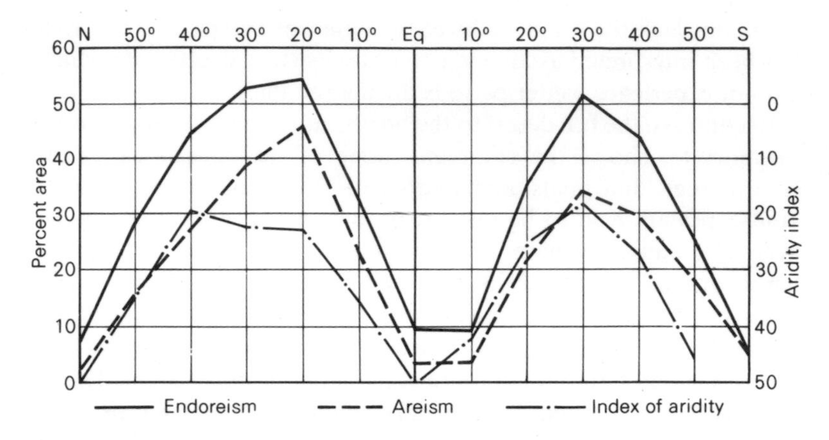

91. Latitudinal variations in the extent of endoreism and areism compared with an index of aridity. From Martonne, 1927.

endoreism expresses a tendency for the tectonic derangement of drainage to persist under aridity, where with a surplus of discharge the rivers would have overflowed and linked to find an exit seawards. Elsewhere, a link between aridity and endoreism may reflect the common factor of distance from the sea, as in the continental desert basins of central Asia or the inner fastnesses of the Sahara. Again, it can express the tendency for topographic barriers additionally to be climatic barriers. Areism may be a function of flat or pervious desert surfaces rather than of extreme aridity, as shown by the Australian examples.

The varying interaction of climatic and relief factors on the organisation of desert drainage is to some extent revealed by comparisons in Table XI. Although the area of endoreism and areism combined varies with the extent of dry climates, their proportions vary significantly. Endoreism is relatively large where strong relief is a determinant of drainage patterns, as in North and South America, and smaller in the flatter shield deserts of Australia and Africa where river systems die out on the open plains for want of replenishment by tributaries.

On a more localised scale, the percentage area of topographic basins occupied by connected drainage, the second of the measures of drainage disorganisation discussed in Chapter VII, has been correlated with lack of rainfall, as by Dubief (1953b) in Table XII. In the drier eastern Sahara effective drainage generally occupies

TABLE XI **Areas of dry climates, endoreism and areism**

	Area of arid and* semiarid climates		Area of endoreism+ (E)		Area of areism+ (A)			
	$km^2 \times 10^6$	% continental area	$km^2 \times 10^6$	% continental area	$km^2 \times 10^6$	% continental area	E+A %	E/A
Australia	6.38	83	1.61	21	3.31	43	64	0.49
Africa	17.94	64	3.45	11.5	11.77	39	50.5	0.29
Asia	16.48	39	4.91	11.5	9.94	23.5	45	0.49
South America	3.01	17	1.07	6	1.45	8.25	14.25	0.75
North America	3.97	16	1.76	4	1.07	4.5	12	1.72

*From Meigs, 1953.

+From Martonne, 1927.

TABLE XII **Relative extent of connected drainage systems in topographic basins in the Sahara* and in Central Australia+**

Basin	Area of connected drainage $km^2 \times 10^3$	% of topographic basin	Annual rainfall (mm)
Niger	1335	63	730
Lake Chad	900	40	378
Melrhir (south Tunisia)	250	36	48
Saoura	11	25	34
Western Sahara	400	17	34
Syrte	60	19	12
Qattara Depression	30	5	12
Finke-Macumba (central Australia)	150	55	160

* From Dubief, 1953b.
+ From M. Sullivan, unpublished.

less than 5 per cent of catchments, whereas values are higher in the west and highest in the relatively well-watered basins of the south. Gautier (1950) has also claimed to see in these differences a contrast between an older eastern Sahara, in which traces of drainage have been extensively obliterated during a long history of aridity, and a younger western Sahara with evidence of Quaternary wetter phases. However, factors such as relief and the extent of aeolian sands are also important. The low values of the eastern

Sahara doubtless reflect the predominance of plains and tablelands, whereas to the west the desert is flanked and broken by mountains from which active wadi systems are generated.

For these reasons such comparisons should be restricted to physiographically similar areas, and in Table XII the catchment of the Finke-Macumba system in central Australia accords fairly well with the figures from the structurally comparable Sahara. Since the efficiency of desert drainage catchments diminishes with extent, comparison should also be made only between basins of similar area. The correlations in Table XII would have been less impressive had the rainfalls not generally increased with catchment area.

Extent of dunefields

Dunefields are popularly regarded as the extreme expression of the desert environment and their occurrence has accordingly been suggested as a measure of aridity. It was shown in Chapter IX that at the scale of comparison between major continental deserts there is no correlation between rainfall and the extent of dunes; nor is the suggestion supported by the distribution of dunefields *within* deserts. Prescott (1936) long ago noted that the Simpson Desert lay in the driest part of Australia and suggested a climatic control. However, correlations between dunefields and rainfall are likely to be associative rather than causal, for important topographic and geologic factors are involved. In most deserts, including Australia, dunefields are situated in basins rimmed by uplands of sand-yielding rocks, reflecting an origin as wind-sorted alluvia. Any climatic link probably arises merely from the fact that these low-lying areas may well be the drier parts of the desert.

DESERTS IN MODERN SYSTEMS OF CLIMATIC GEOMORPHOLOGY

In later years the desert morphogenic system has commonly been treated within broader schemes of climatic geomorphology (Stoddart, 1969).

Synthetic or chorological systems

Stoddart described as the 'synthetic or chorological approach' those schemes which accept the general postulate that climate must

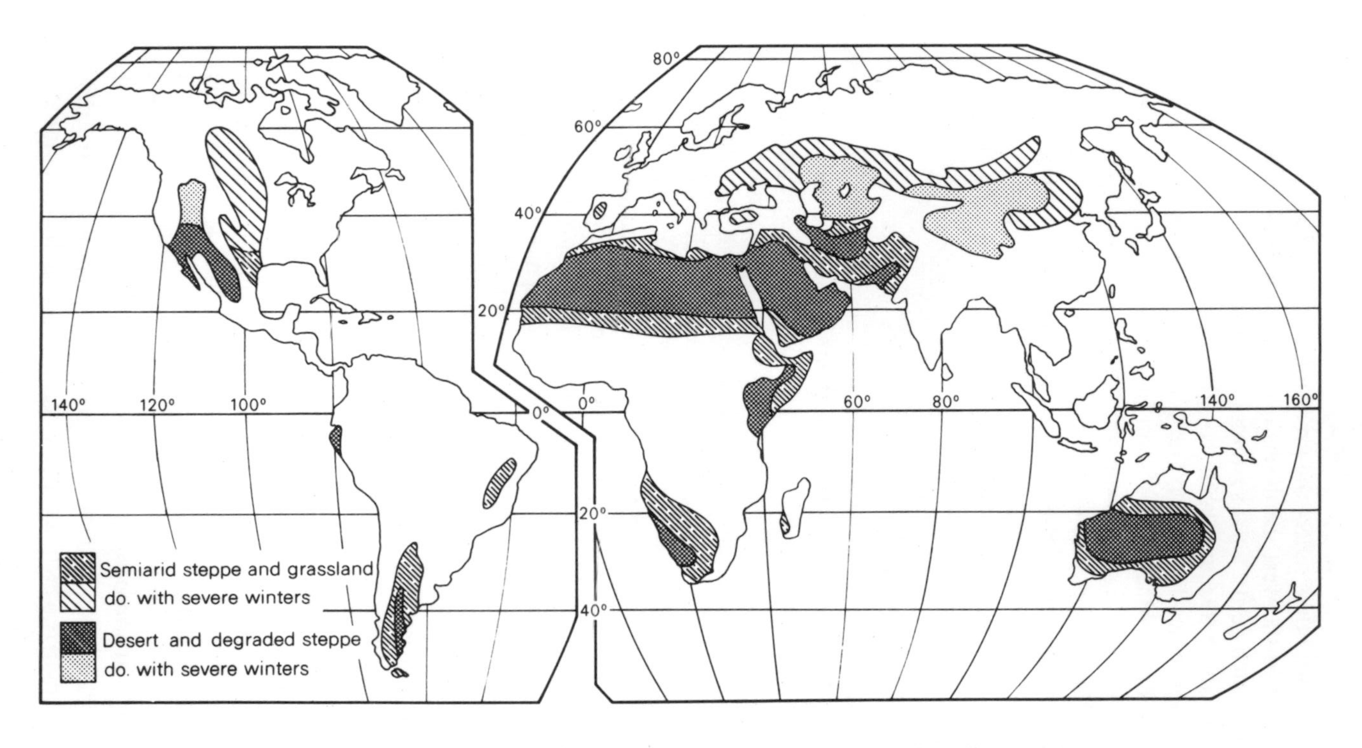

92. Arid and semiarid morphoclimatic regions. After Tricart and Cailleux, 1965.

influence landforms and accordingly choose climatic boundaries which are anticipated to be significant for landforming processes.

The scheme proposed by Tricart and Cailleux (1965) is of this type. It employs the major structural types of vegetation as indicators of morphoclimatic regions and its boundaries are chosen accordingly (Fig. 92). Their map therefore resembles those of similarly based climatic classifications. The dry zones are held to be typified by paucity of vegetation, such that the ground is left open to the direct attack of the climatic agents, and hence to be areas of predominantly physical erosion. Subdivision is by degree of aridity, into deserts and steppes, and on the basis of winter temperature as a determinant of significant frost action. Recognising the susceptibility of semiarid lands to *desertification* by land use, they included degraded steppes within the desert domain, as shown in Fig. 92. In their detailed treatment of these morphoclimatic regions, Tricart and Cailleux (1969) used the following divisions:

Arid and hyperarid deserts and degraded steppes. In the *coastal fog deserts* as in northern Chile, a combination of high atmospheric humidity and salt results in intense superficial weathering. Hillslopes are mantled in weathering products which in the absence of water action are transported by strong trade winds and by mass movement, forming steep straight gravity-controlled profiles. *The tropical deserts with high insolation* comprise large areas such as the Sahara. A range is recognised from hyperarid core areas to geomorphologically more active marginal deserts in which rain is more frequent and where the vegetation is nevertheless inadequate to hinder erosion and transport. There is a corresponding range in rates of weathering, which are predominantly chemical or physico-chemical and hence moisture-dependent, as well as in the efficacy of sheetwash and stream action. It may also be true of wind action, but the authors recognise that dominance by aeolian landforms is confined to areas where sand has accumulated by virtue of other factors, particularly in basins of interior drainage.

Subarid steppes. These are regions in which vegetation cover is sufficient to modify the geomorphic processes, and in which true soils develop. Under natural conditions, transport, erosion and deposition are dispersed and planatory in their action, but these areas have proved vulnerable to human colonisation with its destruction of the vegetation cover, and linear dissection of formerly stable slopes has ensued. Sensitivity to comparable changes is also reflected in extensive terraces formed in piedmonts and valleys during the climatic oscillations of the Quaternary. *Tropical subarid regions*, exemplified by the drier Sahel, experience sheetwash with intense if sporadic warm-season rains. In the absence of frost and with effective chemical weathering, piedmont debris is scarce and low-angle pediments have formed on extensive granitic basement rocks. Active alluvial plains are largely of clay-rich non-saline sediments which resist deflation. *Subtropical subarid regions* such as the Mediterranean fringe of the Sahara are subject to gelifraction at high altitudes, bringing relatively coarse debris to piedmonts generally cut across soft sedimentary rocks. Mantled pediments and terraces with relatively steep concave profiles are widespread in consequence. Cool-season humidity is more effective in the mobilisation of carbonates, and widespread calcretes, relict from past wetter periods, cement these piedmont mantles. Fine-textured saline sediments in structural basins are subject to deflation, or have been in the past. *Continental-temperate subarid regions* are typified by the Eurasian steppes. Severe seasonal freezing here affects lowland soils as well as outcrops on the desert upland rims, and large amounts of comminuted detritus are transported fluvially as silt and deflated by strong winds during the winter and spring. The steppe grasses have proved effective in trapping this air-borne sediment and loess mantles are widespread.

Morphogenic regions

An alternative approach postulates the range of selected climatic parameters in determining the relative importance of the main geomorphic processes and so attempts to establish *morphogenic regions* as defined by the climatic ranges of particular combinations of processes. Peltier (1950) used plots of mean annual temperature and rainfall to define zones of differing intensity of frost action and chemical weathering, and hence weathering regions, and similarly for mass movement, wind action and pluvial erosion as shown in Fig. 93. The ranges of his arid (warm-arid) and semiarid morphogenic regions so defined have been superposed on the diagrams showing the relative effectiveness of the various processes.

Peltier's arid morphogenic region, with less than about 400 mm rainfall, is held to be characterised by weak chemical weathering

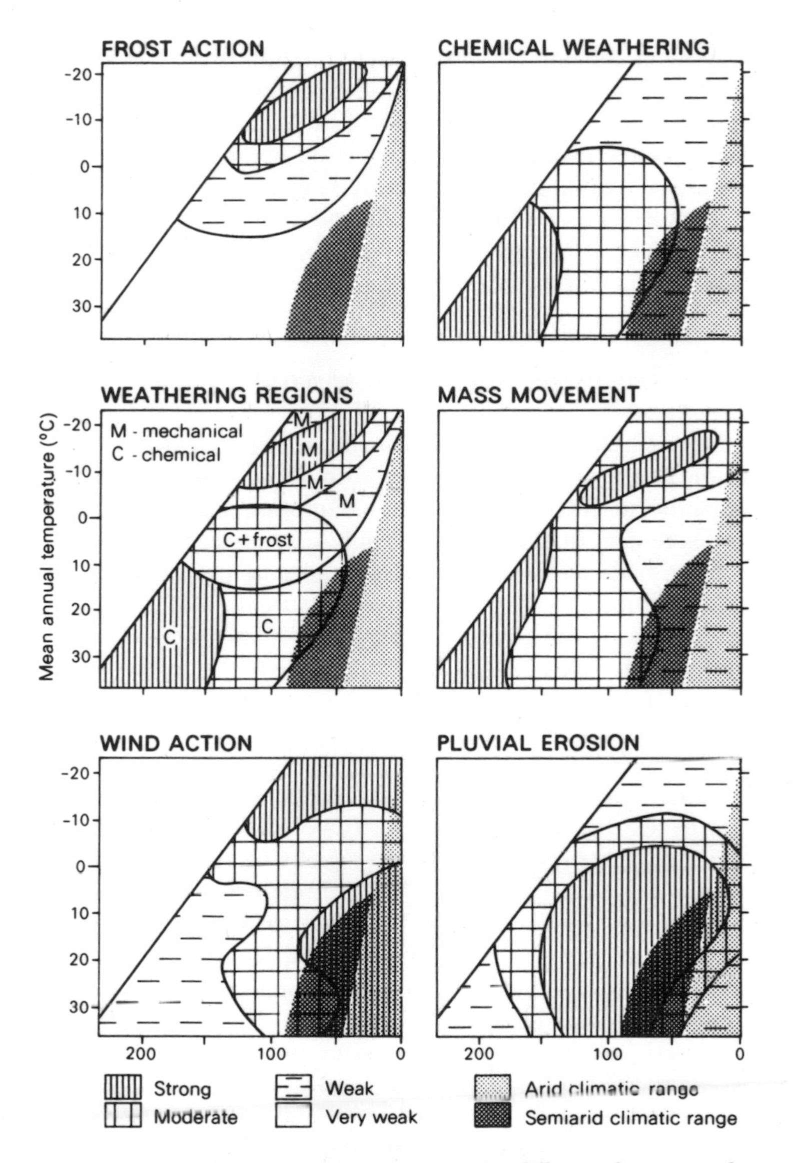

93. Effectiveness of land-forming processes under different climatic conditions, related to arid and semiarid morphogenic regions. After Peltier, 1950.

and insignificant frost action, by minimal mass movement and pluvial erosion, and by strong wind action. The semiarid region, with between 400 and 600 mm annual rainfall, is distinguished by a trend to strong pluvial erosion and by some weakening of wind action towards its humid margin. Peltier's scheme is open to criticism for the broadness of the climatic parameters chosen and for its use of annual means, but more realistic parameters such as rainfall intensity are rarely available, and in any case it is doubtful whether their use would add to the geomorphological precision of what remains a very qualitative generalisation.

VERIFICATION OF THE DESERT MORPHOGENIC SYSTEM

The approaches described above define morphogenic regions in terms of the factors controlling landform development rather than of the landforms themselves, and the assumptions about geomorphic processes remain qualitative and highly subjective. Some attempts have been made to verify the reality and true limits of morpho-climatic divisions, both through the measurement of geomorphic processes and by morphometric analysis of landforms. These are reviewed briefly below, in so far as they relate to the desert morphogenic system.

Verification by measurement of rates of fluvial erosion

Regional rates of fluvial erosion derived from drainage-basin studies by Corbel (1964), given in Table XIII, indicate that the tropical and subtropical deserts are regions of very slow erosion, with rates more than an order of magnitude below those of more humid zones of similar latitude, but that this difference is much reduced in temperate and cold deserts, particularly in mountain-and-basin deserts subject to frost shattering and with snow-fed streams. Differences between erosion rates of uplands and plains in the deserts are held to be in the same proportions as in humid zones, being least in the tropics and greatest in temperate latitudes. Comparable studies by Fournier (1960) and Strakhov (1967) also indicate that the arid zones are areas of slow denudation. However, the establishment by Fournier of the importance of seasonality of rainfall in erosion, as a measure of rainfall intensity and its interaction with vegetation cover, suggests that erosion should

TABLE XIII **Rates of Regional Erosion in Morphoclimatic Zones*** **($m^3/km^2/yr$)**

Climate	Arid under 200 mm rainfall		'Normal' 200–1500 mm rainfall		Humid over 1500 mm rainfall	
	Mountain	Plain	Mountain	Plain	Mountain	Plain
Hot (15°N–15°S)	1.0	0.5	25	10	30	15
Tropical (15°–23.5°N and S)	1.0	0.5	30	15	40	20
Extratropical (Temperature over 15°C)	4	1	100	20	100	30
Temperate (Temperature 0–13°C)	50	10	100	30	150	40
Cold (Temperature under 0°C)	50	15	100	30	180	—
Polar	50	15	100	30	150	—
Glaciated, polar	50		1000		2000	
Glaciated, non-polar	—		—		2000	

* After Corbel, 1964.

increase markedly in the semiarid zones, and this is implicit in the contrast between the erosion rates for arid and 'normal' climatic regions in Table XIII.

It would be realistic to regard such figures as no more than indications of relative magnitudes. Furthermore, the erosional rates cited are derived from medium-sized to very large drainage basins and are for movements of sediment to the sea rather than for local transfers between upland and plain. For this reason the measures are not very meaningful for the disorganised or interior systems of ephemeral rivers in deserts.

A more realistic indication of erosion rates for these localised conditions in upland deserts is given by results obtained by Langbein and Schumm (1958) from measurements of reservoir fills and stream-sediment concentrations in smaller catchments of about 4000 km² (Fig. 94). These indicate maximum sediment yields with an effective annual precipitation* of about 300 mm under temperate conditions, probably rising to 600 mm in the tropics. Although the peaks on these curves may additionally reflect disturbance

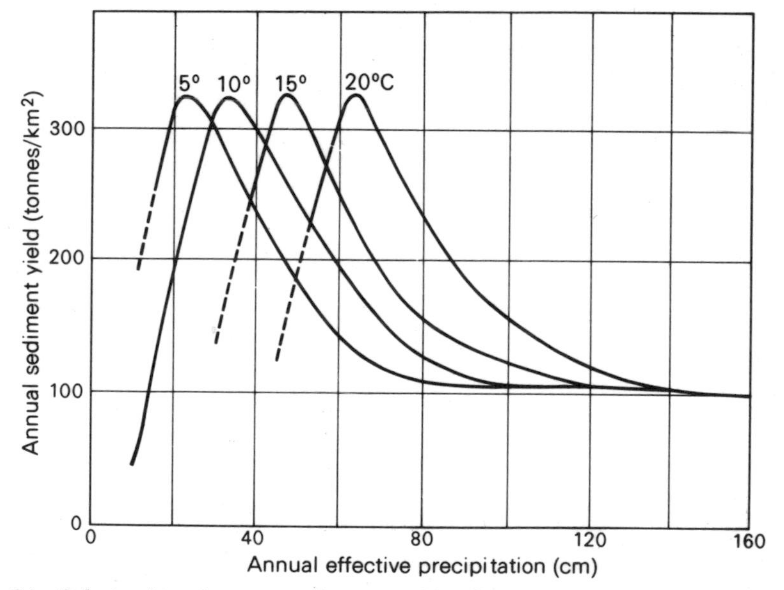

94. Relationships between sediment yield, effective precipitation and temperature. After Schumm, 1965.

* Defined as the annual precipitation required to produce the same runoff, assuming a mean annual temperature of 10°C.

through land use, they rightly demonstrate the natural vulnerability of semiarid regions, where rainfall, while inadequate to sustain a sufficiently close vegetation to protect the surface, nevertheless yields runoff in amounts capable of severe erosion. The deserts, on the left-hand side of the curve, have insufficient streamflow fully to exploit the vulnerability of unvegetated surfaces.

Evidence of landform analysis

Texture of relief. Peltier (1962) correlated mean slope with drainage frequency for his various morphogenic regions by sampling on small-scale maps. His results (Fig. 95) indicate an essential similarity between the dry lands and the moderate (humid tem-

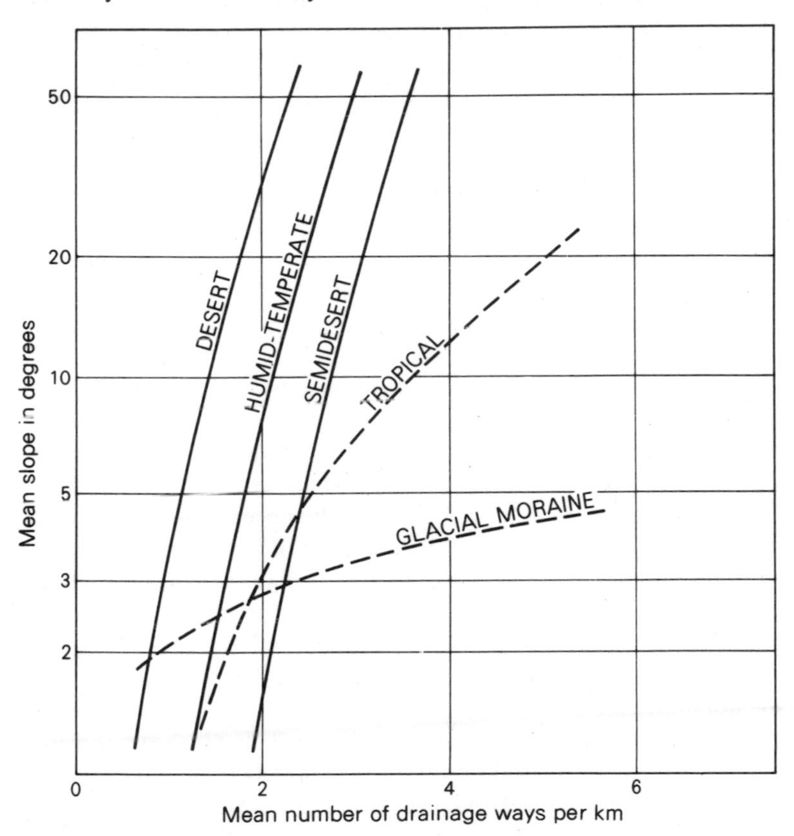

95. Increase of valley frequency with slope in different morphogenic regions. After Peltier, 1962.

perate) regions, indicating the overall dominance of fluvial dissection in the former and a lack of the mass movement on hillslopes which appears to suppress valley formation in periglacial (glacial moraine) zones.

The texture of relief in the deserts is shown to be open, whilst that in semiarid regions is revealed as above-average, consistent with indicated rates of fluvial erosion.

Drainage density. Melton (1957) measured drainage density and stream frequency on topographic maps of 80 upland basins of third to fifth order in the southwestern United States, ranging from arid to subhumid, and correlated the first of these parameters with aridity as measured by the Thornthwaite P-E index, and the second with percentage bare ground. His findings (Fig. 15) show an inverse relationship between drainage density and P-E index, and the significance of the related factor of open vegetation on stream spacing is also demonstrated. Melton found that aridity, infiltration, and rainfall intensity explained 92 per cent of the variability in drainage density, but insisted that these climatic factors act in 'complex ways, through the agencies of vegetal growth, soil formation, runoff and erosion, infiltration, and soil creep' (p. 37).

The relationships established by Melton differ from those claimed by Peltier in that drainage density is shown to continue to increase into the truly arid areas, where Peltier had found the texture of relief to be relatively open. This is explained by the fact that Melton's sample was of upland catchments, whereas Peltier's measurements were from areas of 225 km^2, incorporating uplands and lowlands. The contrast between closely dissected desert uplands and open desert plains from which drainage may be absent has been stressed in the preceding chapters, and no measure of landforms or of landforming processes which conceals this difference can be realistic. A recent examination of drainage density in relation to climate demonstrates that whereas the arid and semiarid zones generally exhibit greater densities than the humid zones, and particularly the humid temperate zones, the ranges of drainage density are also considerably greater in the deserts and overlap with those of wetter regions, indicating the falling off in the intensity of fluvial dissection away from desert uplands (Gregory and Gardiner, 1975).

DESERT AND PERIGLACIAL MORPHOGENIC SYSTEMS: A CASE OF CONVERGENCE

In this chapter emphasis has so far been placed on the uniqueness of the desert morphogenic system, but it is also appropriate to consider affinities, notably between the desert and periglacial morphogenic systems, which resemble each other in many ways. In part this resemblance follows from shared environmental factors such as sparse vegetation. The periglacial regions indeed form a climatic extension of the cold deserts; precipitation is generally small and there may be an excess of evaporation in the short summers, whilst in winter the moisture store, being frozen, is not available to geomorphic processes. Similarities also result from the like expression of different processes, notably between alternate expansion and contraction of soil mantles due to wetting and deep drying in the deserts and those resulting from freezing and thawing in periglacial regions, although desiccation is also involved in the latter.

Mortensen (1930b) was first struck by similarities between landforms in Spitzbergen and in the north Chilean desert, and Meckelein (1965) first reviewed such parallels as a case of 'geomorphological convergence', drawing on his experience of extreme-arid forms in the east-central Sahara, although he later suggested that many of the processes were in fact identical (Meckelein, 1974). He noted the prevalence of salt-hydration weathering in both zones, as shown in granular disintegration and in tafoni formation (Wellman and Wilson, 1965; Calkin and Cailleux, 1962). He also drew attention to features common to periglacial blockfields and desert hamadas, such as pavements swept clean of fine interstitial products by wind action or by sheet erosion, and angular shattered boulders where frost shattering had simulated fracturing due to weathering in the presence of salts in the hamadas. Varnishes of surficial iron enrichment due to limited penetration and capillary return of weathering solutions occur in both settings, for strong summer evaporation can occur under a periglacial regime, and desiccation also accompanies winter freezing. The *Rillensteine* found on limestone fragments in the hamadas of the Negev also occur in periglacial blockfields subject to solution by surface water above frozen subsoils in the cool spring. Restriction of desert weathering to surface crusts by lack of moisture has its parallel in subpolar regions, where constraints set by low temperatures are additionally involved.

A combination of bare ground, loose rock waste, dry frozen surfaces and storminess makes the periglacial regions a zone of wind action second only to the deserts. Active sand dunes are well known from the dry valleys of Antarctica, as they were formerly extensive near ice margins in the glaciated lowlands of Europe and North America. Periglacial regions and nearby cold deserts have been main sources of loess, and frost action on glacifluvial silts has been even more effective in activating deflation than has the effect of salts on alluvium and playa-floor deposits in warm deserts. However, just as deflation of fine-textured desert soils in warm deserts is checked by crusts due to rainbeat or salts, so in the periglacial zone it is restricted by winter freezing. In both zones rillwash is the main agent of erosion on slopes.

Whilst the periglacial regions contain features accepted as typical of warm deserts, the latter contain forms typical of periglacial regions. For example, warm stony deserts exhibit patterned ground on a scale surpassed only by periglacial environments. Here the effects of freeze-thaw are simulated by expansion and contraction due to wetting and drying of base-saturated and often saline expansive clays. Many periglacial patterns have their direct equivalent in the deserts; for example sorted and unsorted polygons are represented by stony and normal gilgai, and sorted steps occur on gentle slopes in both settings. Many reg pavements result from the extrusion of stone from clay subsoils, as in forms of frozen ground, and these may also show stone arrangements on a small scale. In the same way that water is subject to trapping and limited release by frequent frost in periglacial regions, patterned ground in the arid zone forms repetitive hydrologic systems with miniature run-on and runoff components, with inadequate excess water to establish a connected surface drainage.

Both the desert and the periglacial morphogenic systems are zones of landform preservation rather than of marked geomorphic change, with moisture in the one and temperature in the other as the main factors limiting morphogenesis.

PROBLEMS OF CLIMATE-BASED CLASSIFICATIONS OF DESERT MORPHOGENIC SYSTEMS

Survival of palaeoforms

Relict features from morphogenic systems of earlier climates complicate the interpretation of most landscapes in climatic terms, but the problem is particularly acute in the warm deserts where, as discussed in Chapter X, forms produced relatively rapidly in periods of ameliorated climate may survive little-altered under aridity. In these settings the climatic geomorphologist concerned with landforms as well as processes must consider climatic change in the interpretation of landforms. The elucidation of changes of climate may in fact become a major objective of that interpretation. The development of such a climatogenic geomorphology (*klimagenetische Geomorphologie*) is apparent in schemes by Büdel (1963, 1969) which use both fossil and active forms in the designation of morphoclimatic zones.

In the earlier of these schemes arid and semiarid morphogenic systems are not set apart but are incorporated as parts of a tropical zone of planation and a temperate zone of valley formation. The tropical deserts are linked with the savana and humid tropical areas, for Büdel was impressed by the extensive survival in the Sahara and in central Australia (Bremer, 1965) of plains with inselbergs fashioned under previous moister regimes. The extratropical deserts, including the northern Sahara, the Kalahari, and the southern portion of arid Australia are included in the subtropical zones of pediment and valley formation, transitional to the temperate zones of valley formation with which the cold deserts are linked.

In Büdel's later scheme the arid zone as a whole is recognised as one in which tropical planation surfaces (*Rumpffläche*) are preserved and in which pediments (*Fussfläche*) may be both preserved and extended (Fig. 96). Depositional sandy floodplains are also included in the complex of plains characteristic of this desert zone. The geographic grouping of tropical planation surfaces and extratropical pediments doubtless reflects the influence of Mensching (1958), who had studied the morphoclimatic bases of their

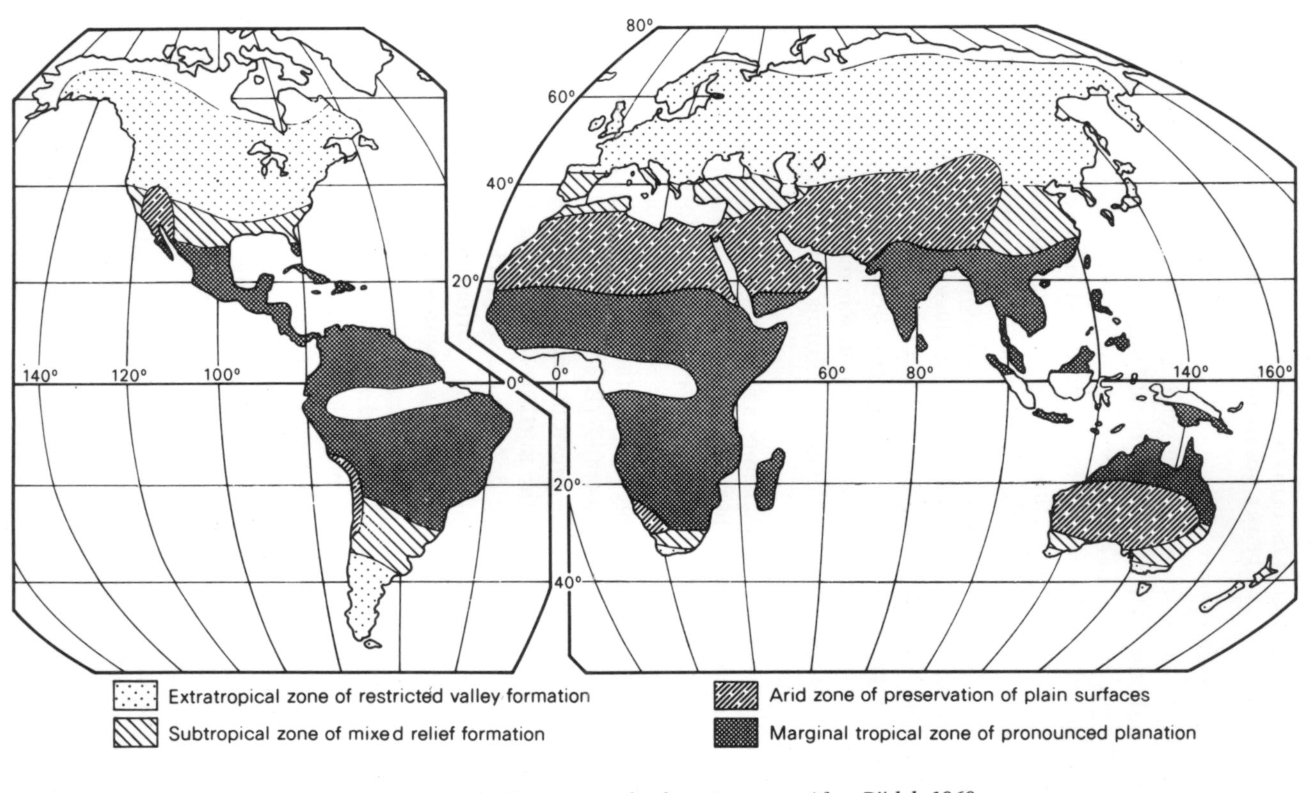

96. Desert and adjacent morphoclimatic zones. After Büdel, 1969.

association in the Saharan and Atlas regions of Morocco, but Büdel continues to regard the two forms as morphogenically distinct.

Control of morphogenesis by structure and major relief

Another obstacle to a morphogenic treatment and subdivision of deserts solely on climatic grounds is the strong expression of structure in desert relief and in the consequent compartmentation of geomorphic processes as discussed in Chapter I. General systems of climatic geomorphology apply most convincingly to tropical shield and platform deserts where stability and a shared tectonic and climatic history over large areas are expressed in uniform and generally low relief. They are less applicable in mountain-and-basin deserts. For example, Gabriel (1964) has demonstrated the inappropriateness of a morphoclimatic zonation, such as that proposed by Mortensen, in the Dasht-i-Lut of southern Iran. In the innermost lowland of extreme-arid climate, dunes predominate and wind action prevails, whereas water action increasingly dominates towards the fringing piedmont with its normal-arid climate. It is perhaps for this reason that many geomorphologists who have worked extensively in extratropical and continental deserts of contrasted relief and marked Quaternary changes of climate, notably the French, have tended to rely principally on morphostructural criteria in the classification of deserts and only secondarily on climatic — and above all bioclimatic — differences (Coque *et al.*, 1972).

BIBLIOGRAPHY

AHNERT, F. (1960). The influence of Pleistocene climates upon the morphology of cuesta scarps on the Colorado Plateau. *Ann. Ass. Am. Geogr.*, **50**: 139–56.

ALIMEN, H. (1958). Premiers résultats de l'étude morphométrique de cailloutis sahariens. *Revue Géomorph. dyn.*, **9**: 161–73.

ALLCHIN, B. and GOUDIE, A. (1971). Dunes, aridity, and early man in Gujarat, western India. *Man*, **6**: 248–65.

ALLEN, J. R. L. (1968). The nature and origin of bedform hierarchies. *Sedimentology*, **10**: 161–82.

— (1969). On the geometry of current ripples in relation to stability of fluid flow. *Geogr. Annalr, A*, **51**: 61–96.

ANTEVS, E. (1952). Arroyo-cutting and filling. *J. Geol.*, **60**: 375–85.

AUFRÈRE, L. (1930). L'orientation des dunes continentales. *Rept. Proc. 12th Int. Geog. Conf., Cambridge, July 1928*: 220–31.

BAGNOLD, R. A. (1941). *The Physics of Blown Sand and Desert Dunes*. London.

— (1951). Sand formations in southern Arabia. *Geogrl J.*, **117**: 78–86.

— (1953). The surface movement of blown sand in relation to meteorology. Pp. 89–93 in *Desert Research, Proc. Int. Symp., Jerusalem, May, 1952. Res. Counc. Israel Spec. Publn.*, 2.

BAKKER, J. P. and LE HEUX, J. W. N. (1952). A remarkable new geomorphological law. *Koninklijke nederlandsche Akademie van Wetenschappen. B.*, **55**: 399–410, 544–71.

BALL, J. (1927). Problems of the Libyan Desert. *Geogrl J.*, **70**: 21–38, 105–28, 209–24.

— (1933). The Qattara depression of the Libyan Desert and the possibility of its utilization for power production. *Geogrl J.*, **82**: 289–314.

BARTON, D. C. (1916). Notes on the disintegration of granite in Egypt. *J. Geol.*, **24**: 382–93.

— (1938). Discussion: the disintegration and exfoliation of granite in Egypt. *J. Geol.*, **46**: 109–11.

BEARD, J. S. (1973). The elucidation of palaeodrainage patterns in Western Australia through vegetation mapping. *Veg. Surv. W. A. Occnl. Paper*, 1.

BEATY, C. B. (1959). Slope retreat by gullying. *Bull. geol. Soc. Am.*, **70**: 1479–82.

— (1963). Origin of alluvial fans, White Mountains, California and Nevada. *Ann. Ass. Am. Geogr.*, **53**: 516–35.

— (1970). Age and estimated rate of accumulation of an alluvial fan, White Mountains, California, U.S.A. *Am. J. Sci.*, **268**: 50–77.

— (1974). Debris flows, alluvial fans, and a revitalized catastrophism. *Z. Geomorph.-Suppl.*, **21**: 39–51.

BEAUDET, G., MAURER, G. and RUELLAN, A. (1967). Le Quaternaire marocain. Observations et hypothèses nouvelles. *Revue Géogr. phys. Géol. dyn.*, **9**: 269–309.

BEAUMONT, P. (1968). Salt weathering on the margin of the Great Kavir, Iran. *Bull. geol. Soc. Am.*, **79**: 1683–4.

BERKEY, C. P. and MORRIS, F. K. (1927). *Geology of Mongolia.* Am. Mus. Nat. Hist., New York.

BERTOUILLE, H. (1972). Étude mathématique des phénomènes de thermoclastie et de cryoclastie. *C.N.R.S. Centre de Géomorphologie de Caen, Bull.*, **12**.

BETTENAY, E. (1962). The salt lake systems and their associated aeolian features in in the semi-arid regions of Western Australia. *J. Soil Sci.*, **13**: 10–17.

BEVERAGE, J. P. and CULBERTSON, J. K. (1964). Hyperconcentrations of suspended sediment. *Proc. Am. Soc. civ. Engrs., J. Hydraulics Divn.*, **6**: 117–28.

BIRD, E. C. F. (1968). *Coasts.* Canberra.

BIROT, P. (1968). *Contribution à l'Étude de la Désagrégation des Roches.* Centre de Documentation Universitaire, Paris.

BLACKWELDER, E. (1928). Mudflow as a geologic agent in semiarid mountains. *Bull. geol. Soc. Am.*, **39**: 465–84.

— (1929). Cavernous rock surfaces of the desert. *Am. J. Sci.*, **217**: 393–9.

— (1931). The lowering of playas by deflation. *Am. J. Sci.*, **221**: 140–4.

— (1933). The insolation hypothesis of rock weathering. *Am. J. Sci.*, **226**: 97–113.

— (1934). Yardangs. *Bull. geol. Soc. Am.*, **45**: 159–66.

— (1948). Historical significance of desert lacquer (abstract). *Bull. geol. Soc. Am.*, **59**: 1367.

BLISSENBACH, E. (1954). Geology of alluvial fans in semiarid regions. *Bull. geol. Soc. Am.*, **65**: 175–90.

BONYTHON, C. W. (1955). The filling and drying-up. Pp. 27–36 in *Lake Eyre, South Australia. The Great Flooding of 1945–50.* Report of the Lake Eyre Committee, Roy. Geog. Soc. Australasia, S. Aust. Branch.

— (1956). The salt of Lake Eyre — its occurrence in Madigan Gulf and its possible origin. *Trans. R. Soc. S. Aust.*, **79**: 66–92.

— (1963). Further light on river floods reaching Lake Eyre. *Proc. R. Geog. Soc. Australasia, S. Aust. Branch,* **64**: 9–22.

— and MASON, R. (1953). The filling and drying of Lake Eyre. *Geogrl J.*, **119**: 321–30.

BOOCOCK, C. and VAN STRATEN, O. J. (1962). Notes on the geology and hydrogeology of the central Kalahari region, Bechuanaland Protectorate. *Trans. geol. Soc. S. Afr.*, **65**: 125–71.

BOULAINE, J. (1954). La sebkha de Ben Ziane et sa 'lunette' ou bourrelet. *Revue Géomorph. dyn.*, **5**: 102–23.

BOURNE, J. A., TWIDALE, C. R. and SMITH, D. M. (1974). The Corrobinnie Depression, Eyre Peninsula, South Australia. *Trans. R. Soc. S. Aust.*, **98**: 139–52.

BOWLER, J. M. (1968). Australian landform example No. 11: lunette. *Aust. Geogr.*, **10**: 402–4.

— (1971). Pleistocene salinities and climatic change: evidence from lakes and lunettes in south-eastern Australia. Pp. 47–65 in *Aboriginal Man and Environment in Australia*, ed., D. J. Mulvaney and J. Golson. Canberra.

— (1973). Clay dunes: their occurrence, formation and environmental significance. *Earth Sci. Rev.*, **9**: 315–38.

— (1975). Deglacial events in southern Australia: their age, nature, and palaeo-climatic significance. Pp. 75–82 in *Quaternary Studies*, ed. R. P. Suggate and M. M. Cresswell. *Bull. R. Soc. N.Z.*, **13**.

BRADLEY, W. C. (1963). Large-scale exfoliation in massive sandstones of the Colorado Plateau. *Bull. geol. Soc. Am.*, **75**: 519–28.

BREMER, H. (1965). Der Einfluss von Vorzeitformen auf die rezente Formung in einem Trockengebiet — Zentralaustralien. *Tagungsber u. wiss. Abh. deutsch. Geographentag, Heidelberg, 1963*, **37**: 184–96.

BROOKFIELD, M. (1970a). Winds of arid Australia. *CSIRO Aust. Div. Land Res. Tech. Paper* **30**. Melbourne.

— (1970b). Dune trends and wind regime in central Australia. *Z. Geomorph.-Suppl.*, **10**: 121–53.

BRYAN, K. (1922). Erosion and sedimentation in the Papago Country, Arizona. *U.S. geol. Surv. Bull.*, **730B**: 19–90.

— (1925a). The Papago Country, Arizona. *U.S. geol. Surv. Water Supply Paper*, **499**.

— (1925b). Date of channel trenching (arroyo cutting) in the arid Southwest. *Science*, **62**: 338–44.

— (1927). Pedestal rocks formed by differential erosion. *U.S. geol. Surv. Bull.*, **790A**: 1–15.

— (1940). Gully gravure — a method of slope retreat. *J. Geomorph.*, **3**: 89–107.

— (1945). Glacial versus desert origin of loess. *Am. J. Sci.*, **243**: 245–8.

BÜDEL, J. (1963). Klima-genetische Geomorphologie. *Geogr. Rdsch.*, **7**: 269–86.

— (1969). Das System der klima-genetischen Geomorphologie. *Erdkunde*, **23**: 165–83.

BULL, W. B. (1964). Geomorphology of segmented alluvial fans in western Fresno County, California. *U.S. geol. Surv. Prof. Paper*, **352–E**.

— (1972). Recognition of alluvial-fan deposits in the stratigraphic record. Pp. 63–83 in *Recognition of Ancient Sedimentary Environments*, ed. J. K. Rigby and W. K. Hamblin. Society of Economic Paleontologists and Mineralogists Special Publn, **16**.

BUTLER, B. E. (1956). Parna — an aeolian clay. *Aust. J. Sci.*, **18**: 145–51.

— and HUTTON, J. T. (1956). Parna in the Riverine Plain of south-eastern Australia and the soils thereon. *Aust. J. Agric. Res.*, **7**: 536–53.

CALKIN, P. and CAILLEUX, A. (1962). A quantitative study of cavernous weathering (taffonis) and its application to glacial chronology in Victoria Valley, Antarctica. *Z. Geomorph.*, **6**: 317–24.

CAMPBELL, E. M. (1968). Lunettes in southern South Australia. *Trans. R. Soc. S. Aust.*, **92**: 85–109.

CAMPBELL, I. A. (1973). Control of canyon and meander forms by jointing. *Area*, **5**: 291–6.

— (1974). Measurements of erosion on badlands surfaces. *Z. Geomorph.-Suppl.*, **21**: 122–37.

CAPOT-REY, R. (1943). La morphologie de l'Erg Occidental. *Trav. Inst. de Rech. Sah.*, **2**: 69–107.

— (1947). L'Edeyen de Mourzouk. *Trav. Inst. de Rech. Sah.*, **4**: 67–109.

— (1953). *Le Sahara Français*. Paris.

— and CAPOT-REY, F. (1948). Le déplacement des sables éoliens et la formation des dunes désertiques, d'après R. A. Bagnold. *Trav. Inst. de Rech. Sah.*, **5**: 47–80.

— and GREMION, M. (1964). Remarques sur quelques sables sahariens. *Trav. Inst. de Rech. Sah.*, **23**: 153–63.

CARROLL, D. (1944). The Simpson Desert Expedition, 1939. Scientific Reports: No. 2, Geology — Desert sands. *Trans. R. Soc. S. Aust.*, **68**: 49–59.

CARSON, M. A. (1971). An application of the concept of threshold slopes to the Laramie Mountains, Wyoming. Pp. 31–48 in *Slopes: form and process*, ed. D. Brunsden. Inst. Br. Geogr Spec. Publn., 3.

— and KIRKBY, M. J. (1972). *Hillslope Form and Process*. Cambridge.

CHURCHWARD, H. M. (1963a). Soil studies at Swan Hill, Victoria, Australia. II. Dune moulding and parna formation. *Aust. J. Soil Res.*, **1**: 103–16.

— (1963b). Soil studies at Swan Hill, Victoria, Australia. IV. Groundsurface history and its expression in the array of soils. *Aust. J. Soil Res.*, **1**: 242–55.

CLARK, D. M., MITCHELL, C. W. and VARLEY, J. A. (1974). Geomorphic evolution of sediment filled solution hollows in some arid regions (Northwestern Sahara). *Z. Geomorph.-Suppl.*, **20**: 130–9.

CLARKE, R. H. and PRIESTLEY, C. H. B. (1970). The asymmetry of Australian desert sand ridges. *Search*, **1**: 77–8.

CLEMENTS, T. *et al.* (1963). *A study of windborne sand and dust in desert areas*. U.S. Army Natick Laboratories, Natick, Mass., Earth Sciences Division, Technical Report, **ES-8**.

CLOS-ARCEDUC, A. (1969). *Essai d'explication des formes dunaires sahariennes*. Études de photo-interprétation, **4**. Inst. Géog. Nat., Paris.

COLE, G. A. (1968). Desert Limnology. Pp. 423–86 in *Desert Biology*, ed. G. W. Brown. New York.

CONRAD, G. (1963). Synchronisme du dernier Pluvial dans le Sahara septentrional et le Sahara méridional. *C. r. Ac. hebd. Séanc. Acad. Sci., Paris,* **257**: 2506–9.

— (1969). *L'Évolution Continentale Post-Hercynienne du Sahara Algérien*. Centre de Recherches sur les Zones Arides, Série Géologie, **10**. C.N.R.S., Paris.

COOKE, R. U. (1970a). Morphometric analysis of pediments and associated landforms in the western Mojave Desert, California. *Am. J. Sci.*, **269**: 26–38.

— (1970b). Stone pavements in deserts. *Ann. Ass. Am. Geogr.*, **60**: 560–77.

— (1974). The rainfall context of arroyo initiation in southern Arizona. *Z. Geomorph.-Suppl.*, **21**: 63–75.

— and REEVES, R. W. (1972). Relations between debris size and the slope of mountain fronts and pediments in the Mojave Desert, California. *Z. Geomorph.*, **16**: 76–82.

— and SMALLEY, T. J. (1968). Salt weathering in deserts. *Nature, Lond.*, **220**: 1226–7.

— and WARREN, A. (1973). *Geomorphology in Deserts*. London.

COQUE, R. (1955). Morphologie et croûte dans le Sud tunisien. *Annls Géogr.*, **64**: 359–70.

— (1962). *La Tunisie Présaharienne, Étude Géomorphologique.* Paris.

—, DRESCH, J. and ROGNON, P. (1972). Géomorphologie et Quaternaire des régions arides. Pp. 107–17 in *Recherches Géographiques en France.* Paris.

CORBEL, J. (1964). L'érosion terrestre, étude quantitative. *Annls Géogr.*, **73**: 385–412.

CZAJKA, W. (1972). Windschliffe als Landschaftsmerkmal. *Z. Geomorph.*, **16**: 27–53.

DAVEAU, S. (1965). Dunes ravinées et dépôts du Quaternaire récent dans le Sahel mauretanien. *Revue de Géog. de l'Afr. occ.*, **1–2**: 7–48.

DAVIES, J. L. (1969). *Landforms of Cold Climates.* Canberra.

DAVIS, W. M. (1905). The geographical cycle in an arid climate. *J. Geol.*, **13**: 381–407.

— (1930). Rock floors in arid and in humid climates. *J. Geol.*, **38**: 1–27, 136–58.

— (1938). Sheetfloods and streamfloods. *Bull. geol. Soc. Am.*, **49**: 1337–416.

DENNY, C. S. (1965). Alluvial fans in the Death Valley region, California and Nevada. *U.S. geol. Surv. Prof. Paper*, **466**.

— (1967). Fans and pediments. *Am. J. Sci.*, **265**: 81–105.

DRESCH, J. (1950). Sur les pédiments en Afrique méditerranéenne et tropicale. *Union Géogr. Inter., Congr. Inter. Géogr., Lisbonne 1949*, **2**: 19–28.

— (1964). Remarques sur une division géomorphologique des régions arides et les caractères originaux des régions arides méditerranéennes. Pp. 23–30 in *Land Use in Semi-arid Mediterranean Climates.* UNESCO, Paris.

— (1968). Reconnaissance dans le Lut (Iran). *Bull. Ass. Géogr. fr.*, **362–3**: 143–53.

DUBIEF, J. (1943). Les vents du sable dans le Sahara français. *Trav. Inst. de Rech. Sah.*, **2**: 11–35.

— (1953a). *Essai sur l'Hydrologie Superficielle au Sahara.* Algiers.

— (1953b). Ruissellement superficiel au Sahara. Pp. 303–14 in *Actions Éoliennes, Phénomènes d'Evaporation et d'Hydrologie Superficielle dans les Régions Arides.* Colloques Int. du C.N.R.S. XXXV, Alger 1951. Paris.

— (1953c). Les vents de sable au Sahara francais. Pp. 45–68 in *Actions Éoliennes, Phénomènes d'Évaporation et d'Hydrologie Superficielle dans les Régions Arides.* Colloques Int. du C.N.R.S. XXXV, Alger 1951. Paris.

— (1971). Die Sahara, eine Klima-Wüste. Pp. 227–348 in *Die Sahara und ihre Randgebiete. I. Physiogeographie*, ed. H. Schiffers. Munich.

DUCE, J. T. (1918). The effect of cattle on the erosion of cañon bottoms. *Science*, **48**: 450–2.

DUNNE, T. and BLACK, R. D. (1970). Partial area contributions to storm runoff in a small New England watershed. *Water Resources Research*, **6**: 1296–311.

DURY, G. H. (1970). Morphometry of gibber gravel at Mt. Sturt, New South Wales. *J. geol. Soc. Aust.*, **16**: 656–66.

— (1973). Paleohydrologic implications of some pluvial lakes in northwestern New South Wales, Australia. *Bull. geol. Soc. Am.*, **84**: 3663–76.

EMERY, K. O. (1960). Weathering of the Great Pyramid. *J. sedim. Petrol.*, **30**: 140–3.

EMMETT, W. W. (1974). Channel aggradation in western United States as indicated by observations at Vigil Network sites. *Z. Geomorph.-Suppl.*, **21**: 52–62.

— and LEOPOLD, L. B. (1964). Downstream pattern of river-bed scour and fill. *Proc. Fed. Inter-Agency Sediment Conf. 1963*. U.S. Dept Agric. Res. Service, Misc. Publn, **970**: 399–409.

ENGEL, C. G. and SHARP, R. P. (1958). Chemical data on desert varnish. *Bull. geol. Soc. Am.*, **69**: 487–518.

ERHART, H. (1956). *La Génèse des Sols en tant que Phénomène Géologique*. Paris.

ESTORGES, P. (1959–61). Morphologie du Plateau arbaa. *Trav. Inst. de Rech. Sah.*, **18**: 21–56; **20**: 29–77.

EVANS, I. S. (1969/70). Salt crystallization and rock weathering: a review. *Revue Géomorph. dyn.*, **19**: 153–77.

EVANS, J. R. (1962). Falling and climbing sand dunes in the Cronese ('Cat') Mountain area, San Bernadino County, California. *J. Geol.*, **70**: 107–13.

EVENARI, M., SHANAN, L and TADMOR, N. (1971). *The Negev. The Challenge of a Desert*. London.

—, YAALON, D. H. and GUTTERMAN, Y. (1974). Note on soils with vesicular structure in deserts. *Z. Geomorph.*, **18**: 162–72.

EVERARD, C. E. (1964). Climatic change and man as factors in the evolution of slopes, in Slope Profiles: a symposium. *Geogrl J.*, **130**: 65–9.

EXON, N. F., LANGFORD-SMITH, T. and MCDOUGALL, I. (1970). The age and geomorphic correlations of deep-weathering profiles, silcrete, and basalt in the Roma-Amby region, Queensland. *J. Geol. Soc. Aust.*, **17**: 21–30.

FAIR, T. J. D. (1947). Slope form and development in the interior of Natal, South Africa. *Trans. geol. Soc. S. Afr.*, **50**: 105–18.

— (1948). Hill-slopes and pediments of the semi-arid Karroo. *S. Afr. geogrl J.*, **30**: 71–9.

FINKEL, H. J. (1959). The barchans of southern Peru. *J. Geol.*, **67**: 614–47.

FLACH, K. W. *et al.* (1969). Pedocementation: induration by silica, carbonates, and sesquioxides in the Quaternary. *Soil Sci.*, **107**: 442–53.

FLINT, R. F. (1971). *Glacial and Quaternary Geology*. New York.

— and BOND, G. (1968). Pleistocene sand ridges and pans in western Rhodesia. *Bull. geol. Soc. Am.*, **79**: 299–314.

— and GALE, W. A. (1958). Stratigraphy and radiocarbon dates at Searles Lake, California. *Am. J. Sci.*, **256**: 689–714.

FOLK, R. L. (1971a). Longitudinal dunes of the northwestern edge of the Simpson Desert, Northern Territory, Australia. 1. Geomorphology and grain size relationships. *Sedimentology*, **16**: 5–54.

— (1971b). Genesis of longitudinal and oghurd dunes elucidated by rolling upon grease. *Bull. geol. Soc. Am.*, **82**: 3461–8.

FOURNIER, F. (1960). *Climat et Érosion: La Relation entre l'Érosion du Sol par l'Eau et les Précipitations Atmosphériques*. Paris.

FRIEDMAN, G. M. (1961). Distinction between dune, beach and river sands from their textural characteristics. *J. sedim. Petrol.*, **31**: 514–29.

FÜRST, M. (1965). Hammada — Serir — Erg. Eine morphogenetische Analyse des nordöstlichen Fezzan (Libyen). *Z. Geomorph.*, **9**: 385–421.

— (1966). Bau und Entstehung der Serir Tibesti. *Z. Geomorph.*, **10**: 387–418.

GABRIEL, A. (1964). Zum Problem des Formenschatzes in extrem-ariden Räumen. *Mitt. Österr. Geogr. Ges.*, **106**: 1–15.

GALLOWAY, R. W. (1965). A note on world precipitation during the last glaciation. *Eiszeit. u. Gegenwart*, **16**: 76–7.

— (1970). The full-glacial climate in the southwestern United States. *Ann. Ass. Am. Geogr.*, **60**: 245–56.

GAUTIER, E-F. (1950). *The Sahara*. 3rd ed. Paris.

GERSON, R. and INBAR, M. (1974). The field study program of the Jerusalem-Elat Symposium, 1974. Reviews and summaries of Israeli research projects. *Z. Geomorph.-Suppl.*, **20**: 1–40.

GILBERT, G. K. (1877). *Report on the Geology of the Henry Mountains*. Washington.

GILE, L. H., PETERSON, F. F. and GROSSMAN, R. B. (1966). Morphological and genetic sequences of carbonate accumulation in desert soils. *Soil Sci.*, **101**: 347–60.

GINZBOURG, D. and YAALON, D. H. (1963). Petrography and origin of the loess in the Be'er Sheva Basin. *Isr. J. Earth Sci.*, **12**: 68–70.

GOUDIE, A. (1972). The concept of post-glacial progressive desiccation. *Oxf. Univ. Dept. Geog. Res. Paper*, **4**.

— (1973). *Duricrusts in Tropical and Subtropical Landscapes*. Oxford.

— (1974). Further experimental investigation of rock weathering by salt and other mechanical processes. *Z. Geomorph.-Suppl.*, **21**: 1–12.

—, COOKE, R. and EVANS, I. (1970). Experimental investigation of rock weathering by salts. *Area*: 42–8.

GREGORY, J. W. (1914). The lake system of Westralia. *Geogrl J.*, **43**: 656–64.

GREGORY, K. J. and GARDINER, V. (1975). Drainage density and climate. *Z. Geomorph.*, **19**: 287–98.

GRIGGS, D. T. (1936). The factor of fatigue in rock exfoliation. *J. Geol.*, **44**: 783–96.

GROVE, A. T. (1969). Landforms and climatic change in the Kalahari and Ngamiland. *Geogrl J.*, **135**: 191–212.

— and WARREN, A. (1968). Quaternary landforms and climate on the south side of the Sahara. *Geogrl J.*, **134**: 194–208.

HACK, J. T. (1941). Dunes of the western Navajo Country. *Geogrl Rev.*, **31**: 240–63.

— (1957). Studies of longitudinal stream profiles in Virginia and Maryland. *U.S. geol. Surv. Prof. Paper*, **294–B**.

HAGEDORN, H. (1968). Uber aölische Abtragung und Formung in der Südost-Sahara. *Erdkunde*, **22**: 257–69.

— (1971). Untersuchungen über Relieftypen arider Räume an Beispielen aus dem Tibesti-Gebirge und seiner Umgebung. *Z. Geomorph.-Suppl.*, **11**.

HALLSWORTH, E. G., ROBERTSON, G. K. and GIBBONS, F. R. (1955). Studies in pedogenesis in New South Wales. VII. The gilgai soils. *J. Soil Sci.*, **6**: 1–31.

HANNA, S. R. (1969). The formation of longitudinal sand dunes by large helical eddies in the atmosphere. *J. App. Met.*, **8**: 874–83.

HARRIS, S. A. (1959). The classification of gilgaied soils: some evidence from northern Iraq. *J. Soil Sci.*, **10**: 27–33.

— (1968). Gilgai. Pp. 425–6 in *The Encyclopedia of Geomorphology*, ed. R. W. Fairbridge. New York.

HASTENRATH, S. (1967). The barchans of the Arequipa region, southern Peru. *Z. Geomorph.*, **11**: 300–31.

HILLS, E. S. (1940). The lunette, a new landform of aeolian origin. *Aust. Geogr.*, **3**: 15–21.

— (1953). Regional geomorphic patterns in relation to climatic types in dry areas of Australia. Pp. 355–64 in *Desert Research. Proc. Int. Symp., Jerusalem, May 1952*. Res. Council Israel Spec. Publn, **2**.

HOLM, D. A. (1960). Desert geomorphology in the Arabian Peninsula. *Science*, **132**: 1369–79.

HOLMES, C. D. (1955). Geomorphic development in humid and arid regions; a synthesis. *Am. J. Sci.*, **253**: 377–90.

HOOKE, R. le B. (1967). Processes on arid-region alluvial fans. *J. Geol.*, **75**: 438–60.

— (1968). Steady-state relationships on arid-region alluvial fans in closed basins. *Am. J. Sci.*, **266**: 609–29.

— (1972). Geomorphic evidence for late-Wisconsin and Holocene tectonic deformation, Death Valley, California. *Bull. geol. Soc. Am.*, **83**: 2073–98.

—, YANG, H-Y. and WEIBLEN, P. W. (1969). Desert varnish: an electron probe study. *J. Geol.*, **77**: 275–88.

HORTON, R. E. (1933). The role of infiltration in the hydrologic cycle. *Trans. Am. geophys. Un.*, **14**: 446–60.

HOWARD, A. D. (1942). Pediment passes and the pediment problem. *J. Geomorph.*, **5**: 3–31, 95–136.

HOYT, J. H. (1966). Air and sand movements to the lee of dunes. *Sedimentology*, **7**: 137–43.

HUME, W. F. (1925). *Geology of Egypt*. Vol. 1. Cairo.

HUNT, C. B. (1954). Desert varnish. *Science*, **120**: 183–4.

— (1961). Stratigraphy of desert varnish. *U.S. geol. Surv. Prof. Paper*, **424-B**: 194–5.

— and MABEY, D. R. (1966). Stratigraphy and structure, Death Valley, California. *U.S. geol. Surv. Prof. Paper*, **494-A**.

— *et al.* (1966). Hydrologic basin, Death Valley, California. *U.S. geol. Surv. Prof. Paper*, **494-B**.

HUNTINGTON, E. (1914). *The Climatic Factor as Illustrated in Arid America*. Carnegie Inst. Wash. Publn., **192**.

Hutton, J. T. (1968). The redistribution of the more soluble chemical elements associated with soils as indicated by analysis of rain-water, soils and plants. *9th Int. Cong. Soil Sci. Trans.*, **4**: 313–22.

— and Leslie, T. I. (1958). Accession of non-nitrogenous ions dissolved in rain-water to soils in Victoria. *Aust. J. Agric. Res.*, **9**: 492–507.

Jack, R. L. (1921). The salt and gypsum resources of South Australia. *Geol. Surv. S. Aust. Bull.*, **8**.

Jennings, J. N. (1968). A revised map of the desert dunes of Australia. *Aust. Geogr*, **10**: 408–9.

— and Sweeting, M. M. (1963). The Limestone Ranges of the Fitzroy Basin, Western Australia. *Bonn. geogr. Abh.*, **32**.

Jessup, R. W. (1960). The stony tableland soils of the southeastern portion of the Australian arid zone and their evolutionary history. *J. Soil Sci.*, **11**: 188–96.

— and Norris, R. M. (1971). Cainozoic stratigraphy of the Lake Eyre Basin and part of the arid region lying to the south. *J. geol. Soc. Aust.*, **18**: 303–31.

Johns, R. K. (1968). Investigation of Lakes Torrens and Gairdner. *S. Aust. Dept. Mines, Geol. Surv. Rept. Investigations*, **31**.

— and Ludbrook, N. H. (1963). Investigation of Lake Eyre. *S. Aust. Dept. Mines, Geol. Surv. Rept. Investigations*, **24**.

Johnson, D. W. (1932a). Rock fans of arid regions. *Am. J. Sci.*, **223**: 389–416.

— (1932b). Rock planes of arid regions. *Geogrl Rev.*, **22**: 656–65.

Joly, F. (1950). Pédiments et glacis d'érosion dans le Sud-Est du Maroc. *Congr. Inter. Géogr., Lisbonne 1949*, **2**: 110–25.

— (1962). Études sur le relief du Sud-Est marocain. *Trav. de l'Inst. scient. chérifien, Série Géol. et Géogr. phys.*, **10**.

— *et al.* (1954). Les hamadas sud-marocaines. *Trav. de l'Inst. scient. chérifien, Série gen.*, **2**.

Judson, S. (1950). Depressions of the northern portion of the Southern High Plains of eastern New Mexico. *Bull. geol. Soc. Am.*, **61**: 253–74.

Jutson, J. T. (1917). Erosion and the resulting land forms in sub-arid Western Australia, including the origin and growth of the dry lakes. *Geogrl J.*, **50**: 418–37.

— (1918). The influence of salts in rock weathering in sub-arid Western Australia. *Proc. R. Soc. Vict.*, **30**: 165–72.

— (1934). The physiography (geomorphology) of Western Australia. *Bull. geol. Surv. West Aust.*, **95**. 2nd ed. rev.

Kaiser, F. (1926). *Die Diamantenwüste Südwestafrikas*. 2 vols. Berlin.

Kesseli, J. E. and Beaty, C. B. (1959). *Desert flood conditions in the White Mountains of California and Nevada.* Natick, Mass., U.S. Army Quartermaster Res. & Eng. Center, Environmental Protection Res. Div. Tech. Rept., **EP-108**.

Killigrew, L. P. and Gikes, R. J. (1974). Development of playa lakes in south Western Australia. *Nature, Lond.*, **247**: 454–5.

KING, D. (1956). The Quaternary stratigraphic record at Lake Eyre North and the evolution of existing topographic forms. *Trans. R. Soc. S. Aust.*, **79**: 93–103.

— (1960). The sand ridge deserts of South Australia and related aeolian landforms of the Quaternary arid cycles. *Trans. R. Soc. S. Aust.*, **83**: 99–108.

KING, L. C. (1947). Landscape study in southern Africa. *Proc. geol. Soc. S. Afr.*, **50**: xxiii–lii.

— (1953). Canons of landscape evolution. *Bull. geol. Soc. Am.*, **64**: 721–52.

— (1957). The uniformitarian nature of hillslopes. *Trans. geol. Soc. Edin.*, **17**: 81–102.

— (1962). *The Morphology of the Earth*. Edinburgh.

KIRKBY, M. J. (1969). Erosion by water on hillslopes. Pp. 229–38 in *Water, Earth and Man*, ed. R. J. Chorley. London.

— and CHORLEY, R. J. (1967). Throughflow, overland flow and erosion. *Int. Assn. Sci. Hydrol. Bull.*, **12**: 5–21.

KOONS, D. (1955). Cliff retreat in the southwestern United States. *Am. J. Sci.*, **253**: 44–52.

KRINSLEY, D. B. (1970). *A Geomorphological and Paleoclimatological Study of the Playas of Iran.* U.S. geol. Surv. 2 vols. Washington.

KRUMBEIN, W. E. (1971). Biologische Entstehung von Wüstenlack. *Umschau*, **71**: 240–1.

KUENEN, Ph. H. (1960). Experimental abrasion 4. Eolian action. *J. Geol.*, **68**: 427–49.

— and PERDOK, W. G. (1962). Experimental abrasion 5. Frosting and defrosting of quartz grains. *J. Geol.*, **70**: 648–58.

LANCASTER, I. N. (1974). Pans of the southern Kalahari. *Botswana Notes and Rec.*, **6**: 157–69.

LANGBEIN, W. B. (1961). Salinity and hydrology of closed lakes. *U.S. geol. Surv. Prof. Paper*, **412**.

— and SCHUMM, S. A. (1958). Yield of sediment in relation to mean annual precipitation. *Trans. Am. geophys. Un.*, **39**: 1076–84.

LANGFORD-SMITH, T. and DURY, G. H. (1964). A pediment survey at Middle Pinnacle, near Broken Hill, New South Wales. *J. geol. Soc. Aust.*, **11**: 79–88.

— and — (1965). Distribution, character, and attitude of the duricrust in the northwest of New South Wales and the adjacent areas of Queensland. *Am. J. Sci.*, **263**: 170–90.

LAUDERMILK, A. C. (1931). On the origin of desert varnish. *Am. J. Sci.*, **221**: 51–66.

LAWSON, A. C. (1915). Epigene profiles of the desert. *Bull. Calif. Univ. Dept Geol. Sci.*, **9**: 23–48.

LEOPOLD, L. B. (1951). Rainfall frequency: an aspect of climatic variation. *Trans. Am. geophys. Un.*, **32**: 347–57.

—, EMMETT, W. W. and MYRICK, R. M. (1966). Channel and hillslope processes in a semiarid area, New Mexico. *U.S. geol. Surv. Prof. Paper*, **352-G**.

— and MADDOCK, T. (1953). The hydraulic geometry of stream channels and some physiographic implications. *U.S. geol. Surv. Prof. Paper*, **252**.

— and MILLER, J. P. (1956). Ephemeral streams — hydraulic factors and their relation to the drainage net. *U.S. geol. Surv. Prof. Paper*, **282-A**.

— and WOLMAN, M. G. (1957). River channel patterns: braided, meandering and straight. *U.S. geol. Surv. Prof. Paper*, **282-B**.

LETTAU, K. and LETTAU, H. (1969). Bulk transport of sand by the barchans of the Pampa La Joya in Southern Peru. *Z. Geomorph.*, **13**: 182–95.

LEWIS, A. (1936). Sand dunes of the Kalahari within the border of the Union. *S. Afr. geogr. J.*, **19**: 25–57.

LITCHFIELD, W. H. and MABBUTT, J. A. (1962). Hardpan in soils of semi-arid Western Australia. *J. Soil Sci.*, **13**: 148–59.

LONG, J. T. and SHARP, R. P. (1964). Barkan-dune movement in Imperial Valley, California. *Bull. geol. Soc. Am.*, **75**: 149–56.

LUSTIG, L. K. (1965). Clastic sedimentation in Deep Springs Valley, California. *U.S. geol. Surv. Prof. Paper*, **352-F**: 131–92.

— (1969). Trend surface analysis of the Basin and Range province and some geomorphic implications. *U.S. geol. Surv. Prof. Paper*, **500-D**.

MABBUTT, J. A. (1955). Pediment land forms in Little Namaqualand, South Africa. *Geogrl J.*, **121**: 77–83.

— (1961). A stripped land surface in Western Australia. *Trans. Inst. Br. Geogr.*, **29**: 101–14.

— (1963a). Geomorphology of the Wiluna-Meekatharra area. Pp. 107–22 in *Lands of the Wiluna-Meekatharra Area,1958*. CSIRO Aust. Land Res. Ser.7. Melbourne.

— (1963b). Wanderrie banks: microrelief patterns in semiarid Western Australia. *Bull. geol. Soc. Am.*, **74**: 529–40.

— (1965a). The weathered land surface in central Australia. *Z. Geomorph.*, **9**: 82–114.

— (1965b). Stone distribution in a stony tableland soil. *Aust. J. Soil Res.*, **3**: 131–42.

— (1966a). Mantle-controlled planation of pediments. *Am. J. Sci.*, **264**: 78–91.

— (1966b). Landforms of the western Macdonnell Ranges. Pp. 83–119 in *Essays in Geomorphology*, ed. G. H. Dury. London.

— (1967). Denudation chronology in central Australia. Structure, climate, and landform inheritance in the Alice Springs area. Pp. 144–81 in *Landform Studies from Australia and New Guinea*, ed. J. N. Jennings and J. A. Mabbutt. Canberra.

— (1968). Aeolian landforms in central Australia. *Australian Geographical Studies*, **6**: 139–50.

— (1973). Landforms. Commentary to map-sheet 'Landforms' in *Atlas of Australian Resources* (2nd series). Canberra

— and SULLIVAN, M. E. (1968). The formation of longitudinal dunes: evidence from the Simpson Desert. *Aust. Geogr.*, **10**: 483–7.

—, WOODING, R. A. and JENNINGS, J. N. (1969). The asymmetry of Australian desert sand ridges. *Aust. J. Sci.*, **32**: 159–60.

— *et al.*(1973). Lands of Fowlers Gap Station, New South Wales. *Fowlers Gap Arid Zone Res. Stn. Res. Ser.*, **3**. University of New South Wales.

McGee, W. J. (1897). Sheetflood erosion. *Bull. geol. Soc. Am.*, **8**: 87–112.

McKee, E. D. (1966). Structures of dunes at White Sands National Monument, New Mexico (and a comparison with structures of dunes from other selected areas). *Sedimentology*, 7: 1–69.

— and Douglass, J. R. (1971). Growth and movement of dunes at White Sands National Monument, New Mexico. *U.S. geol. Surv. Prof. Paper*, **750-D** : D108–14.

— and Tibbits, G. C. (1964). Primary structures of a seif dune and associated deposits in Libya. *J. sedim. Petrol.*, **34**: 5–17.

Madigan, C. T. (1936). The Australian sand-ridge deserts. *Geogrl Rev.*, **26**: 205–27.

— (1946). The Simpson Desert Expedition, 1939 Scientific Reports: No. 6, Geology

— The sand formations. *Trans. R. Soc. S. Aust.*, **70**: 45–63.

Mainguet, M. (1968). *Le Borkou.* Aspects d'un modelé eolien. *Annls Géogr.*, **77**: 296–322.

— (1970). Un étonnant paysage: les cannelures grèseuses du Bémbeché (N. du Tchad). *Annls Géogr.*, **79**: 58–66.

— (1975). Étude comparée des ergs, à l'échelle continentale (Sahara et déserts d'Australie). *Bull. Ass. Géogr. fr.*, **52**: 135–40.

Mammerickx, J. (1964). Quantitative observations on pediments in the Mojave and Sonoran Deserts (southwestern United States). *Am. J. Sci.*, **262**: 417–35.

Martin, J. (1965). Quelques types de dépressions karstiques du Moyen Atlas central. *Revue Géogr. maroc.*, 7: 95–106.

Martonne, E. de. (1927). Regions of interior-basin drainage. *Geogrl Rev.*, **17**: 397–414.

— (1942). Nouvelle carte mondiale de l'indice d'aridité. *Annls Géogr.*, **51**: 241–50.

Meckelein, W. (1959). *Forschungen in der zentralen Sahara. I. Klima-geomorphologie.* Brunswick.

— (1965). Beobachtungen und Gedanken zu geomorphologischen Konvergenzen in Polar und Wärmewüsten. *Erdkunde*, **19**: 31–9.

— (1974). Aride Verwitterung in Polargebieten im Vergleich zum subtropischen Wüstengürtel. *Z. Geomorph.-Suppl.*, **20**: 178–88.

Medlicott, H. B. and Blanford, W. T. (1879). *Geology of India.* Geol. Surv. India, Monograph Publn., 2.

Meigs, P. (1953). World distribution of arid and semi-arid homoclimates. Pp. 203–10 in *Reviews of Research on Arid Zone Hydrology.* Arid Zone Programme 1. UNESCO, Paris.

— (1966). *Geography of Coastal Deserts.* UNESCO Arid Zone Res. Ser., **28**. Paris.

Melton, F. A. (1940). A tentative classification of sand dunes. Its application to dune history in the southern High Plains. *J. Geol.*, **48**: 113–45.

Melton, M. A. (1957). An analysis of the relations among elements of climate, surface properties, and geomorphology. *Dept. Geol. Columbia Univ. Tech. Rep.*, 11. New York.

— (1965a). Debris-covered hillslopes of the southern Arizona Desert — consideration of their stability and sediment contribution. *J. Geol.*, **73**: 715–29.

— (1965b). The geomorphic and paleoclimatic significance of alluvial deposits in southern Arizona. *J. Geol.*, **73**: 1–38.

MENSCHING, H. (1958). Entstehung und Erhaltung von Flächen im semiariden Klima am Beispiel Nordwestafrikas. *Tagunsber. u. Wiss. Abh. deutsch. Geographentag Würzburg, 1957, Verh. Deutsch. Geog.*, **31**: 173–84.

—, GIESSNER, K. and STUCKMANN, G. (1970). Sudan-Sahel-Sahara. *Jahrb. geog. Ges. Hannover 1969*.

— and RAYNAL, R. (1954). Füssflächen in Marokko. *Petermanns geogr. Mitt.*, **98**: 171–6.

MILLER, J. P. (1958). High mountain streams: effects of geology on channel characteristics and bed material. *New Mexico Bur. Mines & Min. Resources Mem.*, **4**.

MITCHELL, C. W. and WILLIMOTT, S. G. (1974). Dayas of the Moroccan Sahara and other arid regions. *Geogrl J.*, **140**: 441–53.

MONOD, TH. (1958). Majâbat al-Koubrâ. Contribution à l'Étude de l''Empty Quarter' ouest-saharien. *Mem. de l'IFAN*, 52. Dakar.

MORRISON, R. B. (1965). Quaternary geology of the Great Basin. Pp. 265–85 in *The Quaternary of the United States*, ed. H. E. Wright and D. G. Frey. Princeton.

MORTENSEN, H. (1927). Der Formenschatz der nordchilenischen Wüste. *Abh. Ges. f. Wiss., Göttingen, math-physik. Kl., NF.*, **12**.

— (1930a). Probleme der deutschen morphologische Wüstenforschung. *Die Naturwissenschaften*, **28**: 629–37.

— (1930b). Einige Oberflächenformen in Chile und Spitzbergen im Rahmen einer vergleichenden Morphologie der Klimazonen. Pp. 147–56 in *Hermann-Wagner-Gedächtnisschrift. Petermanns geogr. Mitt., Ergänzungshaft*, **209**.

— (1933). Die 'Salzsprengung' und ihre Bedeutung für die regional klimatische Gliederung der Wüsten. *Petermanns geogr. Mitt.*, **79**: 130–5.

— (1950). Das Gesetz der Wüstenbildung. *Universitas*, **5**: 801–14.

— (1953). Neues zum Problem der Schichtstufenlandschaft. Einige Ergebnisse einer Reise durch der Südwesten der USA, Sommer und Herbst 1952. *Nachr. Akad. d. Wiss. Göttingen math.-phys. Kl.*: 3–22.

— (1956). Über Wandverwitterung und Hangabtragung in semiariden und vollariden Gebieten. Pp. 96–104 in *Premier Rapport de la Commission pour l'Étude des Versants*. Amsterdam.

MOSLEY, M. P. (1973). Rainsplash and the convexity of badland divides. *Z. Geomorph.-Suppl.*, **18**: 10–25.

MOTTS, W. S. (1965). Hydrologic types of playas and closed valleys and some relations of hydrology to playa geology. Pp. 73–104 in *Geology, Mineralogy, and Hydrology of U.S. Playas*, ed. J. T. Neal. Air Force Cambridge Res. Lab., Environmental Res. Paper, **96**.

— (1970) (ed.) *Geology and Hydrology of Selected Playas in Western United States*. Geol. Dept Univ. of Massachusetts, Amherst.

— and CARPENTER, D. (1970). Geology and hydrology of Rogers Playa and Rosamond Playa, California. Pp.23–65 in *Geology and Hydrology of Selected Playas in Western United States*, ed. W. S. Motts. Geol. Dept Univ. of Massachusetts, Amherst.

MULCAHY, M. J. and BETTENAY, E. (1972). Soil and landscape studies in Western Australia. (1) The major drainage divisions. *J. geol. Soc. Aust.*, **18**: 349–57.

NEAL, J. T. (1965). Environmental setting and general surface characteristics of playas. Pp.1–29 in *Geology, Mineralogy, and Hydrology of U.S. Playas*, ed. J. T. Neal. Air Force Cambridge Res. Lab., Environmental Res. Paper, **96**.

— (1969). Playa variation. Pp. 14–44 in *Arid Lands in Perspective*, ed. W. G. McGinnies and B. J. Goldman. Tucson.

— (1972). Playa surface features as indicators of environment. Pp. 107–32 in *Playa Lake Symposium*, ed. C. C. Reeves. ICASALS Publn., **4**. Lubbock.

—, LANGER, A. M. and KERR, P. F. (1968). Giant desiccation polygons of Great Basin playas. *Bull. geol. Soc. Am.*, **79**: 69–90.

NORRIS, R. M. (1969). Dune reddening and time. *J. sedim. Petrol.*, **39**: 7–11.

OBERLANDER, T. M. (1972). Morphogenesis of granitic boulder slopes in the Mojave Desert, California. *J. Geol.*, **80**: 1–20.

OLLIER, C. D. (1963). Insolation weathering: examples from central Australia. *Am. J. Sci.*, **261**: 376–81.

— (1965). Some features of granite weathering in Australia. *Z. Geomorph.*, **9**: 285–304.

— (1966). Desert gilgai. *Nature, Lond.*, **212**: 581–3.

— and TUDDENHAM, W. G. (1962). Inselbergs of central Australia. *Z. Geomorph.*, **5**: 257–76.

PACHUR, H-J. (1970). Zur Hangformen im Tibestigebirge (Rêpublique du Tchad). *Erde*, **101**: 41–54.

PASSARGE, S. (1904). Die Inselberglandschaften in tropischen Afrika. *Naturwiss. Wochens.*, **3**: 657–65.

PEEL, R. F. (1941). Denudational landforms of the central Libyan Desert. *J. Geomorph.*, **4**: 3–23.

— (1966). The landscape in aridity. *Trans. Inst. Br. Geogr.*, **38**: 1–23.

— (1974). Insolation weathering: some measurements of diurnal temperature changes in exposed rocks in the Tibesti region, central Sahara. *Z. Geomorph.-Suppl.*, **21**: 19–28.

PELTIER, L. C. (1950). The geographic cycle in periglacial regions as it is related to climatic geomorphology. *Ann. Ass. Am. Geogr.*, **40**: 214–36.

— (1962). Area sampling for terrain analysis. *Prof. Geogr.*, **14**: 24–8.

PERRY, R. A. *et al.* (1962). General Report on Lands of the Alice Springs Area, Northern Territory, 1956–57. *CSIRO Aust. Land Res. Ser.*, **6**. Melbourne.

PFANNENSTIEL, M. (1953). Das Quartär der Levante. II. Die Entstehung der ägyptischen Oasen-Depressionen. *Abh. d. Math.-Naturwiss. Kl. Akad. d. Wiss. u. Lit. in Mainz, Jhrg. 1953*, 7: 337–411.

PIAS, J. (1970). *Les Formations Sédimentaires Tertiaires et Quaternaires de la Cuvette Tchadienne et les Sols qui en Dérivent.* ORSTOM Mem., **43**.

PRESCOTT, J. A. (1936). The climatic control of the Australian deserts. *Trans. R. Soc. S. Aust.*, **60**: 93–5.

PRICE, W. A. (1963). Physico-chemical and environmental factors in clay dune genesis. *J. sedim. Petrol.*, **33**: 766–78.

— (1972). Oriented lakes: origin, classification, and developmental histories. Pp. 305–26 in *Playa Lake Symposium*, ed. C. C. Reeves. ICASALS Publn., **4**. Lubbock.

— and KÖRNICKER, L. S. (1961). Marine and lagoonal deposits in clay dunes, Gulf Coast, Texas. *J. sedim. Petrol.*, **31**: 245–55.

QUENEY, P. and DUBIEF, J. (1943). Action d'un obstacle ou d'un fossée sur un vent chargé de sable. *Trav. Inst. de Rech. Sah.*, **2**: 169–76.

RAHN, P. H. (1966). Inselbergs and nickpoints in southwestern Arizona. *Z. Geomorph.*, **10**: 217–25.

RAINWATER, F. H. (1962). Stream composition of the conterminous United States. *U.S. geol. Surv. Hyd. Inv. Atlas*. Washington.

RAYNAL, R. (1961). *Plaines et Piedmonts du Bassin de la Moulouya (Maroc Oriental). Etude Geomorphologique*. Rabat.

REEVES, C. C. (1966). Pluvial lake basins of west Texas. *J. Geol.*, **74**: 269–91.

— (1968). *Introduction to Paleolimnology*. Developments in Sedimentology No. 11. Amsterdam.

REICHE, P. (1937). The Toreva-block, a distinctive landslide type. *J. Geol.*, **45**: 538–48.

— (1950). *A survey of weathering processes and products.* Univ. of New Mexico Publns in Geology, **3**.

ROBINSON, E. S. (1970). Mechanical disintegration of the Navajo Sandstone in Zion Canyon, Utah. *Bull. geol. Soc. Am.*, **81**: 2799–806.

ROBINSON, G. (1966). Some residual hillslopes in the Great Fish River Basin, South Africa. *Geogrl J.*, **132**: 386–90.

ROTH, E. S. (1965). Temperature and water content as factors in desert weathering. *J. Geol.*, **73**: 454–68.

RUELLAN, A. (1967). Individualisation et accumulation du calcaire dans les sols et les dépôts Quaternaires du Maroc. *Cah. ORSTOM Sér. Péd.*, **5**: 421–62.

RUSSELL, R. J. (1936). The desert rainfall factor in denudation. *Int. Geol. Cong.*, **2**: 753–63.

RUXTON, B. P. (1958). Weathering and subsurface erosion in granite at the piedmont angle, Balos, Sudan. *Geol. Mag.*, **95**: 353–77.

SCHATTNER, I. (1961). Weathering phenomena in the crystalline of the Sinai in the light of current notions. *Bull. Res. Coun. Israel. G.*, **10**: 247–66.

SCHEFFER, F., MEYER, B. and KALK, E. (1963). Biologische Ursachen der Wüstenlackbildung. *Z. Geomorph.*, **7**: 112–19.

SCHICK, A. P. (1970). Desert floods. Interim results of observations in the Nahal Yael research watershed, southern Israel, 1965–70. *IASH-UNESCO Symp. on Representative and Experimental Basins, Wellington, N.Z. December 1970*, **3**: 201–15.

— (1971). A desert flood: physical characteristics; effects on man, geomorphic significance, human adaptation. A case study of the southern Arava watershed. *Jerusalem Studies in Geography*, **2**: 91–155.

— (1974). Formation and obliteration of desert stream terraces — a conceptual analysis. *Z. Geomorph.-Suppl.*, **21**: 88–105.

— and SHARON, D. (1974). *Geomorphology and Climatology of Arid Watersheds.* U.S. Army Eur. Res. Office. Tech. Paper DA JA-72-C-3874. Dept Geography, Univ. Jerusalem.

SCHREIBER, J. F. *et al.* (1972). Sedimentologic studies in the Willcox Playa area, Cochise County, Arizona. Pp. 133–84 in *Playa Lake Symposium*, ed. C. C. Reeves. ICASALS Publn, **4**. Lubbock.

SCHUMM, S. A. (1956a). Evolution of drainage systems and slopes in badlands at Perth Amboy, New Jersey. *Bull. geol. Soc. Am.*, **67**: 597–646.

— (1956b). The role of creep and rain-wash on the retreat of badland slopes. *Am. J. Sci.*, **254**: 693–706.

— (1960a). The effect of sediment type on the shape and stratification of some modern fluvial deposits. *Am. J. Sci.*, **258**: 177–84.

— (1960b). The shape of alluvial channels in relation to sediment type. *U.S. geol. Surv. Prof. Paper*, **352-B**: 17–30.

— (1961). Effect of sediment characteristics on erosion and deposition in ephemeral stream channels. *U.S. geol. Surv. Prof. Paper*, **352-C**: 31–70.

— (1962). Erosion on miniature pediments in Badlands National Monument, South Dakota. *Bull. geol. Soc. Am.*, **73**: 719–24.

— (1963). A tentative classification of alluvial river channels. *U.S. geol. Surv. Circ.*, **477**.

— (1964). Seasonal variations of erosion rates and processes on hillslopes in western Colorado. *Z. Geomorph.-Suppl.*, **5**: 215–38.

— (1965). Quaternary paleohydrology. Pp. 783–94 in *The Quaternary of the United States*, ed. H. E. Wright and D. G. Frey. Princeton.

— (1967). Rates of surficial rock creep on hillslopes in western Colorado. *Science*, **155**: 560–1.

— and CHORLEY, R. J. (1964). The fall of Threatening Rock. *Am. J. Sci.*, **262**: 1041–54.

— and — (1966). Talus weathering and scarp recession in the Colorado plateaus. *Z. Geomorph.*, **10**: 11–36.

— and HADLEY, R. F. (1957). Arroyos and the semiarid cycle of erosion. *Am. J. Sci.*, **255**: 161–74.

— and Lichty, R. W. (1963). Channel widening and flood-plain construction along Cimarron River in southwestern Kansas. *U.S. geol. Surv. Prof. Paper*, **352-D**: 71–88.

— and Lusby, G. C. (1963). Seasonal variation of infiltration capacity and runoff on hillslopes in western Colorado. *J. geophys. Res.*, **68**: 3655–66.

Senior, B. R. and Senior, D. A. (1972). Silcrete in southwest Queensland. *Bur. Min. Resources Aust. Bull.*, **125**: 23–8.

Servant, M. (1970). Données stratigraphiques sur le Quaternaire Supérieur et Récent au nord-est du Lac Tchad. *Cah. ORSTOM Sér. Geol.*, **2**: 95–114.

Sharon, D. (1962). On the nature of hamadas in Israel. *Z. Geomorph.*, **6**: 129–47.

Sharp, R. P. (1940). Geomorphology of the Ruby-East Humboldt Range, Nevada. *Bull. geol. Soc. Am.*, **51**: 337–71.

— (1942). Mudflow levees. *J. Geomorph.*, **5**: 222–7.

— (1949). Pleistocene ventifacts east of the Bighorn Mountains, Wyoming. *J. Geol.*, **57**: 175–95.

— (1963). Wind ripples. *J. Geol.*, **71**: 617–36.

— (1964). Wind-driven sand in Coachella Valley, California. *Bull. geol. Soc. Am.*, **75**: 785–804.

— and Nobles, L. H. (1953). Mudflow of 1941 at Wrightwood, southern California. *Bull. geol. Soc. Am.*, **64**: 547–60.

Simons, F. S. (1956). A note on Pur-Pur dune, Viru Valley, Peru. *J. Geol.*, **64**: 517–21.

Singh, G., Joshi, R. D. and Singh, A. B. (1972). Stratigraphic and radiocarbon evidence for age and development of three salt lake deposits in Rajasthan, India. *Quaternary Res.*, **2**: 496–505.

Slatyer, R. O. (1962). Climate of the Alice Springs area. Pp. 109–28 in *Lands of the Alice Springs Area, Northern Territory, 1956–57.* CSIRO Aust. Land Res. Ser., **6**. Melbourne.

Smalley, I. J. and Vita-Finzi, C. (1968). The formation of fine particles in sandy deserts and the nature of 'desert' loess. *J. sedim. Petrol.*, **38**: 766–74.

Smith, G. I. (1968). Late-Quaternary geologic and climatic history of Searles Lake, southeastern California. Pp. 293–310 in *Means of Correlation of Quaternary Successions, Proc. VII Cong. INQUA*, **8**, ed. R. B. Morrison and H. E. Wright. Salt Lake City.

Smith, H. T. U. (1963). *Eolian geomorphology, wind direction, and climatic change in North Africa.* U.S. Air Force Cambridge Res. Lab. Bedford.

— (1965). Dune morphology and chronology in central and western Nebraska. *J. Geol.*, **73**: 557–78.

— (1967). *Past versus present wind action in the Mojave Desert region, California.* U.S. Air Force Cambridge Res. Lab. Bedford.

— (1969). *Photo-interpretation studies of desert basins in northern Africa.* U.S. Air Force Cambridge Res. Lab., Bedford.

— (1972). Playas and related phenomena in the Saharan region. Pp. 63–87 in *Playa Lake Symposium*, ed. C. C. Reeves. ICASALS Publn, **4**. Lubbock.

SMITH, K. G. (1958). Erosional processes and landforms in Badlands National Monument, South Dakota. *Bull. geol. Soc. Am.*, **69**: 975–1008.

SYNDER, C. T. (1962). A hydrologic classification of valleys in the Great Basin, western United States. *Int. Assn. Sci. Hydrologists*, **7**: 53–9.

— and LANGBEIN, W. B. (1962). The Pleistocene lake in Spring Valley, Nevada, and its climatic implications. *J. geophys. Res.*, **67**: 2385–94.

SPRINGER, M. E. (1958). Desert pavement and vesicular layer of some soils of the desert of the Lahontan Basin, Nevada. *Proc. Soil Sci. Soc. Am.*, **22**: 63–6.

STACE, H. C. T. *et al.* (1968). *A Handbook of Australian Soils*. Adelaide.

STEPHENS, C. G. (1971). Laterite and silcrete in Australia. *Geoderma*, **5**: 5–52.

STODDART, D. R. (1969). Climatic geomorphology: review and re-assessment. *Progress in Geography*, **1**: 159–222.

STONE, R. O. (1956). A Geologic Investigation of Playa Lakes. Unpublished Ph.D. thesis, Univ. of Southern California.

— and SUMMERS, H. J. (1972). *Study of Subaqueous and Subaerial Sand Ripples*. U.S. Off. Naval Res. Rep., Arlington, Va.

STRAHLER, A. N. (1950). Equilibrium theory of erosional slopes approached by frequency distribution analysis. *Am. J. Sci.*, **248**: 673–96, 800–14.

STRAKHOV, N. M. (1967). *Principles of Lithogenesis*. Vol. 1, ed. S. I. Tomkeieff and J. E. Hemingway. Edinburgh.

SWINEFORD, A. and FRYE, J. C. (1945). A mechanical analysis of wind-blown dust compared with analysis of loess. *Am. J. Sci.*, **243**: 249–55.

SYMMONS, P. M. and HEMMING, C. F. (1968). A note on wind-stable stone-mantles in the southern Sahara. *Geogrl J.*, **134**: 60–4.

TARR, R. S. (1915). A study of some heating tests and the light they throw on the cause of disaggregation of granite. *Econ. Geol.*, **10**: 348–67.

TATOR, B. A. (1952–3). Pediment characteristics and terminology. *Ann. Ass. Am. Geogr*, **42**: 295–317; **43**: 47–53.

TEISSIER, M. (1965). Les crues d'oueds au Sahara algérien de 1950 à 1961. *Trav. Inst. Rech. Sah.*, **24**: 7–29.

THORNTHWAITE, C. W. (1948). An approach toward a rational classification of climate. *Geogrl Rev.*, **38**: 55–94.

TRICART, J. (1954a). Une forme de relief climatique: les sebkhas. *Revue Géomorph. dyn.*, **5**: 97–101.

— (1954b). Influence des sols salés sur la déflation éolienne en Basse-Maurétanie et dans le delta du Sénégal. *Revue Géomorph. dyn.*, **5**: 124–32.

— (1959). Géomorphologie dynamique de la moyenne vallée du Niger (Soudan). *Annls Géogr.*, **68**: 333–43.

— and CAILLEUX, A. (1965). *Introduction à la Géomorphologie Climatique*. Paris.

— (1969). *Le Modelé des Régions Sèches*. Paris.

TUAN, YI-FU. (1959). Pediments in southeastern Arizona. *Univ. Calif. Publn Geogr.*, **13**: 1–164.

TWIDALE, C. R. (1962). Steepened margins of inselbergs from north-western Eyre Peninsula, South Australia. *Z. Geomorph.*, **6**: 51–69.

— (1967). Hillslopes and pediments in the Flinders Ranges, South Australia. Pp. 95–117 in *Landform Studies from Australia and New Guinea*. ed. J. N. Jennings and J. A. Mabbutt. Canberra.

— (1971). *Structural Landforms*. Canberra.

— (1972). Evolution of sand dunes in the Simpson Desert, central Australia. *Trans. Inst. Br. Geogr.*, **56**: 77–109.

— and CORBIN, E. M. (1963). Gnammas. *Revue Géomorph. dyn.*, **14**: 1–20.

UDDEN, J. A. (1894). Erosion, transportation and sedimentation by the atmosphere. *J. Geol.*, **2**: 318–31.

VANNEY, J. R. (1960). *Pluie et Crue dans le Sahara Nord-occidental*. Mem. Rég. Inst. Rech. Sah., **4**. Algiers.

VERGER, F. (1964). Mottureaux et gilgais. *Annls Géogr.*, **73**: 413–30.

VERSTAPPEN, H. TH. (1968). On the origins of longitudinal (seif) dunes. *Z. Geomorph.*, **12**: 200–20.

— (1970). Aeolian geomorphology of the Thar Desert and palaeo-climates. *Z. Geomorph.-Suppl.*, **10**: 104–20.

— (1972). On dune types, families and sequences in areas of unidirectional winds. *Gott. Geogr. Abh.*, **60**: 341–54.

WALTER, H. and LIETH, H. (1960). *Klimadiagramm-Weltatlas*. Jena.

WALTHER, J. (1891). Die Denudation in der Wüstend ihre geologische Bedeutung. *Abh. math.-phys. Kl. d. kön. sächsischen Ges. d. Wiss.*, **16**: 345–570.

— (1900). *Das Gesetz der Wüstenbildung in Gegenwart und Vorzeit*. Berlin.

WARREN, A. (1967). Dune measurements and their implications in the Nebraska sandhills (abstract). *Proc. Nebraska Acad. Sci.*: 23.

— (1970). Dune trends and their implications in the central Sudan. *Z. Geomorph.-Suppl.*, **10**: 154–80.

— (1972). Observations on dunes and bi-modal sands in the Tenéré desert. *Sedimentology*, **19**: 37–44.

WATSON, R. A. and WRIGHT, H. E. (1963). Landslides on the east flank of the Chuska Mountains, northwestern New Mexico. *Am. J. Sci.*, **261**: 525–48.

WEISCHET, W. (1969). Zur Geomorphologie des Glatthang-Reliefs in der ariden Subtropenzone des Kleinen Nordens von Chile. *Z. Geomorph.*, **13**: 1–21.

WELLMAN, H. W. and WILSON, A. T. (1965). Salt weathering, a neglected geological erosive agent in coastal and arid environments. *Nature, Lond.*, **205**: 1097–8.

WERTZ, J. B. (1964). Les phénomènes d'érosion et de dépôt dans les vallées habituellement sèches du Sud-ouest des États-Unis. *Z. Geomorph.*, **8** *Sonderheft*: 71–104.

WHITAKER, C. R. (1973). *A Bibliography of Pediments*. Norwich.

WHITEHOUSE, F. W. (1944). The natural drainage of some very flat monsoonal lands (the plains of western Queensland). *Aust. Geogr.*, **4**: 183–96.

WILHELMY, H. (1958). *Klimamorphologie der Massengesteine*. Brunswick.

WILLIAMS, G. E. (1970). The central Australian stream floods of February-March 1967. *J. Hydrol.*, **11**: 185–200.

— (1971). Flood deposits of the sand-bed ephemeral streams of central Australia. *Sedimentology*, **17**: 1–40.

— (1973). Late Quaternary piedmont sedimentation, soil formation and paleoclimates in Australia. *Z. Geomorph.*, **17**: 102–25.

WILSON, I. G. (1971). Desert sandflow basins and a model for the development of ergs. *Geogrl J.*, **137**: 180–99.

— (1972). Aeolian bedforms — their development and origins. *Sedimentology*, **19**: 173–210.

WOOD, A. (1942). The development of hillside slopes. *Proc. Geol. Assn.*, **53**: 128–38.

WOPFNER, H. and TWIDALE, C. R. (1967). Geomorphological history of the Lake Eyre Basin. Pp. 119–43 in *Landform Studies from Australia and New Guinea*, ed. J. N. Jennings and J. A. Mabbutt. Canberra.

YAALON, D. H. (1970). Parallel stone cracking, a weathering process on desert surfaces. *Geol. Inst. Bucharest, Tech. & Econ. Bull. (C) Pedology*, **18**: 107–11.

— and GANOR, E. (1968). Chemical composition of dew and dry fallout in Jerusalem, Israel. *Nature, Lond.*, **217**: 1139–40.

— and GINZBOURG, D. (1966). Sedimentary characteristics and climatic analysis of easterly dust storms in the Negev (Israel). *Sedimentology*, **6**: 315–32.

YAIR, A. (1974). Sources of runoff and sediment supplied by the slopes of a first order drainage basin in an arid environment (northern Negev-Israel). *Abh. Akad. Wiss. Gött., math.-physik. Kl., III*, **29**: 403–17.

— and GERSON, R. (1974). Mode and rate of escarpment retreat in an extremely arid environment (Sharm el Sheikh, southern Sinai Peninsula). *Z. Geomorph.-Suppl.*, **21**: 202–15.

— and KLEIN, M. (1973). The influence of surface properties on flow and erosion processes on debris covered slopes in an arid area. *Catena*, **1**: 1–18.

— and LAVEE, H. (1974). Areal contribution to runoff on scree slopes in an extreme arid environment — a simulated rainstorm experiment. *Z. Geomorph.-Suppl.*, **21**: 106–21.

YOUNG, A. (1972). *Slopes*. Edinburgh.

YOUNG, R. G. (1964). Fracturing of sandstone cobbles in caliche-cemented terrace gravels. *J. sedim. Petrol.*, **34**: 886–9.

INDEX

N.S.W. = New South Wales; N.T. = Northern Territory, Australia; Qld = Queensland; S.A. = South Australia; Vic. = Victoria; W.A. = Western Australia. The names of the individual states of the United States of America are employed alone.

Bold face indicates page reference to figures, plates, and tables, but may also include text reference on the same page.